P9-DVZ-107

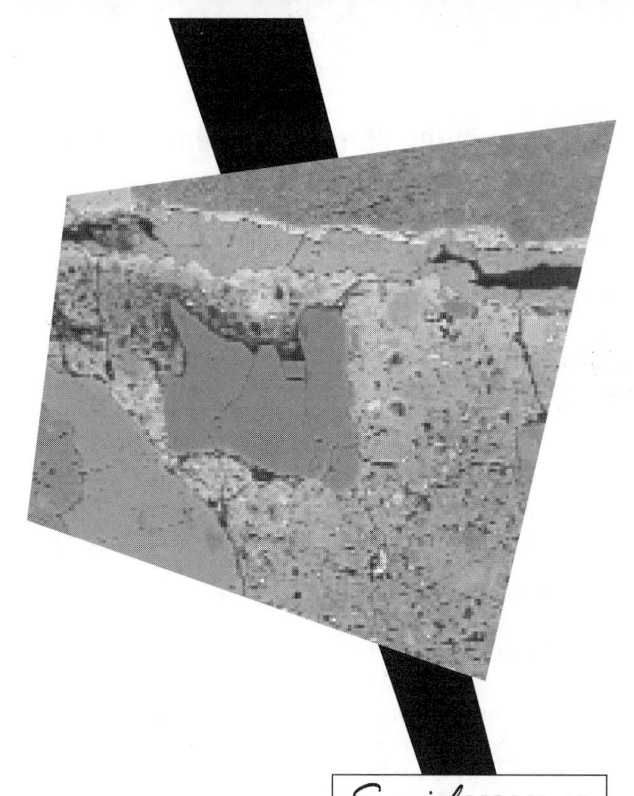

Special VOLUME

MATERIALS SCIENCE OF CONCRETE: **Sulfate Attack Mechanisms**

Other Volumes in the *Materials Science of Concrete* series:

Materials Science of Concrete — Special Volume:
The Sidney Diamond Symposium
Edited by Menashi Cohen, Sidney Mindess, and Jan Skalny
© 1998, ISBN 1-57498-072-6

Materials Science of Concrete V
Edited by Jan Skalny and Sidney Mindess
© 1998, ISBN 1-57498-027-0

Materials Science of Concrete IV
Edited by Jan Skalny and Sidney Mindess
© 1995, ISBN 0-944904-75-0

Materials Science of Concrete III
Edited by Jan Skalny
© 1992, ISBN 0-944904-55-6

Materials Science of Concrete II
Edited by Jan Skalny and Sidney Mindess
© 1991, ISBN 0-944904-37-8

Materials Science of Concrete I
Edited by Jan Skalny
© 1989, ISBN 0-944904-01-7

For information on ordering titles published by The American Ceramic Society, or to request a publications catalog, please contact our Customer Service Department at

Phone: 614-794-5890
Fax: 614-794-5892
E-mail: customersrvc@acers.org

Visit our on-line book catalog at <www.acers.org>.

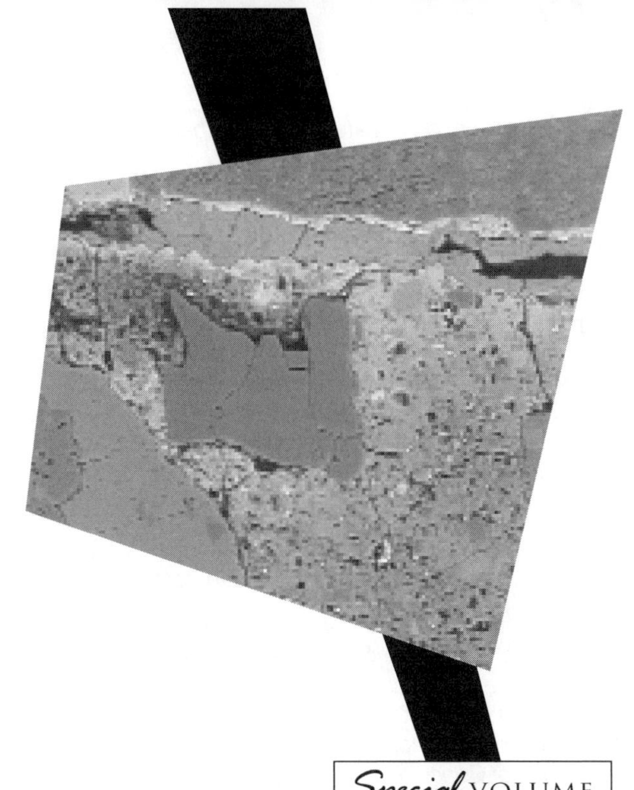

Special VOLUME

MATERIALS SCIENCE OF CONCRETE: Sulfate Attack Mechanisms

Edited by
Jacques Marchand
Jan P. Skalny

Published by
The American Ceramic Society
735 Ceramic Place
Westerville, OH 43081

Proceedings of Seminar on Sulfate Attack Mechanisms held October 5 and 6, 1998, in Quebec City, Quebec, Canada.

Library of Congress Cataloging-in-Publication Data
A CIP record for this book is available from the Library of Congress

For information on ordering titles published by The American Ceramic Society, or to request a publications catalog, please call 614-794-5890. Visit our on-line bookstore at www.acers.org.

Printed in the United States of America.

1 2 3 4–02 01 00 99

ISSN 1042-1122
ISBN 1-57498-074-2

Contents

Preface

This special volume, the second in the Materials Science of Concrete series, represents a collection of presentations given at the Seminar on Sulfate Attack Mechanisms, held October 5–6, 1998, in Quebec City, Quebec, Canada. The seminar was organized as a forum for thorough discussion of some new and old—yet unresolved—scientific questions.

Although it is not considered to be one of the most prevalent mechanisms of concrete deterioration, sulfate attack has received a wide notoriety in recent years, especially in North America, because of costly litigation involving cement manufacturers, ready mix and precast concrete producers, developers, etc. Numerous valid technical questions were raised about both internal (represented by so-called delayed ettringite formation, DEF) and external sulfate attack mechanisms. Issues relevant to the DEF variety of internal sulfate attack have been subject of often semi-scientific, emotional debate for about a decade; for this reason, discussion of DEF-related issues was specifically limited at the above seminar. However, as the mechanistic aspects of internal and external sulfate attack are closely interconnected, it is hoped that the work presented at the seminar, and now published in this special volume, will help explain some of the ambiguous technical issues relevant to both.

We would like to dedicate this special volume to the memory of Dr. Jaromir Jambor, one of the invited speakers, who unexpectedly passed away in May 1998. A well-known Czechoslovak researcher and author, Dr. Jambor's four decades of work on concrete porosity/permeability and various aspects of durability made him a natural, but unfortunately unrealized, choice to coauthor the invited paper on sulfate attack and porosity/permeability issues.

It gives us a great pleasure to thank all authors, coauthors, and participants at the seminar for their active cooperation, friendship, and help in the search for scientific truth. We would like to express our thanks to the reviewers of the manuscripts for timely and eminently professional peer reviews. Special thanks are due to our sponsors: Ash Grove Cement Co., Holnam Builders Technologies Inc., National Institute of Standards and Technology, W.R. Grace and Co.'s Construction Products Division, SIMCO Technologies Inc., and Service d'Expertise en Matériaux, Inc. (S.E.M.).

We also would like to acknowledge the highly skilled cooperation by the staff of the S.E.M.; without their dedicated work, the organization of the seminar would not have been possible. Finally, thanks are due to The American Ceramic Society, and specifically to Mary Cassells, Sarah Godby, and Natalee Sperry, for preparation and publication of this special volume.

Jacques Marchand
Jan Skalny

DISCUSSION OF PRESENTED PAPERS

The following text was transcribed from video recordings made during the discussions that followed the oral presentations, as well as the discussion session held at the conclusion of the seminar. Because parts of some of the presentations were inaudible and could not be recovered, a small portion of the text is incomplete. The editors apologize for this inconvenience.

Discussion of Papers Presented Monday, October 5, 1998

Discussion of Presentation by J. Skalny and J. Pierce

Hooton: I do not think we should discuss minimum cement contents; I think we should discuss restrictions on the maximum water content in concrete. Because you can make a 0.33 water-cement ratio with 200 liters of water and 600 kg of cement or you can make it with 150 liters and 450 cement. That 150-liter water content will give you an order of magnitude lower permeability than the other one even though there is the same water-cement ratio. I do not think minimum cement content is the way to go.

Diamond: Jan, I think you made the point on several different occasions that water-to-cement ratio is at least somewhat associated with permeability or impermeability in a concrete. And, of course, it is true, but to my mind, much—certainly not all—but much sulfate attack occurs in semi-dry environments where sulfate is not washed out of the soil. Curing is, for concrete exposed to such environments, extremely important; as important as water to cement ratio is in trying to get an impermeable concrete. And I would like to see your emphasis amended to ensure that appropriate high levels of curing are mandated so that you do have a reasonable chance of getting impermeable concrete in that kind of environment.

There is a further consideration that I do not think most people think about. When people do laboratory studies of sulfate attack, traditionally they make relatively small samples, mortars sometimes pastes, and they cure them underwater for 28 days traditionally and then expose them to sulfate-bearing solutions. Most concrete in the field is exposed to sulfate-bearing solutions if they are to be exposed immediately. It makes a big, big difference. So water to cement ratio, while very important, is far from a guarantee, I think.

Skalny: Needless to say, Sid, I fully agree with you. As you may have noticed, I mentioned in one of the slides that processing, including proper curing—although most important in DEF—is also important here. I did not elaborate but I agree with you and, of course, not only about the curing but also with the last point you made.

Glasser: I do not want to take issue with anything you have said in your presentation,

Jan, but I think I do want to make a point of philosophy, which I think is important to establish at the outset.

I know that I am probably specially influenced by the work we are doing on concretes for nuclear waste immobilization where the time span for performance is very long or can be very long, but I am concerned by the background to present-day specifications which engineers use. Many of them trace their roots back to a time when performance of concrete structure for 20, 25, 30 years was deemed to be quite acceptable, and almost without noticing the time scale for performance of major concrete structures has expanded and expanded and expanded.

Now, we get lots of people approaching us about structures that were built 40 or 50 years ago and were damaged. These people are often responsible for the maintenance because the structure is not performing and it is needing major repairs or renovation or reconstruction or whatever. And you ask these people: "Well, when you built it, what performance specification did you have in mind?" And they say, "Oh, well, we do not know, we do not think there was any specification laid down and anyway we wouldn't have know how to meet it."

So I think that whenever we are considering the mechanisms to be, we have to consider the time frame, because many of the tests and standards that I see the engineers using seem to me to be adequate only for a fairly short time frame for performance. And so I think, another thing that we constantly have to keep in mind when discussing specifications, changes in specifications and expectations of performance, is the steadily expanding lifetime which we now expect for performance.

Young: I have a couple of comments. One where I will agree with you, Jan and Jim, and I will disagree with you on the second one. But I think your concept, emphasized at the beginning of you presentation, that sulfate attack is a complex series of several processes, is absolutely the right way to think about it.

I think we have had a disservice done, quite naturally because the dominant process that tended to be observed was the formation of ettringite, and we realize that this can be more complicated than that. I think we have to think of the concept that we have several parallel processes going on; which of these becomes the most critical depends on what is the rate-limiting step that is operating in a particular situation.

So you have to always have in mind that any particular situation several possibilities to be the root cause of the initial deterioration will all be there, and you may switch from one process to another as you move through the deterioration process. So it is fairly complicated, and we have to move away from the concept that it is always ettringite that is going to be the problem. I think it is rather parallel, probably, to the ASR concern because at the beginning we thought there were only a few aggregates that were a problem with ASR and as time goes by we see that more and more. My personal view is, if you have silica in an aggregate, you probably have an ASR potentiality of maybe 10 years or maybe 20 years and maybe 40 years, but it will happen. It is along the lines of what Fred Glasser was saying.

But I would disagree with you on that latter part of your talk because I think you are falling into the same trap, that we should be specifying our concrete on a prescriptive basis. If permeability is the issue, by all means have a prescriptive guide for the engineer and the mix desired, but for heaven's sake, we have to have tests that show that you have achieved the permeability that you think you need in this situation.

We should have, essentially, performance specifications along the lines that we are finally moving a little better in the cement specification world. So, if permeability is the problem, let us measure permeability directly with some guidance for the engineers as to how to interpret the data.

Skalny: Thank you, Francis. I fully agree and did not mean to oppose performance-based specifications.

Lagerblad: I'm from the Swedish Cement and Concrete Research Institute. I think we have to realize, when we talk about sulfate attack, that there are two different kinds. One is a surface attack; that means it is more like a mixture of chemical attacks of different variations. And then you have process more like an internal attack, which actually causes the expansion.

I think one has to separate these, because that expansion is more related, for example, to the C3A and things like sulfates in cement. But if you are in the soil and you get all these mechanisms for breaking down of the pyrites, you have an acid environment and magnesium and chlorides and things in it, then that is a bit of another thing.

So I think one has to divide it between these two types of mechanism when one makes a prescription of what you want. If it is a pure sulfate attack, just sulfate water, then it may be that expansion is most dangerous and then you have to look on those restrictions. For example, if you take—say you have a minimum content of cement, but if you have an ordinary Portland cement with a high C3A and put a lot of cement in it, expansion will be worse at higher cement content. But if it is a surface attack from the soil liquids, then you need a lot of cement because one needs a reservoir to buffer it.

So one is dividing between those two, and I think it is important in these talks that we do not mix all these mechanisms because there are two types. One is the expansion, and one is a surface attack that will act differently on the sulfate in the mixture of solutions.

Skalny: Thank you very much for the comments. I am, of course, aware of the complexity of the simultaneous processes occurring during sulfate attack, and that is the reason it is so difficult to separate them from each other. Later today we will have two papers discussing the various mechanistic aspects of sulfate attack, and the above two issues, I believe, will be discussed.

Discussion of Presentation by E. Hill

Marschall: Regarding concrete pipe, I would just like to point out that concrete pipe that is made today, that is manufactured today using modern vibrating processes, is very dense and very impermeable. So I do not want anybody to become alarmed that there is a problem with concrete pipe because that is not the case today.

Hill: Well, concrete pipe has performed very well in services. My colleague was in the concrete pipe business for many years and he commented rather recently to me that in a certain manufacturing process, "In a rainstorm, never take with refuge inside a concrete pipe!" That is for a pipe of a certain process.

Marschall: My guess is that certain process isn't even being used today throughout North America. I would like to know what that process is.

Hill: Packerhead.

Marschall: Packerhead pipe is still being made but differently from what it used to be. They make it in a manner now that is watertight.

Hill: Thank you.

Haynes: Gene, on the topic of history, you know when I look at the literature—and I go back to the beginning of this century—and you read articles where there were real failures of concrete due to sulfates, they did not really know what was going on, they jumped on it to try to figure it out, and they found out it was primarily a C3A problem. In the early '30s they came up with a cement type which was resistant, and then you really do not seem to encounter those stories of failures that appear to occur before then. Now, that gives me an indication that the C3A finding was really significant. I was wondering if you had that same kind of feel or not?

Hill: Well, it is an interesting question, Harvey. Until the last few years, interest in sulfate-resisting concrete tended to be centered with a few people who made it and a few people who though it was an interesting subject, of not very great application, not very wide application.

When I got involved in the ACI committee on durability dealing with this issue, I felt a little bit like Will Roger's comment that said: "You know, certain jobs are like the fellow that was going to be tarred and feathered and ridden out of town on a rail." And he said, "You know fellows, if it weren't for the honor, I'd just as soon walk!"

But I asked Jim Pierce, who is with me here, and I hope that I am quoting him properly. Jim told me that he had asked the Bureau people to review their sulfate-resistant concrete requirements. As a public servant, he felt a very strong obligation to offer the most economical as well as durable concrete for public structures and their review, which I think was conducted some time before the ACI-201 effort started, indicating that they were not engaging in an overkill and that their standards were as they should be. Jim, is that a fair description?

Pierce: Yes.

Hill: I think that the Bureau has done a good job in addressing these problems, but I do not think they have gone away. And we are seeing evidence in some areas...with an awful lot of disputes and there is some distress. Now, there are arguments with respect as to how bad the distress is. But I think it is still a problem, and I have periodically seen Steve Gibler or others and they would say, "Well, I am working on another water tower and of course the sulfate is chewing up the water tower." I think they have been more of a problem then we thought. By the same token, I was convinced that 20 years ago alkali silica reactivity was a special problem. And you know it did not occur all that much; it did not occur with that many aggregates.

And earthquakes only occurred in California—and I felt sorry for those folks—and now we have earthquakes in Ohio, we know we have had them in Boston, in Charleston, South Carolina, and we are seeing more ASR. So I think these problems are more widespread than we have recognized.

Mehta: You have just shown this graph, with the y-axis representing the average rate of deterioration?

Hill: I cannot answer that specifically but, generally speaking, the Portland Cement Association adopted a series of specimens which represented different degrees of deterioration and I think that they...I think in their report they certainly used an evaluation based on visual observations. In later work I know we have used sonic modulus, we have used weight change, we have used expansion, and I have a feeling that PCA used all of those.

Discussion of Presentation by J.R. Clifton, G. Frohnsdorf, and C. Ferraris

Mehta: Thanks for the excellent presentation, to start with. Now, as regards to specifications in standardization: it is so difficult to do it, and always there is the need to simplify it as much as possible. However, life is a little bit more complicated, as you are aware of.

Clifton: I fully agree.

Mehta: I feel that we have to distinguish—and Skalny and Pierce talked about it—between at least two separate mechanisms: magnesium sulfate attack, which is driven basically, ultimately, by decalcification of the C-S-H, which is the primary mechanism of the deterioration of the paste itself, the binder itself, and sodium sulfate attack which is driven by expansion. So as the amount of tricalcium aluminate that is essential for the sodium sulfate attack, obviously then you can reduce the tricalcium aluminate content, you can measure the expansion, and you get a good sense of the rate of deterioration, or a good sense of prediction rate deterioration.

However, for example, if you are talking about something else, and if you are introducing acids, you will lower the pH. So by lowering the pH, you will get the same deterioration—well, it will be faster, at shorter period of time. So again, to sum it up, I feel that there is a need to classify it, to say: OK, right now, if this is the main type of deterioration, then you have that kind of set of rules. If there is a different kind of deterioration, another set of rules applies, one that doesn't apply to the first type.

Clifton: Good point.

Bonen: To sum up the comments made earlier today, I think it has been well put that there are many factors of permeability, water-to-cement ratio, curing that are very important. And I think we have learned a lot about paste and mortar and the mechanism of sulfate attack through stories about paste and mortar. We can say that a mortar or paste will be resistant, but I do not think that will guarantee that the concrete will be resistant. On the other hand, we can say the reverse, the paste is not resistant but can we state that concrete will not be resistant?

My question is: Do we know enough about the link between this paste and mortar resistance and concrete, and what kind of work do we need to do in that regard?

Clifton: Of course, recently we did not use concrete in our studies. You could take something stronger or with larger amount of cement in it, and the time period for, let's say, a standard test method would probably be unacceptable using cement, a large amount of cement for a series of test.

But I agree with you, for service conditions we should be looking at concrete test because the behavior depends on a holistic approach to the whole material, how they interface, and the transition zone between aggregate and the cement content are important.

We have evidence that the principal migration path for sulfate attack is through the interfacial zone. So I agree, because that comes from a fracture in a concrete; deterioration is at the interfacial zone. So I totally agree with you.

Diamond: Just a further comment on Dr. Bonen's comments, who suggests that you decide whether the primary mechanism driving a case of sulfate attack is sodium sulfate or magnesium sulfate and test accordingly. Unfortunately, we have a fair amount of experience with the composition of ground waters in at least one part of the country. And do not quote me on this but the numbers that come to mind, the typical molar concentration of cations might me something like 50 percent sodium, 35 percent magnesium, 15 percent calcium. How do you decide what the primary driving force is, magnesium sulfates, sodium sulfates? You cannot. You cannot.

Clifton: Well, thank goodness, I do not have to do that, that is what Paul Brown is going to talk about next. So I have kept away from talking about mechanisms on purpose. We have to understand the mechanism for developing a test. But clearly, I hope that throughout this workshop we have a better understanding of the mechanisms. We have our ideas, sure, but other people have been working more on this, so I am going to defer that to Paul Brown and Hal Taylor.

Diamond: Right. The point of my comment is that you cannot distinguish separate mechanisms. In practical exposures there is a combination, and not a linear combination, there is a lot of interaction going on. It is all one complicated process.

Clifton: Well, any standard, any good standard test method should take that into consideration, and if you are going to be serious you would use that composition for that location in your test method.

Young: I want to avoid the issue that Sid just raised, but I think that in the photographs in which you were showing spalling, that was with sodium sulfate.

Clifton: Yes.

Young: And the model of Atkinson and Hearn, I believe also assumes the spalling mechanism.

Clifton: Yes.

Young: The question is: Do you necessarily have to have this process of deterioration in sulfate attack? It seems to me that it is more a question of this particular form of deterioration probably due to, say, rate-limiting penetration of sulfate.

Clifton: Yes.

Young: And so I think... there can be other forms of deterioration in sulfate attack depending on particular situations.

Clifton: Well, you can have a good old sulfate attack that you have in stone where you have a peeling off, like onionskin. That is when you see...you can have spalling. You may get some discoloration from rust where concrete show spalling. Yes, you can have different, complex types of deterioration to occur.

Glasser: I wanted to pick up from where Sid Diamond left off because I do agree with him. Now, you talked about a test in which the cement or concrete —I am not sure which—was put into an iso-pH unit, where the pH was stabilized by adding sulphuric acid. Now, I do not know all the details of the test, but that test can be realistic for some environments but very unrealistic for others. For example, there is a test widely used in Germany for measuring the resistance of concrete pipes to sewage waters where, of course, sulphurous and sulphuric acid are produced in the flowing water through the pipes, particularly when the flow is slow.

So under those circumstances, if that is your special interest in developing cements and concretes which are durable to sewage acids, then titration with sulphuric acid is, I think, an acceptable basis on which to begin constructing a test. But that is a rather special case and most of the cases that engineers are concerned with, you're not effectively getting a titration by sulphuric acid.

If you are in seawater or brackish water, for example, the huge volume of water maintains the constancy of pH at around 8.

Clifton: Yes.

Glasser: So I think that you have got to make sure that not only are the engineering requirements for the test achievable, but that the chemical requirements for the test are achievable. I think that with these different ranges in environments, you may not be able to achieve the Holy Grail of finding one single test method that is applicable to concretes in the range of service environments that we actually encounter.

Clifton: Well, I think that there are different ways of overcoming this. You can adjust it up to 11, maybe 11.5. So you can adjust the pH or sulfate ion concentration limits. By adding (inaudible) acid is controlling the pH but also it is raising...putting sodium hydroxide back to sodium sulfate. That is the whole idea to keep the sodium sulfate concentration at a certain level. That is really the chemistry of it.

I think that by developing a proper test arrangement we can probably handle that. But I contend, as...the way we're doing now and...we have no control. The pH is not controlled, the sulfate ions concentration is not controlled. We need better understanding of basics at this time. Things are not perfect, I agree. None of these...but we are trying and strive to make our tests better and more representative of nature.

Tavares: Jim, thank you for your comments. A question. I agree with you that we should be taking a look at service life in concrete *per se* and take a look at sulfate resistance and other durability issues. But my question really is: We keep talking about C3A, and most modern concretes today contain supplement materials other than, you know, just plain ordinary cement. We should be talking about slags, fly ashes, or other materials.

How do you see what you are doing, and how can we develop new test methods because the present test methods do not give us really adequate results re sulfate resistance in concrete?

Clifton: That is a yes. The only thing I know is, C 1157, which is maybe the Holy Grail of all cements now, hydraulic cements in the United States, they might all come underneath that. Portland cement in C 150 may disappear and become a part of C 1157. At the present time we are requiring C 1012 as a test method.

So I am saying yes to what we have and I ask the question: Is that the best method? I am trying to propose other methods for discussion. I presume this morning is for discussion. I think there have to be better test methods available and how do we do it. I would hope that this workshop will give us information about mechanisms that will help us to develop a better test. So I want to answer your question, but that is what it is about the best answer I can give you. I mean, I do not know.

Lagerblad: We are all trying to figure out what happens when you lower the pH, or increase the pH, because when you get expansion, it has to do with sulfate ions that have to get into the concrete. It is not a surface attack; it is a material attack.

Clifton: Yes. It is a moving boundary attack. The inner face is moving inward.

Lagerblad: Yes, the broad problem is to get the sulfates in there and what does that do to the pH? The only thing I can see about it is that if you want SO_4 to get into the concrete, you have to get some other anion to get out of it. In might be the reason if they have a lower pH outside; that means it is easier to make exchange OH ions without quicker (inaudible) toward having a lower pH. That would mean it would be easier to balance it; that means the SO_4 will "win."

It doesn't have anything to do with magnesium and things like that, it just has to do with an ion exchange for the diffusion it will get in. Because in the case of ASTM 1012, that test is a total diffusion control because the concrete is totally immersed. The situation will be quite different when you have it in soil. There we have sulfate attack and different environments with different levels of saturation.

Clifton: But it is if you're in a saturated zone; [in an] unsaturated zone, it really depends where you are at in the soil profile.

We developed a model, as you know, called FORSIGHT where we are looking at the soil effects. I will discuss that here tomorrow. But I am primarily discussing what test methods we have at this time. And that is a model dealing with underground concrete vaults where it is not saturated.

Discussion of Presentation by P. Brown and H. Taylor

Taylor: I would like to add a couple of aspects, things that are actually in the written paper, which Paul Brown didn't mention.

First of all, I think two separate things. One is the accepted wisdom that ettringite forms early in the hydration of a cement paste, but then all the sulfate is used up and you still have C3A reacting, so you get monosulfate. That was no doubt true in the 1950s when this was first put forward and cements were low in sulfates; but as Wally Clermont and Larry Adams, among others, I think, have said fairly recently, it just is not true of present day cements. It is a much more complicated situation, which we do mention briefly in the paper, and ettringite cannot persist indefinitely in a cement paste. And, in fact, way back in 1960, Lou Copeland and others showed that even with the cements of that time. And because small crystals are unstable relative to big ones, that ettringite will recrystallize if it gets a chance and, therefore, seeing ettringite in a concrete does not necessarily mean that ettringite has caused the stress. And I think this point is important.

Now, even if ettringite has caused the stress and you can see big crystals of ettringite, it is not necessarily the big crystals that have caused the damage. It may be very much smaller crystals that you do not see in the paste that have caused the damage, and the big ones you can see by petrography or for that matter with the SEM could be secondary products of recrystallization. They may have added to the damage, they may have not. There have been arguments about this, which I will not go into.

The other aspect suggested: I want to draw your attention to work done originally by two groups of Japanese workers, one was Kondo and others, and the other was Nakamura and others, reported at the Cement Symposium in 1968, where they made the point that where the ettringite is formed is extremely important.

They were talking about essentially Type-K expansive cements, but I think it also applies to external sulfate attack. And they noted the observation that if there was calcium hydroxide there, then you could get expansion, but if there was no calcium hydroxide you can control an awful lot of expansion—even in a hardened paste—and not get expansion.

And their interpretation of this, which I believe to be correct, was that if there was calcium hydroxide there, the aluminate could not migrate easily away from the surfaces of C3A or C4AF, and for that matter, monosulfate and another AFm phase; ettringite was formed in conditions of very high supersaturation where it exerted considerable stress, or could do. If there was no calcium hydroxide, it formed all over the place and did not cause expansion.

And this has recently been said also by a couple of Chinese workers whose names, I am sorry, I have forgotten, but it was in CCR, *Cement and Concrete Research*, pretty recently.

And I think it just reinforces the point which Paul made and which Jim Clifton, I think, also made, that you would not expect expansion to be related to the amount of ettringite that is formed; where it is formed is extremely important, perhaps more important.

Beaudoin: Paul, that was a very interesting paper. I do have a couple of questions, if I may. One of them relates to the expansion, and the other one to a point made about the topochemical growth, crystal growth.

On the expansion side, if expansion is not due to crystal growth impinging on the boundaries of a limited space, to my knowledge, this would imply that the expansion is either due to the free energy difference resulting in a length change or some kind of inner collation phenomena within the ettringite crystal itself. Now, it would seem to me that I would rule out the first type of expansion because the surface area, the nitrogen surface area of ettringite, would be not large enough to account for the amount of expansion that you would observe.

And in terms of an inner collation phenomena, we have worked with AFt phases trying to measure its sorption isotherms and this sort of thing, length change isotherms, and we just have not been able to measure the kinds of length changes that would be associated with that. Could you comment on that?

Brown: Are you asking the question with regard to the bridging theory or... ?

Beaudoin: Well, perhaps I misunderstood, but I thought your premise, or one of your conclusions, sort of dismissed the crystallization growth pressure concept.

Brown: It doesn't *per se*. The point I am making is, the pressure even though the crystals are not physically touching the crystal pore walls in a region, when the growth surface of the crystal pore walls here. ...What develops in this region is an osmotic pressure, OK? The pressure here is higher than at the bulk associated with the concentration of ions.

Beaudoin: So it is due to a pressure gradient. That is what I meant by the intercalation kind of. ... It is a structure where you can ... I am trying to visualize it. It is a channel. ...

Taylor: You cannot put extra water. ...

Beaudoin: No, that is my point. Yes. So that is why I dismissed intercalation. Well, it is an osmotic-related mechanism. When you say osmotic pressure, you are saying there is a pressure difference.

Brown: Pressure difference between the bulk solution in this region here, yes. I was not making any argument. ...

Clifton: Could I just get my two-part ... the second part to my question?

Beaudoin: Yes, OK, I will be brief. On the topochemical growth in experiments we have done with C3A and C3A plus gypsum, compacted systems, low-porosity systems, I believe we could get direct formation of the AFt phases. These experiments were done at slightly elevated temperatures—very little water in the system initially. And I always thought that was a kind of topochemical formation.

Brown: I do not ... I do not accept your explanation. When you say "direct formation," you mean directly on the C3A?

Beaudoin: Yes.

Brown: I do not think so.

Diamond: I do not know exactly how to express my comment. Oh, leave that slide on, please. Sometimes you win the battle and lose the war. I followed your argument, and George Scherer made the same argument, as you know, in Banff, six months ago, and I was pleased to see that you were completely honest and said that at the concentrations we are talking about the closeness, the overwrap of the double layers, are such that you have to have a—what did you say— 20 Å?

Brown: Yes.

Diamond: In your diagram the AFt crystal was five microns long maybe, three microns, four? You are talking about bridging of void; the void may be five microns. If it is an air void, it may be 50 microns or 100 microns. You are talking about gap remaining between the growing ettringite crystal on the opposite pore wall of 20 Å, so I would say you are growing just a little out of proportion. More to the point ... more to the point, a 20 Å distance of separation on a relatively rough surface like a pore wall, and also I suspect it to be a relatively rough surface on a scale of angstroms of a growing ettringite crystal, does not depart, to my mind, very much from contact. You are so close to being in contact that I think it is a kind of a self-restraint to say it doesn't have to contact the other wall and ...

Brown: The problem I have is: conceptually, how do we get the nutrients to that growing surface for growth to continue? The point of this is that … which is in the interface; which is in the solution?

Diamond: All right. But the interface itself is rough on a scale at least as large as the gap. There may very well be channels trough it but ... all right.

Brown: I think someone needs to model this in a way that I am not capable of doing so we can understand the issues associated with ... probably stress.

Bonen: Do you suggest that the potential expansion is about the same whether it is through solution mechanism or topochemical expansion replacement?

Brown: Well, I do not believe it is topochemical reaction. So I think it is things that may have some sort of topochemical overtones to them, but I do not know the exact answer to that.

Sahu: After so many smart and intelligent questions, let me ask, maybe, a more than naive and stupid question. I am not from the university side but from the industry. What is so special about ettringite? I really do not understand, because if you accept the theory of crystal pressure, crystal growth pressure, or pressure on the walls, why is it not ...

cannot be applied to other hydration mechanism, like, let's say, for C-S-H? Why not? Is it just because of the quantity, because ettringite has a much greater volume-expansion or ... what is the reason why we cannot apply these theories also to C-S-H, I mean, for hydration of calcium silicates?

Brown: I am not so sure that we need to apply them to C-S-H, but I think that you are right. When we talk about other crystal ... or hydrates in the cementitious systems ... why does it not apply to this? I do not know the answer.

Berliner: Well, let me expose my ignorance a little bit more and ask you if you would comment on the state of knowledge of thermophysical parameters characteristics of ettringite. I have a couple of concerns. One, as this crystal grows, as you say, it is going to be storing up a lot of elastic energy. I do not know that that is in your crystal growth formula. The second is, if these crystals are really under a lot of stress, is there any evidence for dispersion lattice parameters measurements? I mean, you should see this if you do a careful diffraction experiment.

Brown: You mean a shift in the lattice parameter?

Berliner: Exactly.

Brown: I am not aware of any measurements. Hal, would you comment?

Taylor: I would just like to comment briefly on that, if I may, and also I should get who it was who said that—it was Sid Diamond—why the scale of things is so different between the space and the size of the ettringite crystal.

If you were to make lattice parameter measurements, they would tell you almost exclusively about the biggest ettringite crystals, which may not be the ones that you are interested in. I do not know how much Paul agrees with me on this, but my own feeling is that the ettringite crystals that actually produce the expansion are minute—they are sub-micron in size—and you are not going to pick that up by X-rays, and also it puts a different light on the scale question that Sid was talking about.

Mindess: I am going to terminate the discussion here just so that we can keep on time. Please, remember, we have all tomorrow afternoon to attack each other. So, thank you very much, Paul.

Discussion of Presentation by F.P. Glasser

Young: You mentioned, Fred, the binding capacity of the cations as sulfate, but you did not mention the fact that the sulfate also can undergo binding reactions mainly with the calcium hydroxide to form gypsum. And I think that is the other possibility we should look at it in terms of improving sulfate resistance. And I think one of the reasons why blended cements could be good in this respect is reduction of calcium hydroxide, which can affect the binding capacity of gypsum and the formation of ettringite binding capacity of sulfate, and therefore the formation of ettringite, because you have to have lime to form ettringite.

Glasser: Yes, sure. I think there is another aspect to that. And I agree with what you said, obviously. And if I had longer to spend I would have gone into that; it is mentioned in the paper.

But I think another important aspect about blended cements, if I can just follow that up for a moment—I have the microphone they cannot take it away from me—is that the alumina in blended cements often gets tied up in phases that are not convertible to ettringite at normal sulfate ground water and seawater concentrations. For example, if you use a slag cement, a lot of the alumina gets tied up as a hydrotalkite-like phase. If you use fly ash or slag, some of the alumina gets tied up as a stratlingite and that is not readily convertible to ettringite. So there are a number of reasons why the equations that you need to take into account change when you go from Portland cement to blended cement.

<u>Discussion of Presentation by N. Hearn and J.F. Young</u>

Brown: ... did you say "complete hydration" at w/c at 0.38 ? I thought it was much lower than that, I thought it was 0.26.

Hearn: .38.

Glasser: Well, I was concerned about the question that you asked about where this ratio came from because I did some calculations about special cement types and I thought I would just go look up the number in the book for what the theoretical water requirement was to hydrated Portland cement. And the author of the book, I will not mention his name, but the author of the book is here present, a book on cement an concrete chemistry, and there is a huge range of values given in book but without comments. So where did you 0.38 number come from?

Hearn: I think that is the original Power's calculation.

Glasser: OK. I did want to make an observation about the conclusions that you have come to. The vast bulk of research on cements and concretes in seawater particularly is done in northern countries where seawater temperatures are typically seven, eight degrees—possibly as high as 14 or 15—but if you compare the experience of behavior of concrete in aggressive waters, with papers that have been reported in places like Saudi Arabia and Nigeria and tropical countries, there is a world of difference. The increase in temperature drastically foreshortens time frames for deterioration, and I think if these studies are based on relatively low-water temperature exposures, you do need to add another caution in that they may not be representative of the sorts of reactions and rates of reactions that you might experience in tropical or sub-tropical climates.

Hearn: I think, when you look at the temperature effects, again, we come back to the relative volumes of their open or capillary structure. Under higher temperature conditions we may have coarsening of the pore structure, so we may have increased permeability and we may have increased rates of deterioration. So I think that is probably paramount. And then, secondly, what happens to the pore structure, to the continuous pore structure? And then, how does that pore structure get affected by the aggressive elements? And that is where, I think, the temperature comes in.

Young: If I could just add a little bit. More than that, I think you also have to be concerned about the issues of quality of construction and whether you get higher internal temperatures within the concrete that could lead to some cracking. So the real issue perhaps is not what the ideal permeability might be, from the calculation we learned from the textbooks, but what indeed might be the actual permeability of the field concrete.

Hill: I cannot comment on the effects of water temperature in the Arabian Gulf, but I can comment that in the Portland Cement Association long-time study on the exposure site at Tread Island, Maine, which of course is very cold weather, the Type 2 concrete performed best in seawater. At St. Augustine, Florida, which would certainly be semi-tropical, the cement type that performed best had C3A level was ranging from about 9 up to 10.3. So there, certainly in that point in time you have a different effect. Now, part of the distress that you were seeing in both cases was the cracking due to corrosion reinforcement, but the temperature effect is significant.

Hearn: Based on the observations of the works that I looked at, most of the water-cement ratios that were investigated were 0.5 and above. And if you look at those studies, yes, the cement type is important. If you want to eliminate that issue altogether, then you go to lower water-cement ratios.

Hill: Well, the concretes that they used were four-sack, five and a half-, and seven-sack mixes. Now whether they did that on the piling in Florida, I rather suspect it was rather close to that, and I am not sure that a seven-sack mix would not be rather close to 0.4 or 0.5 water/cement ratio. I cannot remember those numbers.

Bentur: Looking at the issue of temperature, and what we are doing here is looking at a concrete versus a uniform environment, but in practice the problem of temperature is not the temperature itself but it is the wetting and drying. Most of the problems that you get in hot concrete is really where you have a concrete structure intersecting the soil, which is a sulfate, or some other kind of soil, and you are getting the capillary suction in drying, and this cycle is most hazardous. And from our practice, you cannot deal with it by just making good concrete. Even if you make the best concrete, the pumping cycle there is so intensive that you have to look into the structural detail and at this junction do something else. So it is not the average properties of the concrete, the permeability and so on, which are going to be affected by temperature, which is the issue there, but only capillary suction in drying.

Haynes: Most of the studies that are done on permeability were done under pressure; you are pushing water through the material, whereas the capillary action of pulling water through is quite different. I am wondering: Is anybody doing research on looking at the quantities of water that go through concrete due to the capillary action? There is a yes here by Doug Hooton.

Hearn: There is work that is being done, basically on wicking action and diffusion. I know Doug Hooton is doing some; I have been doing some work in that area as well. Many studies have gone away from doing a Darcian type of flow experiments where you have a flow under pressure, and you are actually looking at percolation through the material because that is where you are dealing with aggressive ions, so you are dealing more with Fick's law rather than Darcian flow characteristics.

Discussion of Presentation by S. Diamond and R. Lee

Beaudoin: ... The decalcification of C-S-H, can it be in itself an expansive process?

Diamond: Well let's put it this way. I see no apparent evidence for such a thing. Where we see decalcified C-S-H, we do not generally see extensive cracking, which to me is an indication of what was local expansion. But cracking is more commonly observed where you have heavy ettringite deposition.

Beaudoin: OK. Thank you.

Mehta: ... Very interesting works in the '80s, as ettringite have ... inserted sulfate into C-S-H system as such, are they more common compared to those 80 years where you found deposits of ettringite around the coarse aggregates? What is the relative proportion of? ...

Diamond: Yes, I think I understand what you are trying to get to. Yes, I would say, in my judgement at least. Yes, the areas where you have rim deposition of ettringite, as you see in DEF, are uncommon. Areas where you see ettringite inserted in the paste, in my experience at least, seem to be far more common.

Mehta: And is that, again, a true solution formation of ettringite compared to topochemical?

Diamond: Obviously you cannot tell from a static moment in time, but I would be very surprised if it were anything but a through solution of precipitation.

Hawkins: Sid, you mentioned that you commonly see a bottom zone that is attacked and at the top zone, and the middle section is typically relatively unattacked. You see carbonation at the top and the bottom, and you thought that rather than carbonation from the bottom coming from bicarbonate, it was CO_2 coming from the top because the system was drying out and the water front bouncing down. Would you not expect to see carbonation in the middle section, if that was the case, rather than from the bottom?

Diamond: No, I do not think that CO_2 is coming from the top, I think it is CO_2 coming around to the point of contact of the soil with the bottom of the structure. As things dry out, as you well know, the soil shrinks away from the contact, and I think a certain carbon dioxide diffuses in, and when things get wet it dissolves and you get some carbonation, maybe, because of that.

Taylor: Sid, a very minor point. You mentioned these rims around voids are high in magnesium and alumina. Do you think it could be a hydrotalcite-type of thing?

Diamond: Could be ... from the composition, yes, it could very well.

Taylor: The other thing, carbonation certainly destroys ettringite when the pH gets low enough, but there is an intermediate stage where, if Seligman and Kuzel were right, and I think they were, it actually forms it because carbonate plus monosulfate gives ettringite plus monocarbonate. And I think your first ettringite, it may be only a very thin zone where this happens, but then as carbonation gets more intense you destroy it.

Diamond: Oh, absolutely. I not only think so, I think that the only way you can think about this situation, if I am right, and you have periodic periods of wet exposure to the sulfate attack bearing-solution alternating with dry exposure where some carbonation potential material seeps in, that you do have to have a conglomeration or a sequence in time entering, either being created and then perhaps being destroyed by the carbonation. And they have been back and forth on various occasions, yes. Yes?

Lagerblad: A comment on these structures, because I have seen that quite often in concrete that has reacted to water for a long time, and I have sometimes caught a single ... rate of ... when you form a concrete you have a better rate of hydration and the system is not really in thermodynamic equilibrium, all the crystal are too fine grained. You entrap a lot of calcium hydroxide into the calcium silicate hydrates.

When you see some of these phenomena, it is really that you try to lay out energy of the system by having crystals grow coarser, getting calcium out of it, of the calcium silicate hydrate, and you get coarser crystals growing and bubbles and things like that. So it is a reorganization of the material to lower its energy by forming bigger crystals because that is what nature wants to do.

And some of these phenomena, I think, are associated with that one. But also, if you have sulfate you will loosen the system and it will reorganize and get a coarser structure. I think that is very much what we are seeing really in these textures.

Diamond: Well, clearly it's a reorganization process, and if we believe in thermodynamics, it is only an equilibrium situation. As the potentially equilibrium situation changes over time, and it does, I do not, in the pictures that I have shown and in the concrete that I have examined, I do no see much evidence of the coarsening of the C-H-S. There are no crystal ...

Lagerblad: But you get a coarsening of ettringite, you get a coarsening of the structure.

Diamond: Ah, clearly, that is correct.

Lagerblad: And you actually get a lower calcium silicate ratio of the calcium silicate hydrate.

Diamond: We do get coarse ettringite being deposited—at least in the affected zones where there is no calcium hydroxide, it is gone—and we see what seems to be locally massive alterations of the C-H-S chemically. But I do not see anything that I could call a coarsening of the C-S-H structure. Maybe I am not looking at it properly.

Lagerblad: It is difficult to say, but we have looked at very old concrete that has been in water for a very long time, and you will get a lowering of the calcium silicate hydrate and get more calcium hydrates for some reason. And I think it is part of the process when it tries to lower its energy by coarsening itself. I do not think it is thermodynamically stable to have the calcium silicate hydrate like that. It will try to reorganize itself into a more stable structure, slowly but surely.

Diamond: May I ask a question? You have worked with a fair bit of old concrete; have you ever seen altered residual cement grains such as I have shown?

Lagerblad: Yes, I have. There is another interesting thing there because if you leach a concrete that has not been leached for 100 years and look at it, you can see this reorganization quite nicely. Actually we expected all the portlandite to go out during this leaching process. But it doesn't do it, the calcium silicate hydrate reorganizes itself, it lowers the contents but the portlandite is still remaining there. It is something we will never find in a leaching test but we will find it in very, very old concrete.

So you get different things. Everything has to do with exactly what we can see in geology. It will try to force itself, it will try to reorganize itself into a more stable system. It is really: when you form a cement base, it goes very fast. Just in a day you will form quite a lot of it, it is a changing composition with a clinker or a hydrate in different sequences, you have an over-saturated solution where you form all of this. You actually form ... imbalance in the system from the beginning, and it will reorganize itself with time.

Bonen: Sid, I wonder if you can rate the degree of alteration according to the microstructure. Usually, from my experience at least, when you see larger ettringite and still can identify calcium hydroxide, this denotes a slight alteration. As it progresses, you do not see the calcium hydroxide any more, and then you start to see a smeared structure, whether it is due to carbonation or to complete decalcification. Did you get the same notion?

Diamond: Well, not clearly at least. There is some sort of zonation obviously inherent in the process but this is, as we discussed earlier, this is in concrete not cement paste, and the physical presence of the sand grains and the aggregate tends to break up the parallelism of the zonation. And if you have a certain amount of excessive transport around sand grains, as you very likely do, that is a further complication.

So at least I have not been able to classify these things into clear degrees of reorganization, let's say. It may be possible to do so, but I have not ... I do not think my colleague Dr. Lee has either, to this point.

Discussion of Presentation by M. Moranville

Taylor: I have no doubt that the composition of U-phase found originally, I think by Dosh, which you gave, is at least approximately correct, but I have never been convinced by this admission of aluminium, which seems to me crystal-chemically a bit unlikely. How well established is the exact composition? Is there a possibility that, in fact, what you have is the full amount of aluminium and you have more anions into layers as well as the calcium? Is there a possibility of checking the structure further by X-ray, structure determination or spectroscopic method or whatever?

Moranville: I agree, Hal. I was also of the same expectation about this crystal structure, so I think some young scientist should do the true structure. I agree with you.

Taylor: Has anybody crystals big enough to do X-ray with?

Moranville: No, I do not think so. I do not remember. I think it was... I do not remember where I found this crystal structure, but I was also a little astonished by this. But I reported it; I did not know myself.

Mehta: Micheline, two questions, very short ones. First of all, I am very much intrigued that these XRD spacings of your U-phase are very similar to what I called phase "M" in my doctoral dissertation. Not metaphase, but because of the similarity to monosulfate hydrate. Because it happened to be a monosulfate hydrate with 18 molecules of water. So I wonder whether there is a... whether there is any coincidence or whether it is actually sodium substituted?

Moranville: Jan told me, "Please, no DEF at this seminar!" So I said, maybe I will show this diagram if somebody asks the proper question. Here, on this graph, you have a normal Portland cement with the monosulfate, with 12 water molecules. And when you add silica fume you will change the alkalinity of the pore solution, and you will get a mixture of ettringite and monosulfate. If you enrich the pore solution with sodium hydroxide, you can see here the U-phase appearing because you are richer in sodium ions. And if you use, with the same alkaline concentration, silica fume, again you will decrease the sodium and you will have only ettringite. And then for 1.8, you still have the U-phase, but the silica fume, I think it was at 10%, was not enough to void the U-phase formation. So I agree, you have different kinds of monosulfates, but in this case I would believe that it was the U-phase.

Mehta: And my second very short question is: AFm or monosulfate is not considered expansive, what makes you think or state that this is expansive?

Moranville: Because we think that we have an intake of water into the structure. Because we did the experiment with no solution and with water after we stored the samples in water, we found expansion, substantial expansion.

Glasser: I just wanted to say that we have had a U-phase preparation which, I think from memory, is in 0.5 molar sodium hydroxide, and we have had it at 20 degrees for, I think, six years. Unfortunately when we last looked at it, although it was a reasonably pure phase, and probably good enough for analysis, there were no single crystals ... not big enough for X-ray diffraction.

Discussion of Presentation by Jan Olek

Allow me to briefly review some important facts regarding microscopic examinations. While observing microscopic features and noting that they are there, it is very important that we also have the ability to interpret what we see correctly. In other words, we need to understand the origin of the features that we see. We should know: why did they form, how did they form, when did they form, and how that formation leads to the destruction or deterioration of this micro-section.

This is not as simple a task as one can imagine, and all of you are surely aware of that. In order to answer some of those questions that I just posed, we often supplement the microscopy with some other techniques. In other words, we rarely use the microscopy alone. Typically, the best results of any sort of determination of this kind are achieved when you combine the macroscopic, visual observations of the structure that is being affected with more detailed microscopic observations of the core taken from the structure. If available, collect as much information as you can about the mix design and type of cement and water/cement ratio and so forth; also collect the information about the environment.

Obviously, if you do not have sulfate present in any form in the vicinity of the structure, it is hard to argue that you have external sulfate attack. So if you have some characteristic features that point to some sort of a sulfate damage, then the damage was probably caused by a form of internal sulfate.

There are various techniques used for microstructural or microscopic observation of concrete and they can be simply divided into three categories: optical microscopy, scanning electron microscopy, micro X-ray analysis using microprobes.

I will concentrate today only on the optical microscopy and scanning electron microscopy. However I will mention a couple of points about the other techniques as well. I also want to point out that in order to supplement the information that we obtain by above techniques, we also use chemical analysis, X-ray diffraction analysis, chemical imaging or chemical mapping, and using energy dispersive X-ray analysis; and, of course, we could use thermal and thermogravimetric analyses and scores of other techniques.

As far as the optical microscopy goes, we can do it either in the transmitted light or in the reflective light. Transmitted light being the most common of the two, thus I will spend more time talking about that. And, as I have already mentioned, I will talk a little bit about scanning electron microscopy and very briefly about micro X-ray analysis.

As far as the optical microscopy goes, the standard that one can use to evaluate concrete using the optical microscopy is ASTM C-856, which is the standard covering the petrographic examination of concrete. Some of the proponents of optical techniques argue that there are somewhat better than the scanning electro microscopy techniques because, among other things, they allow for observation of a larger field of view, and the other point is that samples are not exposed to high-vacuum environment that the sample will be exposed to when in the stage of the SEM, therefore we do not have as much of the other facts as we would with some of the other techniques.

Discussion of Presentation by J. Olek

Mehta: I think Dr. Skalny did a very good job in asking you to fill in for Niels Thaulow, because you have really done very well in a short time. You gave us an excellent review. Thank you.

The question I have is: most of your studies from the Minnesota highways, microscopy, both optical and SEM, show that ettringite either was in the pores, see the last slide, or some was in partially filled gaps around aggregate. Have you also found ettringite dispersed within C-S-H?

Olek: Not in this particular case. Actually, the problems related to this particular highway are

only partially related to that excessive sulfate; there is a freezing and thawing problem and primarily ASR in those pavements. So I use it mostly as the illustration of the capabilities of a technique rather than a diagnostic tool for this particular deterioration problem.
Constantiner: Just a general comment ; I do not know if you are the right person to answer this, but we know that especially in polished sections of steel or other very dense materials, the information that we obtain from X-rays is commonly from the few microns below the surface. Do you know or does anybody here have any idea from how deep does the information come from when we are looking at polished sections?

Olek: Varied. Actually, the information that we get is from a very shallow layer. I would say not more than three microns. Would you agree with that? So it is not very deep.

Open Discussion of All Monday Presentations
Hill: This morning there was discussion about the testing error in the C-265 test, which is the amount of soluble remaining after 24 hours, and I would like to offer a couple of comments on that. Our practice at the IDA Laboratory was to determine an optimum SO_3 for a cement, regardless of what the specification said and then add about half a percent extra SO_3 over that optimum and then run C-265. The feeling was that we could always encounter that much manufacturing variation, we wouldn't be proud of it but it could happen, and we wanted to make sure we were not over-sulfated. Never, when we did it, did we ever come close to the maximum on C-265. Now, C-265 is a more severe limit than the mortar bar expansion test. It sets you at a lower level, but we didn't have any problem with it.

Now, our lab participated in developing that test and the temperature limits on the test were too severe, they were ridiculous. But I am convinced that a great deal of the reported testing error is due to the fact that laboratories do not pay any attention to the procedure.

I realize that I am preaching to the choir but I do not want to kill the procedure simply because nobody follows it. I think it is a far more useful thing, and I think we ought to be thinking more about it.

Hawkins: I agree with you, Gene. We recently have been—with respect to the precision of the test or the stated position of the test—we recently have been doing a fair bit of work with C-265 adding various increments of SO_3 to the cement and doing C-265 not just 24 hours, but curing it for also three and seven days and doing the same test. And once you get —if you plot out $S0_3$ of cement against concentrations you have extracted—very shortly after it is upturned, you get a straight line. And our standard deviation, we found, we were using one person to do each step of the way, each time was of the order of plus or minus 0.01 gram per liter as opposed to, I think Doug showed us plus or minus 0.39, which is very much different.

It certainly shows that you can get good precision with it. But the other point is that the method calls for, after you have cured the mortar —from grinding it into the mortar and pestle —we found that as you went from one to three to seven days, the slope of the line was increasingly more shallow as you went to the higher ages, which suggests that the particle size distribution is changed—of that ground mortar is changing—and probably getting a little coarser as the pace gets harder and the age gets longer.

I think particle size distribution is one of the problems when you compare various laboratories' data, and even within the laboratory. I think, choosing a narrow size distribution from that overall powdered mixture may well improve the test or the precision.

I believe there are a few things being undertaken in that method. We have also just very recently, and I think it has just been brought in now, we have broadened the temperature range over which you need to. ... the temperature range that you had to adhere to was ridiculous, and I think we have broadened that to be reasonable. By calculation, it makes very little difference. So I think you can at least expect you are going to see some improvements in that method, and I think it is a very good method.

Hooton: I guess the point is, and Peter mentioned the fact, with one technician in one laboratory doing it over and over again and getting very good precision ... and as long as you are not near the limit.

Hawkins: Right.

Hooton: What was your precision like when you get close to the limit? I think that is where the problem is, it has spread.

Hawkins: We went to, for our solutions with our alkaline concentrations up to, I think, a little over two grams per liter and the scanner was the same all the way through.

Hooton: But the big problem in that disputed case is you have two laboratories doing it with two different people.

Hawkins: Right. But if we can narrow down ...

Hooton: Yes, if you can do something to improve the precision, then I think it would improve it drastically.

Constantiner: This is a general comment regarding the use of admixtures. I do not know if the information that we have and we looked at and has been presented and reviewed today is relatively old, but there have been significant changes. One of them is the use of admixtures that may significantly change how cement reacts, disperses.

The use of admixtures is not only a water-to-cement ratio issue, but it disperses cement particles, it has significant effect on how fast cement hydrates, etc. And I am just wondering if anybody here has an idea of whether this really plays a role on the way we make concrete?

Glasser: A comment regarding U-phase. I am not quite sure how to tie up the argument about U-phase with blended cements. I think we started out with one thing and went to another, but I think, if I pick up the threads right and try and find a common theme in this discussion, it is the alkali content of a cement, which of course has the interplay with the stability of the other phases, including ettringite.

Now, one of the problems with blending materials is that they typically bring more alkali into the system, on a chemical basis, then you get from the cement. It is not uncommon to find cements with 0.7 to one weight percent Na_2O equivalent, but fly ashes with perhaps several weight percent alkali, and even blast furnace slags where steel makers know that alkalis in blast furnace slags harm their refractories you find those with quite a bit of alkali in them too.

Now, we have been interested in this question of where alkalis go in cement and, of course, we are not the only ones to be interested in that. If you go right back to the earliest pore fluid expression experiments that were done by Longway and his colleagues they, of course, really, I think were the first to show conclusively that alkalis concentrated in the pore fluid.

And quite a lot of people in this room, e.g., Sid Diamond has published, we have published on pore fluid expression values from cements. And I think that Hal Taylor—and I am trying to keep a rough chronology to this—Hal Taylor did some work on collecting pore fluid analysis in trying to determine a partition coefficient —"B" value I think you called it, Hal, wasn't it?—to effectively a partition coefficient between cement paste substance and alkalis.

Now, we haven't submitted this paper yet, although I have written it in manuscript form. We looked specifically at the alkali sorption of C-S-H. Hal had to work with a cement paste, which is a mixture of substances, we just used synthetic C-S-H and aluminous C-S-H. But we have made a few quick recognizances - if that is the right plural. There are so many French speakers here, I have to be careful of what I say when I use French words taken into English!

We have had a look at, I mean, U-phase study is a part of a longer term project to look at the stability of AFm phases, but we have had a very quick look at some of the other AFm phases. I think that when we come to looking in detail at the alkalis in cement we are going to find quite a few surprises, which AFm phases will pick up alkalis and whether they are selective for sodium and potassium.

So I think the U-phase issue is most interesting. A little window has opened up and we can look and see how AFm phases are responding to increased levels of alkali in cement. And we must remember that if AFm is forming in preference to ettringite, or indeed other AFm-type phases —I mean stratlingite is basically an AFm-type phase—these all affect the distribution of alumina in the system. And alumina that is tied up in some phase, which under those particular conditions is more stable than ettringite, is alumina that is not able to form ettringite.

So it is a complicated picture but we are getting insights into it. And this is what I referred to this morning in my summary when I said that if we are going to have good engineering in cements and concretes we need to back it up with good science as well. We need to know more about these phases and how they interact with each other and with the minor constituents in cement paste. I hope this is going some way into answering your question, I know it is a bit roundabout.

Mehta: With your permission, I would like to return to another paper, Paul Brown's and Hal Taylor's, which I believe was very, very stimulating as far as an overview of sulfate attack-related mechanisms.

I think Paul very convincingly demonstrated that some of the old myths we believed in, like solid-state formation of ettringite, are probably out. Most of us now believe—he demonstrated the on the basis of his thermodynamics, but even scanning electron microscoping observations tell us so—that ettringite is forming through-solution rather than topochemically or solid state-formation.

And the question of crystal growth pressure upon pore walls was a very, very interesting observation, because many others in the past also have commented very often that crystal growth pressure is not really the key in enlarged expansion in many phenomena in nature. It has to be something related to phase changes, just like water changes to ice, or the alpha-beta quartz phase change, similarly thenardite and menablite in sulfate attack. And the observation that they have made—that it is not the large crystals of ettringite but the small ones, and they happen to be in areas where calcium hydroxide is or was—is quite important because this raises the issue of how we can explain the mechanism of expansion associated with very small crystals.

Now, Thorvaldson's theory was based on osmotic pressure or double layer associated with adsorption of water, or imbibing of water. So, basically, swelling of ettringite by water adsorption. And there are others thinking that maybe it is very similar to some of the expansive clays, where high surface area and surface forces are involved in attracting water molecules and causing a double layer-type of repulsion. This theory was more or less confirmed by field observations from structures where we found two very important things. One is that always somehow the loss of cohesion and strength was observed before you could see any expansion and cracking.

Even in laboratory testing it takes several weeks. We may use very small specimens and very, very porous specimens of paste, but it still takes three weeks, four weeks before you can see any expansion. And this is because there has to be some loss of cohesion, dropping of elastic modulus before the expansive force generated by water absorption can manifest itself.

So up to this point, I think, more or less, there seems to be general agreement with what Hal Taylor and Paul Brown are saying, and what I have also commented in some of my observations. The more important point from here onwards to resolve seems to be how do these small, very small crystals of ettringite cause expansion by water absorption?

And my initial thinking was based on Lee's book, in which he mentions that C-S-H, although it has high surface area, very low crystalline state, like a gel, but due to some chemical bonding between the layers it has a kind of limited swelling capability.

I first thought that maybe this is another case with ettringite, although the surface area is not as low as the C-S-H, but maybe there is no chemical bonding between ettringite layers and, therefore, it is able to absorb a lot of water and expand just like in case of clays. And it seems to me that Paul Brown or Hal Taylor's thinking is that there may be a water-starved starting ettringite structure. Rather than a 32-mole ettringite maybe it is a 15- to16-mole ettringite when it forms in spaces where there is lack of water, and later on it will pick up water and expand.

Taylor: I really do not buy the idea that ettringite crystals will be attacked by water molecules. Kumar quoted something from Lee's book about C-S-H having chemical bonding between layers. Well, I do not know what this was based on. Frankly, I think that C-S-H is a far looser structure than an assemblage of ettringite crystals and I cannot see why ettringite crystals or assemblages of ettringite crystals should absorb water.

I do think it is just possible that when ettringite is formed it forms in a locally water deficient environment so you do get a dehydrated ettringite or a partially dehydrated ettringite and expansion could occur when that gets some water. I think it is also possible that ettringite forms as ettringite and it draws its water from surrounding C-S-H and expansion occurs when that C-S-H gets its water back.

But both those theories involve an assumption that you have a water deficient situation locally —I think it has to be strongly queried whether that is a realistic suggestion. I do not have strong views as to whether expansion is as a result of water being absorbed in one or the other of those two ways or whether it is crystal growth pressure on the whole encline to crystal growth pressure, I do not think we know.

Lagerblad: I would think about the consequences of the U-phase, because the U-phase, you put a lot of sodium sulfate into the concrete mix. But in the normal case, if you just put alkali, the more alkali you have in the pore solution you actually decrease the stability of ettringite. So that means it is a very specific situation to get this U-phase. And I wonder, when do we get it in a real situation?

It has to be concrete that is very rich in alkalis, you have an ingress of sulfate of some kind but it has to be a very large increase of sulfates before you could actually get a U-phase in a normal circumstance.

Moranville: We also found this U-phase in steam-cured concretes because we had a temperature of 80°, so we release the alkalis, then you have the U-phase. ...

Mehta: You went out of the stability field of ettringite?

Moranville: Yes. And we found this U-phase coexisting with the CH, calcium hydroxide, and then after a long time we have again a release of the alkali, at low alkali concentration, and we got it as a delayed ettringite formation. So we can have the U-phase in normal concrete. We also found U-phase in concrete with fly ashes, because we have a high amount of alkalis in these fly ashes.

Clark: Actually, I hoped Paul was going to say the same thing. In our research we have looked at tricalcium aluminate and gypsum, where we have looked at one to one molar concentrations of the gypsum in the tricalcium aluminate and we are finding U-phase formations at 0.25 molar sodium hydroxide solutions, and there are some indications even down to 0.1 molar. And so to say that it's high concentrations of sodium, I do not think that is quite right.

Brown: We seem to see cracking in concrete exposed to sulfate associated with AFm phases. And I am wondering if we are forming U-phase, if it is possible to form a U-phase in concretes that have been exposed to external sources of sulfate, and if that phase is not expansive to the point where it can actually disrupt concrete? So this is more of a question than a comment.

Moranville: We found this U-phase in a dam.

Brown: Exposed to sulfate?

Moranville: We have, just on the face exposed to the water. But we used a super-sulfated cement. So we have sulfate ... internal sulfate, but we found the U-phase.

Brown: With expansion?

Moranville: Yes, with expansion, yes. Cracking.

Discussion of Presentations on Tuesday, October 6, 1998

Discussion of Presentation by J. Marchand

Diamond: Well, my first comment as usual when I listen to Jacques is "Wow!" But I have a couple of specific questions relating to details, and sometimes details are important, and sometimes they're not so important.

One detail that I think is important is the water distribution set up through the slab under a given set of boundary conditions. And you calculated this and you "normalized"—I think is the proper expression for the calculation—with some respect to some actual data as a function to water/cement ratio in the concrete, which makes perfect sense, of course.

But the question is: Were you assuming for that concrete, water-cement ratio of 0.45, water-cement ratio 0.65, that that concrete had itself been properly cured to a reasonable hydration condition before you attempted to formulate the water distribution? To the extent that you did and that the real concrete that you're trying to measure was not cured, the transport coefficients would be even greater, I assume.

Marchand: Yes. And while all the simulations were, I mean all the results presented this morning are for a piece of concrete which is assumed to be fully hydrated. So, of course, if we—and I mean for those of you who were at the Banff Conference—I guess Mike showed a few slides on the effect of hydration on the transport properties of various concrete mixtures, and that is certainly a significant parameter that should be accounted for in a model.

And you can easily account for that fact in the model but you need an equation, I mean, you need a trend. And, I mean, in most cases it is quite difficult to evaluate in a reliable fashion, in a reliable way, the evolution of the transport coefficients for early-age concrete mixtures. I mean, young concrete mixtures. After a certain while you tend to have ... the system will reach a certain level after which the evolution of transport properties will vary but very slowly with hydration.

Diamond: But in ...

Marchand: But we have very little information on that aspect of the problem.

Diamond: But certainly the direction of the change is clear?

Marchand: Yes. Oh, yes.

Diamond: The other is a more technical question of detail. You mentioned that your model accounts for the behavior, in essence, the precipitation-dissolution of specific phases such as ettringite and gypsum and of course, calcium hydroxide in the normal components in cement. But when we examine real concrete subject to sulfate attack, we see monosulfate.

Marchand: Yes.

Diamond: ... we see brucite, we see all sorts of other things. Is the absence of specific mention about these things an indication that the model does not consider their stability or is it something that is implicit in the model, whether you have talked about it or not?

Marchand: In all the slides I presented this morning I was limiting myself to a simple case of external sodium sulfate attack.

Of course you can modify the model to account for the presence of magnesium and the formation of brucite as an external. I mean, this is an additional ion that can be added to the model. As regards monosulfate, it is a nonstable phase within the system, so it is mainly the effect of monosulfate on the amount of ions, sulfate ions, which are bound to the solid is accounted for in the interaction isotherm. So we do not really distinguish between; in fact, we do not really make that clear distinction between ettringite and monosulfate. We are mainly accounting for the presence of these two phases in the interaction isotherm.

Diamond: All right. In trying to express this in words of one syllable that I can understand, either you figure sulfate is bound as ettringite or you figure the sulfate in solution ... because these are the stable conditions.

Marchand: Yes.

Diamond: And the fact that metastable sulfate happens to be bound in a monosulfate does not enter into your equation?

Marchand: Yes, that is right.

Bonen: I would like to comment on the degradation front, and maybe I will continue from where Sid stopped. If you start ... once you are going from the ettringite phase of precipitation to the gypsum stage of precipitation then brucite has to be precipitated as well, otherwise you cannot count all the ions. Now, once the brucite starts to precipitate it forms kind of impermeable barrier ...

Marchand: Possibly.

Bonen: ... to an extent until there is spalling and so on. Now, if you're talking about paste, then I assume that it is very ... it is quite well to imagine a kind of uniform front of degradation. However, in field concrete usually what is seen under careful examination is patches. One patch where it is terribly affected and complete alteration where, next to it—and this on a microscopic scale, and basically Sid also showed it the other day—you see that it is not so bad.

Now, I know that it does not make life simple, but how do you say in it

Quebec—"C'est la vie!"—so instead of going with kind of uniform fronts, maybe it is better to complicate it further more, or to continue to complicate it and slice it into sections. I can give you another example: the rate of degradation in saturated brines, high brines. According to the decalcification of the paste, the rate was about, I think, two millimeters per year, which is much more—in places, no longer a uniform front —which makes it much more, a greater rate of decomposition simply because you're not going along a front.

Marchand: Well, two comments. First, on brucite formation: the effect of brucite formation on the transport properties of your concrete can be accounted for in the damage equation because while you are forming brucite you are in fact affecting the capillary porosity of the system and that you can account for in the damage equation. So, when we are dealing with magnesium ions we are in fact accounting for that effect in the model by using the damage equation, the last equation I showed you.

Second comment, I want to emphasize just the fact that we are playing at the material scale, we are averaging the material properties over a certain volume. So, of course, if you were to look at the concrete at the very specific location and then looked at the concrete at another location and see microstructural differences, that is not accounted for by the model because we are working at a different scale.

So what you really need to do, if you want to validate the model on the basis of actual observations, is to average what you see at that scale and come up with an average picture, which is, of course, not such a reliable picture of what is really happening at the microscopic scale. You cannot play with ... what you receive from one site and give from another site. And if you want to play with the material at the microscopic scale, you can do that but you need to change a little bit the focus of your model. I appreciate that's "c'est la vie."

Lagerblad: Now, I would think about this at a microstructural rate … we increase the porosity. But it does it only if you have that chemical system, because you would take some components out of the system. But then, have a look on the volume of the product being formed into that system. If you have an ettringite being formed it will actually expand, so that means that it will decrease the porosity, and that will change all these concepts.

Marchand: The simple answer to your question, or to your comment, is yes. But then you play within the damage equation. So if you really know how the dissolution or the precipitation of certain phases is affecting locally the porosity of your system, if you have that knowledge, then you can input that information into the damage equation.

I appreciate that it is probable that ettringite formation is not affecting the transport properties of your system in the same fashion as portlandite dissolution or C-S-H decalcification, but that you can play with in the damage equation.

Lagerblad: Why I made that comment is because you use the same equation when it is leaching; you have about the same way of modeling it. You use very often the vertical shrinking called model, they use for having speed or ...

Marchand: Yes.

Lagerblad: And, therefore, as I said yesterday, we were looking at very old concretes that had been possibly leached for a long time. And those are cases that give a much too high dissolution depth, diffusion control than what you get in reality. That presumably is due to the ... type of alteration that actually compensates for the loss of material by volume changes.

Marchand: Yes. That is the most difficult to account for; yes, I agree.

Discussion of Presentation by W. Ju and S. Mindess

Beaupre: Could you put back the slide—the distribution throughout the cylinder that was cast in the laboratory? Was there a mold when you cast at the bottom of that cylinder when you casted?

Ju: There was a mold, yes.

Beaupre: OK. Could you just put it back, please? How do you explain the difference between the bottom and the top? How do you explain the reduction?

Ju: I believe that is due to some degree of aggregate … (inaudible) …in fact.

Beaupre: Or bleeding?

Ju: Yes.

Beaupre: OK. Have you tested similar … have you done a similar test on a cylinder that was cast on a granular base or on sand? Because then you will have water absorbtion through the bottom and you might find that the distribution is quite different.

And you may find that the maximum is not at the bottom or that there might be slightly less difference from the bottom to the top. I agree with the result you show on the core from the slab, but to take that as what you should expect from one that was cast on a granular basis and make the comparison to show that there was more than—I do not know the number—but 80% or 20% reduction, I am not sure you can make that switch so easily because on a different test program we saw a big difference from the quality of the surface if you cast it in a mold than if you cast it on a granular base. So I just wanted to notice that point.

Ju: I should tell you that when we measure a core, I show that the ... a diamond-blade cutting machine, because we have to do direct tension test to assess the tensile strength. So all we do is, we actually cut off the bottom of the core, so we do not even measure the true bottom part of the core. So let us say the bottom one-centimeter layer is gone for a field core, so I think what you mentioned would not play a role.

Beaupre: Are not most of your cores cut horizontally?

Ju: Well, for slabs they are vertical cores. For foundation they could be at an angle, let us say a 30° angle from the horizontal line. In some cases probably it is more horizontal but in some cases it may be more vertical. So we have examined all types of angles.

Discussion of Presentation by M. Pigeon

Diamond: People perhaps wonder why the calcium hydroxide has what seems to be disproportionate mechanical effect. And I think the answer to that question is implicit in the fact that when you look at ordinary concrete, you see calcium hydroxide primarily associated with the sand grains and the coarse aggregate grains. They are sitting there, precipitated on the surface, as of course other deposits of calcium hydroxide would in the bulk of a hydrated cement paste.

But an awful lot of it is really in the form of a bonding agent between the surface of these aggregates and the bulk cement paste. If you remove that bonding agent and you do not put another one in its place, you have lost contact and that makes it certainly a satisfying explanation for the importance of maintaining that contact for the mechanical properties of the system.

Pigeon: I would just like to say that I partly agree with that remark because, yes, it is true that you have often lots of portlandite at the bond between the aggregate particles and the cement paste. But for having played around with the electron microscope for a little while, very often it is not that much. It depends on the density of the paste matrix and if you are using pozzolans and so on, it depends on various aspects, and I think it is part of the answer.

But I would think that the portlandite that you would lose, for instance, that is microcrystalized in other parts of the paste would also have, I think, some importance.

Taylor: I am delighted that both you and earlier Jacques stressed the importance of decalcification of C-S-H. It was generally accepted until fairly recently that this only happened with magnesium sulfate but, in fact, we found some time ago that even with a low water-cement ratio paste in sodium sulfate, it is detectable although, of course, it is nothing as massive as with magnesium sulfate. And I absolutely agree with what you have both said on this.

The other thing about C-S-H decalcification, I think it will contribute to an increase in porosity. It is tempting to say: Oh, it is just losing silica, the body of the stuff is still there, it is going to be weak maybe, but it is not going to make real holes in the material. But, in fact, all that we think we know about the structure of C-H-S suggests that it will not happen that way, that the volume fraction of lime in it will stay approximately constant, and there will be a reorganization so that the silica is more concentrated in it, and that is going to contribute to an increase in porosity.

It is true, you are forming ettringite or gypsum, which is having the opposite effect, but are not necessarily formed in the places where the C-S-H—or for that matter the calcium hydroxide—was lost.

Pigeon: Yes. What I could add to this is that if you look here in this diagram, they had lost all the calcium. It was fully degraded all over, and you had some strength left just, you know, with a silica gel in fact, or the silica gel, and it amounts to about 20%. And Young's modulus, from their calculations, is about of a fully degraded C-S-H. Young's modulus is about something like 16% of the original.

Discussion of Presentation by P.K. Mehta

Taylor: I totally agree with your comments! The reason, incidentally, why I did not say why you get ettringite below water zones—well, you do get ettringite, but you do not get expansion—was that I did not know, and I have never worked on this myself; all the work I quoted on seawater attack was from other people.

And I think that your explanation that the chloride is replacing the hydroxide is a very reasonable one. And, of course, as you yourself have shown, there are many other cases showing where ettringite is formed in ...you do not get expansion. Now, we may have to disagree on the mechanism of that, whether it is what you said or whether it is a matter of where the ettringite is formed. I do not believe it is worth arguing about this because we are going to settle this not by argument but by more experimental work. And it is my belief that microstructural work with the SEM and probably also with the TEM on ultra thin sections is what is needed.

Thaumasite, you also quoted, I think and can I just say that I agree 100% with everything you said on that. I too have been very worried about this idea of putting in ground limestone for just the reasons that you said and we know that causes thaumasite formation, and thaumasite formation destroys C-S-H.

But at the same time I have to recognize that in France they have been doing this for quite a long time, and they seem to have gotten away with it. I do think there is something here that needs more work in order to see whether your comment that there has not been any trouble yet, you know, expresses a justifiable fear or not. I do not believe we know.

Discussion of Presentation by M. Thomas

Hooton: Mike, did you see thaumasite in fly ash concrete along the aggregate interface?

Thomas: I didn't see any thaumasite at all in the fly ash concrete, no. And I think there is a reason for that—but in this audience I am not going to say what I think about that until I have confirmed it!

Hearn: Mike, in your opinion, what is more significant with additional fly ash: the change in the chemistry or the change in the pore structure?

Thomas: I think both are important. And a comment that Michel Pigeon made that if you do not see calcium hydroxide in concrete, you have a problem. I like to see concrete with calcium hydroxide or at least with as much of it as possible, it has been reactive. Of course it wants to be consumed by pozzolanic reaction and not leached out, but I think both are very important. I think one of the main things, obviously, is the reduced diffusivity of the fly ash concrete. I think that is critical.

And some of the research that we have done on chloride diffusion shows that even at equal water-cement ratios of the fly ash and Portland cement concrete, the diffusivity are fairly similar in the early ages. So you get order of magnitude decreases in diffusion in the fly ash concrete with age. And after about 30 years it may be as much as a 100 times less diffused than it was at 28 days, and I think that is very important.

Hearn: So it is related to the water-cement ratio?

Thomas: Yes, it is. But it is not just the water-cement ratio; different materials have very, very different durabilities at the same water-cement ratio.

Discussion of Presentation by Lagerblad

Haynes: What about the concrete expanding … right underneath your points? At the very end … your measuring points.

Lagerblad: But it will, because you have the outside expansion here, you will stress this one and you will get a stress...(inaudible)... of the concrete ...(inaudible)....

Haynes: Well, I agree that it will. Just the measuring gauges themselves are sitting on concrete; how deep do they go in the concrete, the measuring gauges?

Lagerblad: ... they were put there, they are sitting there; they are not glued on.

Open Discussion of All Monday Presentations

Skalny: Welcome to the afternoon session! Before we start discussing the presented papers, I asked Dr. Gunnar Idorn to present in about 15 minutes his views on durability, particularly on sulfate attack. As you know, Gunnar is an admirer of past American cement and concrete research but, at the same time, he is very critical of the present American research, and I think he is basically right.

After the Idorn presentation, we will have an open discussion session with a coffee break in the middle for about an hour and 45 minutes. For the conclusion I have invited two gentlemen from the audience to say a few words about whether we had accomplished anything. Hal Taylor will represent the scientific side, Jim Pierce will represent the engineering community. So, without further delays, I would like Dr. Idorn to take the podium.

Discussion of Presentation by G.M. Idorn

Haynes: I would like to take the opportunity of saying a few words following you, Dr. Idorn, because your idea of doing field work; I really agree with that. I would like to make the point that, sitting here for the two days, there is definitely a momentum that is going on. We are here to talk about the problem of sulfates, and everybody is doing that, and it looks like it is a very large problem. And I at least, would like to be a voice saying I observed there is quite a difference between what comes out of the laboratory, and what I read in research reports, and from what I see in the field. I really do not see much damage in the field.

I have talked with just a few other guests here, and they do not report much sulfate attack in the field either. One of the things we do see in the field is "salt attack" as opposed to the "sulfate attack." That is one of the reasons I like having a distinction between the two instead of trying to blend the two. And I ask myself the question: Why is there this seemingly big difference between what we see in the laboratory and what we see in the field? I would like to think it goes to the holistic view that there are things going on in a field that are different than the reductionist view of studying a problem in such detail in the laboratory. And I will mention a few of those differences.

In the lab, very high sulfate contents are used. It is common to use 5% sulfate solution. Well, that's 50,000 parts per million of sodium sulfate. That is about 30,000 parts per million of the sulfate ions. In the field, when we get a soil sample analyzed, we are more looking at 2,000 to 3,000 parts per million. Once in a while this number might jump up higher than that but, you know, that is in the bad areas.

Now, I am from California, and most of my attention has been related to residential foundations, which is really a problem that has led to this symposium. There is no doubt that there are other examples in industry where, you know, sulfate attack can be significant because of the industrial processes. But related to residential foundations we are dealing with what are the sulfates in the soil.

Another item that is different in the field is that there are other ions in the soil than just sodium sulfate or magnesium sulfate; there are chlorides. And in seawater we understand that there is a mitigating effect when chloride ions are present. Well, why does not that happen when our concrete is in the soil, and there are also chloride ions present?

Another effect is that in a laboratory we are dealing with small specimens. Some specimens are an inch-by-inch cubes, others are two-inch diameter, four-inches long, and this phenomena of surface attack is a surface effect, and that surface effect has quite an a substantial influence on the size of the specimen. But if I get out in the field, and I am dealing with a concrete foundation, and there is sulfate attack, you know, an eighth of an inch on the surface of a foundation, one quarter of an inch does not have anywhere near the significance of what you see on small specimens.

Another difference is in the field there is a limit of water availability, but in the laboratory we are dealing with submerged specimens, abundant water.

I will pretty much stop with those few items of noticing there is a difference but, you know the problem as it has come across here has been very eye-opening, and I hope I am a little bit of a taunter that my experience that it is not as eye-opening as it sounds.

Mehta: Thank you, Gunnar, for very stimulating remarks. You are right that we not only have some chaotic situations as far as phenomena are concerned, all of them are interconnected, interrelated, but I think there are many other issues which make the whole situation much more chaotic. And I think one of the things to which you referred to is after World War II, when Denmark and Europe were able to ...that there was a national effort to improve things.

You also mentioned the importance of very meticulous laboratory studies which simulate field conditions as well as, equally important, are the field studies. Now, all those things in my days, during my stay in the U.S., where I came as a student, graduate student, are going down the hill. The budgets, the laboratory support is not there, not too much support available to do the field studies, and there is no spirit of national cooperation that we need to solve these problems in a time-bound economical manner.

Do you have any suggestions? How do we start the process, because I find it difficult. Even in the educational institutions where the process should start to build future generations, I find it hard to get anywhere.

Skalny: Thank you, Kumar. I would like to add that I don't think that it should start at educational institutions; it should start with real world, and the universities should listen to the industry and react to what is needed rather than the other way around; but that is just a difference of opinion.

Hooton: It might be a minor point, but you showed your traffic marker, the traffic sign, the little marker on the road, the sign; you showed the top of it being undamaged to a large extent and the bottom being damaged. And you said the mortar bar from the top would appear to be fine, but that is not true.

The mortar bar would fail in both cases. What you had there was a different environment because you had wicking or capillary action of water rising, the environment at the bottom of the sign was different. So the lab test can only tell you the potential of a combination of materials to do something. You then have to go study the environment and realize the environment the structure is in, and part of it may be subjected to an environment that is going to cause damage.

So I do not think that your statement that a mortar bar at the top of the sign would not fail is true.

Idorn: Are you saying that I should have said the mortar bar would be worse? What I really meant was that it is an illustration of the complexity of the problem.

Hooton: Right. And I agree with your philosophy; one has to understand the whole picture.

Open Discussion

Thomas: I would like to pose a question to the cement chemists here. I think a comment was made either by Fred Glasser or somebody else as a result of his presentation that pozzolans and slag do not explain the fact that the permeability or the diffusivity of a system changes, but maybe they are altering the calcium aluminates by forming hydrotalcite or stratlingite?

I do not think there has been enough talk on pozzolan and slag because I really think they are probably one of the most effective ways of dealing with the problem, and I wondered if any of the cement chemists here would like to comment on whether you do form more durable aluminates?

Taylor: I would like to say something briefly and mainly—if not entirely—about slag, to say that the effects can be good and they can be bad as far as resistance to external sulfate attack is concerned. On the one hand, they undoubtedly decrease permeability, and that is clearly a good thing. On the other hand, they may increase the amount of available alumina, and I think it was Locher, if not it was somebody else from Dusseldorf, who showed quite some time ago that if you add, say, 40% of slag, if it was low in alumina, it improved sulfate resistance but if it was high in alumina, it worsened it.

So you cannot really generalize about the effects of slag, and I think it is probably true as well about pozzolan and fly ash, Class-A fly ash, for example. I think I am right in saying it will decrease permeability but it will also put more alumina into the system. And I think, really, that is all I want to say about this. The point about what happens to the alumina; I would like to stress the formation of hydrotalcite-like phases, which happen with slags and that, of course, does gum up some of the alumina; this is a plus. I do not know whether there is any evidence of high magnesium in the slag is a positive factor, but I would rather expect it to be.

And, incidentally, if we are talking of sulfate resisting Portland cement, the effect there is very largely that a very high proportion of the alumina gets tied up in ways that seem fairly harmless. Some goes into the C-S-H, some goes into hydrotalcite, some probably goes into hydrogarnit phases, and together with the small content of alumina you have, this, I think, largely accounts for the beneficial effects. You have to look at the formation, certainly, of things like hydrotalcite in that case.

Incidently, somebody this morning—I cannot remember who it was—said that hydrocalonite was essentially an AFm phase with hydroxide; it is not. It is an AFm phase with chloride; it is either the same as Freidel salt or something pretty like it.

Glasser: Well, I certainly agree that the problem of putting pozzolan into cement is a complicated one with regard to assessing an improvement or otherwise of their resistance to sulfate. Now, you first have to look, I think, at the principal that I invoked of the reactive fraction. If you are going to put in an unreactive fly ash or an unreactive slag, to take an extreme example, you might just as well put in some extra fine sand for all the difference it is going to make. It will increase somewhat the tortuosity of diffusion paths but that is about all.

If you are then going to put in a more reactive material, you are also going to have to cure it properly. You are not going to get very good performance out of a pozzolanic cement if the pozzolan has not been allowed ... reasonably reactive pozzolan, and you allow good practice and good curing, then there are potential benefits to be gained.

There are, as Hal Taylor says, and I agree with him, potential demerits. The advantages include not only the postponement of any possible attack from external agencies owing to refinement of the pore structure and the slowing of diffusion generally throughout the mass—something which, of course, will only be obtained with good cure—but also mainly the phases that you produce will themselves have sulfate-binding capacity.

For example, if you put a hydrotalcite like phase into contact with simulate seawater, it will uptake quite a lot of sulfate, you will effectively get a sulfated hydrotalcite. Indeed, I think there is a whole series of Japanese patents on the use of hydrotalcite-like materials to remove sulfate from lightly contaminated sulfate waters. It will work reasonably well even in fairly neutral pHs. So that is a possible benefit to be obtained.

There are other possible benefits to be obtained, but I think some of them at least are mentioned in the paper, but just to give one of the possible demerits. We have looked at a series of cements and concretes that had been cured for a long time with blending agents in them, such that they developed a lot hydrogarnet.

And it is interesting that you can, in certain blended cements, get a problem that is analogous to the old conversion reaction in high aluminous cements. So with time, as hydrogarnet develops, it is much denser than the phases that it replaces. Hydrogarnets have no binding capacity for sulfate; they don't take sulfate into their structure at all. But you could conceivably, if you are in the wrong areas of compositions, actually get a cement or concrete formulation whose microstructure deteriorates with time.

Now, that brings me back, first, to something that Kumar said with his charge that we were not taking a holistic view. Well, I would plead that I am taking a holistic view of the situation, but I was charged with you, Jan, to present only a specific aspect of it rather than try in 15 minutes to deal with all issues totally globally. But I want to point out while I have the microphone in hand the importance of the modeling studies. Now, I thought we had an excellent example of the way to go in the talk that was given by

Marchand and his colleagues. I very much appreciated that contribution.

And I am very pleased to see, as one who has been a voice in the wildness for some years, on the modeling side my colleagues in the nuclear field accept that this is the only way of modeling the performance of cements and concretes over hundreds or thousands of years, but civil engineers have not paid the slightest bit of attention to me.

I am pleased to see, therefore, that engineers are picking this up and are doing it, because the very act of modeling forces you to take a holistic view of things. If you have left chemical factors out that are important, your model will not solve. If you are missing important data with which to make solutions, your model will tell you that; it will stall and say: give me data for, or whatever.

So, OK, I will stop at this point, and we may return to that in the discussion. But I think I have given us as reasonable and comprehensive answers as I can about the blended cement issues, good and bad. It is a complicated problem and we always have to have good practice if we are going to realize the benefits from these materials, but there are also debit sides. And I have also said a bit about, I think, the way to go forward on the chemical, mineralogical side of things, which is the modeling.

And what I would say is that you guys were talking about field evidence and field experience and field concretes, you have to get your act in shape, too, so that you can glean from your practical observations the right input to make into the modeling studies, so that the modelers sharpen up their model. And there is a two-way duty of communication, not only for the modelers to tell you what they can now do and what they cannot do, but also to get not a mass of garbage from you guys who are doing field studies, but a sorted out mass of information so that the really relevant facts can be put into the model.

Skalny: Before I let Sidney to say something, I just would like to explain that when I asked some of you not to talk about DEF, it was not because it is not an interesting and relevant issue; to the contrary! I just wanted to avoid revisiting some of the nonsense discussed during the past five years at various conferences where we didnt talk much about science. I wanted to avoid deviations from the topic of external sulfate attack and that's why I said no DEF at this conference. So do not take it personally, please.

Diamond: I am going to talk about DEF!

Skalny: Go on, Sidney.

Diamond: The reason I asked for the floor in connection with the effects, beneficial, harmful sometimes of pozzolanic material and other blended components was to relay a little background and a few results that Jan Olek and his students are having about a particular illustration of this.

There are concerns in the United States, particularly in some of the States in the Midwest, about the potential for problems induced by using high calcium fly ashes. In particular, both Hal and I think Fred mentioned aluminum that you are bringing into the system by adding various kinds of blending materials. And certainly there is a lot of extra aluminum your are bringing into the system if you add a significant amount of most high calcium fly ashes. You are also bringing some sulfate into the system.

The question arose quite seriously: When you do this in field concrete, are you setting yourself up for delayed ettringite formation? You have the ettringite-bearing components being added to, introduced deliberately, so to speak, for the benefit of the pozzolanic action that you are going to get out of using high calcium fly ashes.

In Fred's terminology, these are "reactive species," these are not inert; reactive fraction is high. Well, we, again, and students I help from a distance, did a reductionist kind of approach to the examination. Various combinations of cements, sulfate contents in the cements and proportions of several different high calcium fly ashes to see how long the mortar bars expanded linearly and also to determine how much ettringite was being formed. We did do that and a number of auxiliary investigations.

To make a complicated story relatively short, the appearance seemed to be that we were not contributing to DEF because, in spite of the presence of the extra alumina we were introducing, the major product of reaction in the system seemed to be not ettringite at all but over a period of time either monosulfate or strattlingite. And presumably strattlingite is innocuous, and presumably the monosulfate in the middle of the concrete, that was reasonably secured by barrier layers against other problems, would be fine.

However, to the degree that the barriers did not exist—here we go holistic again—to the degree that the concretes were placed with high water-to-cement ratio conditions, high water-binder ratios, or were disrupted by freezing activities, and to the extent that sulfate externally was able to penetrate the concrete, there is a potential hazard that you can indeed convert some of that monosulfate to ettringite and maybe the ettringite might be destructive.

So you go up, down, sideways, and it comes back again to the usual consideration that you really have to understand the details of what is governing the response in a particular concrete, in a particular exposure.

And to my mind, despite my admiration for what Gunnar Idorn said and my agreement with his philosophy, that is not something that you can get by fuzzy logic, that is something you have to get by examining the particular situation in detail; end of philosophy. Thank you.

Mehta: Since we are on chemistry, I would like to ask Micheline a very short question; something I do not know and I want to learn. U-phase, Sid also raised a very important possibility; we are using type-C fly ashes, many of them are very high in alkalis, but more particularly I was interested because at the Paris Symposium, you had presented some very interesting papers on sulfate attack making a difference between what orthorhombic C3A would do versus the cubic C3A.

Now, we are increasing the alkali contents in some cement and clinkers because of many other environmental issues, does that have any implication to U-phase?

Moranville: I do not know exactly if we have a high alkali clinker if we will get U-phase. But I can tell you that you have U-phase in a concrete with fly ash high in alkali.

Mehta: In concrete with high alkali cement?

Moranville: I do not know exactly the alkali concentration in the clinker. I do not know; I cannot answer. We will look it up.

Lagerblad: I just want to make a comment about the pozzolan. If we look at concrete in a real environment, you get expansion at a certain period but then, also it involves some kind of acid attack, and you get ion exchanges, and you get a lot of things that actually cause softening of the product. So I think it is very important to see actually what is the reaction product that you get at the surface, is it strong or is it weak?

And from my experience, we put a lot of slag also in silicate, you get a stronger product because you have less portlandite, and portlandite is the one that reacts to form gypsum, and that makes it very weak, it makes it fall off. So you could put a lot of pozzolan in it, you get a stronger …(unaudible) ... and the rest by itself then would make the damage less severe than it would do in pure Portland cement.

Barger: Just a very short comment on the question of alkali, high alkali clinkers. You must invariably find, that the alkali in clinker with high alkali content is balanced by sulfates that will occur as soluble sulfates; it is really only under specific conditions that you get alkalis going into the aluminate phase.

Idorn: I just want to say that if you have enough alkalis, then you do not need Portland cement in order to get make concrete.

Skalny: We have several topics we did not discuss yet and maybe we should. I think we should spend some time on the mechanisms of expansion. There were several ideas presented here, and I wonder whether we can maybe clarify what is really going on. Some of the extreme statements, to the best of my knowledge, were that ettringite is the sole reason for expansion, and that secondary ettringite can open a crack ... some ideas imported from DEF. Some other people say expansion has nothing to do with ettringite ... maybe it is the calcium silicate hydrate that is swelling. So I think we have some chemical and mechanistic issues here to discuss.

Taylor: Well, as you have said, Jan, there are these various hypothesis going around. Now, one that has been very widely held is that it is just the increase in solid volume and, frankly, you can forget about that one right away because the increase in solid volume, if you form ettringite from—for the sake of argument—C3A gypsum and water or almost any reasonable starting material, is just about the same as when your form C-S-H and calcium hydroxide from C3S and water or, for that matter, just about any other hydration reaction, which is what this is. So it may be a necessary condition that you have an increase in solid volume, but it certainly is not a sufficient condition.

Now, people have talked about topochemical reaction; it was originally Lafuma. I never know what this means, and I think when this was put forward nobody really had—or Lafuma, at any rate, and other people in the field—had very little idea of crystal chemistry, and there were some funny ideas that C3A could turn into ettringite without going through a solution. Well, it is about as likely as a monkey turning into an elephant!

So if what they mean by it is that an interface moves into the C3A before that interface gets there, there is C3A and then afterwards ettringite, OK, fair enough, but I think that in that case there does have to be some sort of a film of liquid in which various ions can move around, reconstitute themselves and so on. So if it means, rather more generally, simply that the ettringite is formed closer to a C3A surface, OK, fair enough, we will come to that in a moment.

Crystal growth pressure. A very, very widely held view, you can use simple thermodynamics on this and calculate a growth pressure, but that is only a maximum that you can possibly get, it is not the value you do get, which is always less than that.

And this was described very thoroughly by Georges Shearer, from Princeton, of the main Engineering Foundation in Banff recently, and one of several factors that determines the actual pressure is connected with the surface energies of the crystal that is growing, the thing is pushing against and the interfacial energy between the solids, that has to be greater than the sum of the other two; otherwise the crystal will actually come into contact with the wall, and there is no way the ions can get there to make it grow.

Now, an important thing, I think, about the crystal growth theory is the question of where the crystals are forming is probably extremely important. And I think I mentioned it yesterday, work by some Japanese, two Japanese groups in the '60s and also by ... I have forgotten, two Chinese workers, much more recently saying the same thing. If you have calcium hydroxide there, the aluminate ions cannot migrate very easily, the solubilities are even less than they would be otherwise, so the ettringite is formed close to the aluminate surface; you get very high supersaturation and can get expansion. If there is no calcium hydroxide, it is formed all over the place, in places where it can fit in and you do not get expansion.

And I know this is a controversial thing. I know Kumar thinks it is much more a matter of the size of the aluminate crystals. My own feeling is very much that this other idea is a good one, and that if ettringite forms close to an aluminate surface, C3A, C4A3S, or, for that matter, even monosulfate, it can be expansive, but it is formed scattered around the place away from those surfaces or, for that matter, in voids or cavities, it is not going to be so expansive.

And there are other views, there is one Paul mentioned, which is that you have the possibility that ettringite crystals grow in cavities, which are not constant in size, but they are constantly changing for various external reasons. So there is the possibility the ettringite crystal grows while the cavity is large, and it stays grown when the cavity tightens up again, and this continues indefinitely. And this has also being called "dirt filling."

It has been suggested by Thorvaldson, who has been mentioned several times today, that ettringite formation is a total side issue, that there are reactions that happen to produce ettringite, but the really important thing is that they alter the C-S-H and make it much more likely to expand when it picks up water.

And merging into that theory is Kumar's idea, which I will not repeat because I think everybody knows it and he can add to it himself if he wants, that colloidal crystals of ettringite attract water molecules, and you get expansion by some sort of imbibition process. I cannot really find this reasonable in its original form, because I do not see why colloidal size crystals of ettringite, even if they were sub-micrometer, either singly or in aggregates, would attack water molecules any more than C-S-H would. In fact, less because C-S-H is a gel that could very easily pick up water, go in between layers, wherever, and there is no way you could put more water into an ettringite crystal, and I do not see why it should be strongly attracted to them.

But there are other possibilities described here as "expansion caused by reversal of local desiccation." As I said, formation of ettringite is essentially a hydration process. You are doubling a solid volume by taking up what has to start as, well, liquid water or water from somewhere else. And there are just two things one could think of: one is that you have local deficiency of water, that the ettringite requires water faster that it can be supplied from outside, and it gets it from the surrounding C-S-H, and the expansion actually happens when the C-S-H, by fair means or foul, gets its water back. And the other variant on this is that you can form a dehydration product of ettringite—we know it can be done by heating or evacuation—in which under certain conditions it stays crystalline, but the axis shrinks as the columns lose water and fuse together. With the small crystals that you get in a cement paste or, indeed, any synthetic ettringite, and it does not stay crystalline, it goes amorphous, but something of the same sort may happen, though in a more confused way.

So there is just the possibility, too, that the ettringite forms; what you get initially is not ettringite but this dehydration product, and expansion happens when that acquires more water and forms ettringite. The term "amorphous ettringite" has been bandied around for ages. It is a monstrosity of a term because ettringite is known for a crystalline phase, so it is an absolute contradiction in terms.

But if you mean this "amorphous dehydration product," then it is not necessarily what is formed but, at any rate, it is not stupid. Well, there are pros and cons to just about all these theories, and perhaps the first one I mentioned, which I think is a little con, but I guess I have said enough to set the ball rolling.

Brown: I agree with everything you said, except I do not believe a dehydrated ettringite can form by through a solution mechanism.

Taylor: I agree with you.

Glasser: I want to show an overhead that I put up earlier. Now, Hal says that if you mix the components of ettringite and water, there is not a net expansion in volume. Have I quoted you correctly?

Taylor: I said that there is an increase in solid volume.

Glasser: Yes.

Taylor: What I did not say and could have said is that there is a small contraction in total volume capturing water.

Glasser: OK. That calculation is unconvertible. It is easily enough done, you can go away and check it for yourself, and I agree totally. But the way in which you get expansion with ettringite, and this is patented and it works, you can prove it in the laboratory, is, if you take a cinder cake of C3A and gypsum—and I have deliberately put them in a ratio of one C3A to three gypsum - and stick it into a cement such that water can then diffuse in, it is widely expansive. You could easily achieve expansions of 100%.

It is like one of these novelty tricks you get in a shop, you can take a cement bar and unfortunately, it does not grow quite quickly enough to do it as a party trick, you have to wait a day or so, although I dare say you could fiddle around with it and probably make it into a party trick, but you can actually make it double in its length. Sorry to be back to length again but ...

Now, that is a bit ... and if you do this calculation, if you take the solid volume of C3A and three gypsum and diffuse it in necessary water and calculate the volume it is before and the volume after, there is of course a big expansion. And the expansion is partly due to the low density of ettringite itself, but it is actually mainly due, I think, to the fact that the packing of water, molecular water in ettringite, is more efficient than it is in liquid water.

Now, if you want to do the calculation again for some hydrates, there is a table of molar volumes in the paper, and you can go away and do the calculation for any combination you want, and you do not even have to assume that these are densely packed.

There is specimen calculation done in the paper where these are assumed to be 60%, the rest of the space is filled with water, so you do not need to diffuse in 15 H_2O, you need to diffuse it in a lot less—I cannot remember—and that is still expansive.

So this is the mechanism that is, I think, in many cases directly responsible for the expansion. And that is, you have one place in the cement, an aluminate source, and in another place you have a sulfate in water source. In the sulfate, I suspect since the sulfate in water is faster diffusing than the aluminate, that the sulfate in the water diffuses to the aluminate and that is the center for expansion.

But it does not need to be submicron size or nanometer size, you can make these little grains about 100 microns or so and that produces lovely expansion. So I think that is the...

Taylor: You are not saying anything different from what the Japanese people that I mentioned are saying. The product is formed first in the aluminates, as in your case, and there could be C3A or it could be AFm.

Glasser: Yes, it could be. As I say, there is a table of molar volumes, including AFm in the paper, and you can take hypothetical bottles of this and bottles of that off the shelf, mix them together in the proportions ...

Taylor: You have to ask, I think, what the microstructure is of that starting material, let us say, C3A and gypsum, and you have to ask what the microstructure of the product is. There is the question of how much of the water has come in from the outside. I mean, you have put 26 H_20 there, but unless that contact that you start with has zero porosity, some of the water is inside the yellow circle. If all the water came from the outside, then the change in total volume would be completely irrelevant. But on the other hand, if it is all in the yellow circle to begin with, it is highly relevant.

Glasser: Well, these are simplified examples. Now, I am not saying that regions of this sort or that sort really exist in a cement paste. What I am saying is, there is a table of data in the paper and you can calculate yourself. And I do a calculation where I assume that only 60% of this volume is filled with solid—it is an arbitrary number, but it is not an unreasonable number—60% of this is filled with solid, 40% with water, so you do not need 15 H_2O diffusing, you need a lot less than that, but it is still expansive by a big margin.

Taylor: You will still, if it is the proportions you said, you will still have to bring in roughly half the water you need from the outside. So the total volume changes.

Glasser: I have calculated the minimum density of these solids you would need to have in order to have a dimensional neutrality. But I suspect that this would have to be about 80% water and 20% solids by volume.

Taylor: No, it would have to be roughly 50-50, I think. I have to work it out.

Glasser: No, because the densities in solids are higher, you see, it is less than that. Go away and do the calculation. But ...(inaudible) ... the low density ...(inaudible)... in solids is quite a little water. As long as you still need some water ...(inaudible)... the reason it will form, it is expansive.

Taylor: But I do not think we are arguing about anything, I think ... what you are saying is exciting.

Brown: The way I see this is ... the cement hydrates ... in the paste fraction of the concrete hydrates, produce AFm throughout the paste. Sulfate intrudes from an external source; we come back five years later and the AFm has converted to AFt. And I am wondering what the origin of the semipermeable membrane you are invoking is?

Glasser: The origin of the semi-permeable membrane is already hydrated cement. Remember that this expansion is not significant unless the cement has hydrated enough to gain some strength.

Brown: So it is not significant unless the ettringite has something pushing against it?

Skalny: Thank you. I would like you to interrupt this discussion for a second, excuse me. I asked Jim Pierce to say a few words at the very end of this session. He has to leave for the airport soon, so he will make a few comments about the conference now. Jim?

Pierce: My apologies, Fred and Paul and Hal, but I do want to get back to Denver tonight. First of all, a number of "thank yous." First of all, to Jan and Jacques for putting this program together, organizing it with the central theme of "let us get to basics, let us get to facts, let us recount what we have known for some time, let us look at how we can predict things for the future through modeling." And I think they have done an outstanding job in that regard. Certainly, thanks are due to Jacques' staff, which made things happen, to the sponsors who made it possible to happen, and to you the participants.

As Gunnar Idorn said, I was particularly delighted to see several newer and, may I say, younger faces in the crowd. I notice now, each year it is easier to find younger faces in the crowd! And thank you, Jan, for inviting me to meet them.

The office I represent, the Bureau of Reclamation in the U.S. government, has been concerned with sulfate problems for a long time. I echo what Harvey Haynes has said: "We do not see problems in the field," ... but I believe that is because we have taken the necessary precautions. As Francis Young said: "If it is my tax money you are spending, please be very, very cautious, but if it is your tax money, I do not care!" But for his tax money, we do use low water-to-cement ratios, we do pay attention to what the cement type is, we do use low water contents, we do insist upon quality construction, which means consolidation and good curing. And, therefore, we do not see problems today in our sulfate areas.

And certainly, those of you who know the western U.S., which is the only place the Bureau of Reclamation works, it is arid, semiarid, a lot of—as we call them—alkali flats, sulfate soils in the ground water, so we have a lot of potential problems there. But, in reality, because we have taken these precautions, conservative though they may be—but when I am spending Francis' tax dollars, I need to be—and we do not have those problems.

We have done a fair amount of research over the years dealing with external sulfate attack, and our work goes back a number of years now, certainly as far back as Miller and Manson, but when I came aboard in the mid-, well, early 1960s, we were doing things then ... in the Thorvaldson seminar or symposium was in late sixties.

My first exposure of a crowd of this nature was at the Verbeck Symposium in 1980, and I just sometimes think how much smarter we might be if we had some of those speakers still with us today: certainly Georges Verbeck, Catherine Mather, Bob Philleo, Bill Dolch. If we had them still contributing, as they did in those days, how much smarter we might be. We would not have to worry about the complexity of things because those guys were smart enough to figure out a lot of the complex things.

We have done things. Back in 1975 we came out with the bold statement "fly ash improves ... or increases the resistances of sulfate attack." And sure enough, the data that we looked at were great but, in fact, when we looked at more data we found out that that was not always true, and most of that is related to the high-lime/low-lime fly ashes.

But in all of our concrete, in addition to those other things I mentioned today, we are using Class-F fly ashes. Class-F fly ash is in just about all the concrete that is being used in Reclamation projects today. So we have a fair amount of experience. We tried, at one point, to sort out the cations. And what we encounter in the western U.S. are the three basic ones: sodium, magnesium, and calcium sulfates, and we could not find, we were not smart enough, we did not have enough brain power perhaps, but we could not find a simple method for differentiating the cations with a simple field test, so we are still doing belts and suspenders to make sure that we do not have problems with sulfate attack.

Even when President Clinton came aboard, he talked about how it was very important to maintain and rehabilitate the infrastructure of the U.S. We had trouble with semantics back then because we said, "Certainly, we can do that for you." And we put together a very sizable research program. Jeff Ramstroff was involved, Tom Pasko of Federal Highways, and a number of people through the Civil Engineering Research Foundation. So, in fact, we presented them with something that he wanted to do and, once again, he had trouble clearly stating what his needs were or what it was that he wanted, so we did not get money.

There is now an effort within the American Concrete Institute, called the Strategic Development Council, in which they are facilitating collaborative research. It is an opportunity for those with needs and those with funds to get together on a basis of collaboration to try and find answers to the questions we have.

It is the necessary way, apparently, in the U.S., to fund research because the government is not doing as much as it used to. I know in my own agency when I talk about concrete research they say: "Oh, you found the answers to alkali aggregate reaction, you found answers to sulfate attack, why do you need to do any more research?" Structural engineers.

But it is difficult, it is not glamorous, it does not catch the eye and, therefore, you do not see the quick payback and yet we, in this room, know that there are lots of needs and I think we have talked about them today.

Just keep in mind: we do need your efforts in the standards development community; how better can we transfer our technology than through standards? So although they may be tedious at times, slow in developing, it is the people in this room that we need participating in the standards development organizations, and in the U.S., principally ACI and ASTM. And as we have talked about, the problem is concrete so, in fact, maybe C-1 ought to be a part of C-9. With that, I close.

Many thanks to the organizers; very good program. You did, in my mind, meet the target, and it is some old stuff, some new stuff, but better that we remember what we have learned long ago enmeshed with what we have learned today so that we can provide concrete that resists the aggressive environments that we have been talking about. Thank you.

Skalny: Thank you, Jim! Let's return to the discussion. We talked about chemistry, and we have only about 25 minutes to go; is anybody here, for example Sidney, who would like to say something more about the unusual changes in morphology, changes in microstructure, etc?

I have a particular question to Prof. Diamond: you presented a picture where you showed that during sulfate attack, the original clinker minerals can be completely restructured and, of course, we showed also at the 1998 RILEM meeting in Haifa, that not only belite but also alite can be restructured. You see an alite crystal; it looks like a nice unhydrated crystal ... everyting looks normal. But when you pinpoint the beam on it, you see pure silica or magnesium silicate. What is in your view the mechanism of this restructuring? How does it happen?

Diamond: Well, lots of different mechanisms. ... In some cases you are simply dissolving the whole grain. We have presumably, who can tell, you are looking at one spot in time, but it looks like in some cases you dissolve the whole grain, leave a hole and subsequently something like monosulfate precipitates in the hole.

In other cases you are clearly stripping away the cations. Sometimes you see low calcium silica, a far lower ratio than belite should have. In other cases, we see no calcium, all you see is the silica and the gray texture suggests that it is quite hydrous. So, presumably, in that kind of response you have removed the calcium, left the silicate in place and brought in some water.

Then, in other instances, you see that magnesium has moved in, and what seems to be a low magnesium-to-silica ratio, hydrous, magnesium silicate hydrate has entered that space. In other cases, you see a high magnesium content, a high magnesium silicate hydrate.

The impression I have is that sometimes there is a complete dissolution, other times the silica remains behind and it is the cation that moves. But all this is inference, and I have zero evidence.

Taylor: Yes, I agree with what you said, Sid. I think you could add something to it: If you do your microanalysis with standards and apply ZAF corrections and so on and get absolute values, you find, in fact, that even though you cannot find anything there but silica, and you assume that there is oxygen to balance this, you find that it does not add up to anything like 100%; it is nowhere near that.

And it is a characteristic of X-ray microanalysis and this applies equally to back-scattered electron imaging, that if you have porosity on a scale comparable with the interaction volume, which is about a micron for microanalysis and less for back-scattered imaging, then the X-ray output goes down and the back-scattered electron output goes down. And I do not think anybody really knows why this is, but it is a fact. So ... (inaudible) ... the silicate phases are sometimes ...they work out with a 30% silica or something like that from microanalysis calculations.

Glasser: I just wanted to ask Sid a question about the photographs he showed, the micrographs. Really two questions. One is: the magnification is showed comparable in many ways with optical micrographs; are we looking at features on a comparable magnification to what you see in optical microscopy? When you showed us those zones, they appeared as strips running across the photograph. What is the relationship of those to the external face? And do you have sections in other directions to show that they are sheet-like units.

Skalny: He is talking about gypsum.

Glasser: Gypsum, was it? Yes, yes, yes.

Diamond: Those pictures were all taken normal to the surface of contact ...(inaudible)... those were just about at the surface ...(inaudible)... bottom of the ...(inaudible)...

Glasser: I missed that ... yes. OK. Thank you.

Lagerblad: You know, thinking about that, the microphotograph I could see was fairly... is that really an ettringite or a sulfate attack with expansion? Because I think you can look upon it, when you have a sulfate attack you start by spalling off the surface, you will form cracks and things like that; you start to break it down.

But what you see in this micrograph seems like a multiple chemical attack of some kind, a dissolution and breaking down of the minerals and things like that. Is it really the sulfate attack or the secondary consequences of a sulfate attack that we do really see in those late stages that you show on the outer surfaces of a sulfate attacked concrete?

Diamond: Well, you know, an elephant is an elephant if you define the elephant in the proper manner. The attack we see is the consequence of sulfate solutions moving through the concrete and interacting with the concrete and causing a degradation of the original structure, which I hoped was fairly obvious for everybody to see. People see the consequences of sulfate attack differently in different exposures. The word mush was used extensively in somebody's particular description.

The question of expansion and the consequent cracking versus alteration of structure leading to a reduction in the strength and elastic behaviour, internally that is, conversion to a mushy from a hard strong structure, vs. exfoliation, which also occurs partly at top surfaces where you get high concentrations of the salt crystallizing just below the surface and causing lifting and exfoliation. Those, the relative importance of at least three different phenomena in sulfate attack, vary from place to place.

I think it is nice to agree with Gunnar Idorn, and I particularly appreciate his condemnation of the idea of equating damage to expansion in the linear mortar bars. In sulfate attack, we have had this syndrome very seriously. People stick mortar bars in sulfate solutions, and they wait a while, and they measure the expansion and if there is extensive expansion, yes, it's a serious sulfate attack, and if there is not much expansion below a certain level, there is no sulfate attack. I think that is shortsighted. This does not reflect the complexity of things that actually happen in real holistic sulfate attacks in the field. So, yes, the answer to the question, from a very long way around, yes, to me, that is sulfate attack.

Skalny: I would like to repeat that what we call "sulfate attack" is not only the reaction of tricalcium aluminate with sulfate to give ettringite; it is much, much more complex. First of all, as we all know, we do not have a solid material like calcium sulfate or sodium sulfate in the solution, we have various ions in the solution. So whatever enters the concrete within the liquid phase and whatever reactions happen due to this sulfate and corresponding cations, whatever the other components are in concrete ... I call it all sulfate attack.

Lagerblad: That type of sulfate attack, you could even have a sulfate-resistant cement because if you have high enough sulfate content, you get gypsum and you will break it down, and then you get all the other components coming in, the magnesium, the chloride, and whatever coming in there. Of course, it is a sulfate attack, but it is not the same sulfate attack. I think one has to divide a little bit in between the ones with expansion and those more like a multiple chemical attack where sulfate is a very important part of it.

Taylor: I just wanted to add to Sid's comment, which I agreed with. I think Bjorn Lagerblad's original question as to whether the damage was a direct cause ... a direct result of ettringite formation or indirect, is a good one. And I would answer very briefly by saying that in many cases it is an indirect effect.

The thing that turns the cement paste into a "mush" is the destruction of C-S-H. And that, of course, happens particularly with magnesium sulfate. But it is a fact that even in cases where you have magnesium sulfate attack, and the main effect is destruction of C-S-H, the damage is the worse the more alumina you have around. So I think the ettringite formation damages the microstructure that paves the way for further damage which is directed caused by the magnesium.

Skalny: Thank you very much, Hal. I think we all basically agree...we just use different words.

Diamond: Fred, I think, spent some time discussing the pore solution chemistry inside of concretes, and we are aware that despite the fact that you may have in a sulfate attack relatively free communication of fluid from one part of the concrete to another, you do get zones of accumulation of particular salt, dissolved salts, and so on.

So it is hard to tell by squeezing the pore solution out of a core what the zonal distribution might be. But at least you can determine something about the internal chemistry of the pore solution from collecting and analyzing pore solutions out of a full-depth core material.

And we have done a little of that, and as you might expect, we find that the alkalinity is reduced. The hydroxide ion concentration instead of being 13.2, 13.3, 13.5 is, in the cases we have examined, flirting with the 12.5 margin. And that is reasonable because the middle of the concrete still has some calcium hydroxide in it, the top and bottom do not but the middle still does.

In some cases, in a few cases where there seems to have been more deterioration, the ion concentration is significantly below 12.5, it has picked up in a few places, 12.1, 12.2. In cases of extreme damage, and we have a couple: one case of extreme damage, 20-some odd years old, believe it or not, the pH of the pore solution squeezed out of a concrete core is about 7.2.

What we are doing, in part at least, is replacing the original pore solution by ions and water moving into the core from the outside. Original pore solution, we expect to see mostly potassium as the cation with greater or lesser but still minor content of sodium. In even these early stage pore solutions, eight, 10 years old, not incompletely reactive, we find almost no potassium. The main cations are clearly sodium. You know, where is the potassium gone? Remember the counter-diffusion idea that Jacques illustrated? I think—I cannot prove anything—but I think the potassium is gone down into the soil and what has moved up is sodium.

So we have sodium-bearing solutions of modest pH—depending on how much lime we have left—some magnesium has moved in; it is hard to keep magnesium in solution inside of a concrete environment, but there is a little magnesium in solution that squeezes out. We have a lot of sulfate, far more sulfate in solution that any unaffected concrete ever has in its pore solution. And if there is chloride in the ground water, we pick up chloride in solution in the concrete as well.

So there is a way, sort of, of tracing the changes taking place, not in the concrete solid structure, but in the solution of the concrete and inside the concrete and following the ongoing processes of change by following the solution chemistry. And that is all I wanted to say.

Skalny: Thank you, Sid. I am experienced enough to realize that half of the audience has reached the point of saturation ... this was a very intensive session. So why don't we agree to adjourn? However, before we do so, I would like to ask Professor Taylor to very honestly and openly tell us what he liked and what he did not like about this seminar. I think it is much more important, Hal, to tell us what you did not like than what you did like. So why don't we spend a few minutes ...

Taylor: Well, I am not good at thinking out things impromptu, and I really cannot think of all that much that I did not like, so I do not know whether I am going to be able to meet Jan's request in that respect.

Like Jim Pierce and Gunnar, I have been in this field a long time, in fact, just over 50 years. I never had the pleasure of meeting Le Chatelier, but I was actually born before he died. During that 50 years, there have been enormous advances in many, I would say, most aspects of the chemistry and material science of cement and concrete, but I think external sulfate attack has—as I think was said earlier in this meeting—been an exception to that, though always with the reverse exception of the Thorvaldson Symposium that Jim mentioned this morning and that was in Saskatchewan, and it is really appropriate that this meeting, which I think is the first of any substance that has been held on the science, the first on the technology of sulfate attack should be held in Canada.

And it has also been a very timely meeting because there has been a lot of progress on external sulfate attack in the last six, seven to eight years, so it is a fine time to take stock of what we know, what we do not know, what we ought to be doing.

I thought there were three things that were outstanding at this meeting. The first one, which has been mentioned by many, many people, is the complexity of external sulfate attack. And I think one reason why it was not studied much was everybody says: well, there is really not much to it, you have monosulfate, it reacts with sulfate from the outside, and it gets some lime from somewhere, I do not know where, and that forms ettringite and it expends and it cracks, and that really is all there is to it.

So, enormous amounts of work, sticking cubes in tanks containing sodium and magnesium sulfate solutions and during, I am inclined to say "ritual expansion measurements," and from a fundamental point of view, I think it is fair to say that nothing at all worth speaking of came out of it.

But we now know, and it has been emphasized by many people at this meeting, perhaps most clearly by Kumar Mehta this morning, that you have a lot of closely interrelated processes involving not just formation of ettringite, formation of gypsum, destruction of C-S-H, very often formation of magnesium compounds and the physical effects are not just expansion, may not even be expansion but the softening, general destruction of material resulting most directly from the destruction of C-S-H.

The interrelations between all these effects were shown particularly carefully, elegantly and in a way that, to my mind, was totally in accordance with what microstructural information we had by Jacques Marchand this morning, and I think his modeling paper was one of the highlights of that meeting, and I am not just saying this, Jacques, because you organized it.

The second theme is connected with experimental approaches. We can certainly learn a lot from chemical studies on pure systems, studies on equilibria, perhaps on kinetics, other things, even from studies at high water-to-cement ratios, but they are a necessary beginning to understanding what happens in concrete, but they are nothing more than a beginning.

And studies on paste, especially microstructural studies, are a very major step in the right direction, but if we are going to understand what happens in real concretes, there is absolutely no alternative to studying real concretes, particularly by microstructures. Perhaps I am keen on that because it is one of my major interests, but I do think that it is a vitally important way of approaching things.

And I would like to highlight what I though was another outstanding paper, the one by Sid Diamond and Richard Lee, which I think was an absolute model of how such studies such be carried out.

It is no intended criticism of their work or anybody else's to say that work on field concretes, though absolutely necessary, does suffer from certain limitations. In particular, all the necessary data are not available, or at least are not available to the people doing the work, and one way in which I think further research might progress is by trying to simulate what happens in real field concretes in all its complexity, which is manifold, with regards to not just making them but as regards their surface, service life, and studying that in the lab varying the various conditions in a controlled way, just as you would doing the same methods of study, you would with a field concrete. It is a major task. Do not underestimate it, but I think it is something that is needed.

The third thing that I think comes out of the meeting concerns the question of how to minimize attack from external sulfate reactions, and I think some very useful things were said here. Many of the people speaking out emphasized the extreme importance of making good concrete: a suitably low water/cement ratio, suitable contents of cement, good concrete in every possible respect, and no amount of making new cements or improving the properties of existing ones, or choosing the right kind of cement will compensate for bad concrete.

One aspect of this, which was also touched on in the meeting, which I thought was very important, was that of concrete standards. Jan Skalny mentioned it in his introductory talk, and also Jim Clifton and his colleagues from NIST, and I think others have added to it in more detail, and this does seem to me to be something very, very desirable.

Of course, from the point of view of academic researchers like me, it would be a very poor show if all concrete was good, because nobody would fund any research! But looking at it from a rather less selfish more general point of view, I think one does come to the opposite conclusion.

In conclusion, I do just want to add my personal thanks to Jacques Marchand and Jan Skalny, and everybody connected with the organization and running of this meeting, which has been outstanding and nice. Quebec is a lovely city; it is grand to be here. But Jacques and his colleagues added those various extra touches that have made it not just a very useful occasion but a very pleasant one. So, thank you.

Skalny: Thank you, Hal. Thank you all for active participation. The afternoon session is adjourned. Jacques?

Marchand: As a concluding remark, I would like to thank you all for attending the seminar. I know that some of you came from Europe, California, Japan, and other distant places. Quebec City is probably not the most obvious or easy place to reach by plane, and I know that there could be some problems.

Thank you again for coming, thanks to the sponsors for supporting the organization and some of the travel expenses, and I wish you a safe trip home ... and all the best for the future. Au revoir!

SULFATE ATTACK ISSUES: AN OVERVIEW

Jan Skalny
Consultant, Holmes Beach, FL 34217, U.S.A.

James S. Pierce
U.S. Bureau of Reclamation, Denver, CO, U.S.A.

ABSTRACT

Sulfate attack represents a complex set of chemical and physical processes, and thus cannot be fully characterized by a single mechanism. These processes decrease the durability of concrete by changing the chemical and microstructural nature of the cement paste. Depending on the conditions of projected use (concrete quality, severity of sulfate exposure, environmental factors), the defense against sulfate attack should include design for low-permeability concrete (primary defense), use of sulfate-resisting cementitious materials, and proper processing.

INTRODUCTION

During the past decades, sulfate attack-related deterioration has resulted in expensive rehabilitation of concrete structures in several countries, and has led to numerous litigations in Canada and the United States. This development, in conjunction with changes in the composition of concrete making materials [e.g., 1, 2] and processing approaches, gaps in our knowledge [3], and possible misinterpretation of the spirit of existing standards and relevant tests, has revived interest in sulfate attack, whether caused by internal or external sulfates.

The ongoing discussion of what is "right" or "wrong" with sulfate attack-related issues involves a diversified group of professionals, including engineering and scientific personnel, specialists interested in the development of standards and test methods and, for good or ill, attorneys and lobby groups. We do not find this situation disturbing: the way to progress is complex and often frustrating.

The issues of internal and external sulfate attack are intimately interrelated and overlapping but, in line with the overall topic of this Seminar, we shall concentrate our discussion on issues relevant to the ingress of soluble sulfates into concrete from external sources, such as soils and ground water. Internal sulfate attack represented, among other mechanisms, by the somewhat controversial *delayed ettringite formation* (DEF), was recently reviewed in numerous publications [e.g., 4-6].

Let us briefly highlight some of the external sulfate attack-related issues that should, in our view, be considered for discussion at the Seminar:

Materials:

(a) cement composition (relative amounts of C_2S to C_3S, C_3A and sulfate contents, role of aluminate and ferrite ions from C_4AF, form of sulfate, and fineness);
(b) mix design (cement type and content, focus on strength or durability, water-to-cementitious materials ratio, impermeability,);
(c) use of less-than-perfectly characterized mineral additives;
(d) effects of chemical admixtures;
(e) use of lesser quality (sulfate-containing) aggregate;

Processing Practices:

(a) mixing procedures;
(b) curing conditions;
(c) concrete placing and consolidation; and
(d) maintenance and repair.

Environmental Exposure Conditions:

(a) chemical environment (soil chemistry and mineralogy, ground water composition, form and concentration of sulfates, presence of chlorides, etc.);
(b) atmospheric conditions (temperature and relative humidity, frequency and range of changes, precipitation, other sources of sulfates);
(c) other (soil properties, drainage, etc.).

Available Knowledge and its Application:

What is the proper definition of "sulfate attack", considering that it *does not* represent a single or simple chemical reaction? Is the available knowledge on sulfate attack understood by the field engineers and is it reflected in their practice? Is the classical knowledge comprehensive enough to enable prevention of the damaging attack? What is the best measure of concrete failure due to sulfate attack?

Standards:
Are the present standards and test methods adequate to prevent concrete deterioration by sulfates? Is our present interpretation of existing standards correct in view of the above changes in materials and processing? Do we "cut corners" to remain competitive? How does current practice affect the *long-term* durability of structures exposed to sulfates?

We would like to believe that the purpose of this Seminar on external sulfate attack is to restore the technical sanity of the discussions and to disseminate the applicable knowledge. This should be based on **scientific facts** rather than undocumented claims and assumptions, the goal being to upgrade the ways of constructing cost-effective but, at the same time, durable concrete structures. Quality and cost need not be antagonistic!

SELECTED ISSUES

It is obvious that we cannot cover all important sulfate attack-related issues in this presentation. Neither do we want to do so; the speakers that will follow us will surely have much more to say about the mechanisms, physical chemistry, and mechanical consequences of the individual reactions that characterize "external sulfate attack".

What is Sulfate Attack? What Damage Characterizes Sulfate Attack?

Some accepted facts:

- sulfate attack is not a single or a simple chemical reaction;
- it is a complex sequence of physical and chemical processes resulting in chemical and physical (micro-structural) modifications of the cement paste matrix;
- restructuring of the matrix is eventually evidenced by several possible modes of deterioration, and leads to loss of mechanical and physical properties expected of any particular concrete structure;
- sulfate attack is not fully characterized by any one of the many possible reactions between the sulfates and cement paste components (e.g., presence of ettringite, formation of gypsum, dissolution of calcium hydroxide, or decalcification of calcium silicate hydrate);
- relationships between the degree of chemical, physical or microstructural change caused by sulfate attack reactions and the degree of mechanical damage are complex and non-linear.

The meaning of the above is as follows:

Deterioration of concrete mechanical properties, whether measured visually, by decreased strength, volume expansion or other parameters, is the **consequence** of chemical and microstructural changes caused by sulfate attack. At the time such deterioration is observed the damage is already done! To quote from the conclusions of a 1988 WES study [7]: *"Measurable deterioration of mortar bars exposed to sulfate solutions occurs **before** excessive expansion occurs."*

The prevailing mechanism(s) of deterioration will depend on:

- the concrete quality (e.g., type of binder used, type of concrete structure, chemical and physical stability of the concrete as delivered, placed and cured, concrete impermeability, and damage due to other mechanisms),
- the exposure conditions at the site (e.g., soil conditions, concentration of available sulfates in soil and groundwater, form and distribution of the sulfates, water/moisture transport opportunities), and
- the environment of use (humidity changes, range of temperatures, frequency of changes, exposed surfaces, etc.).

It is now accepted that strength, especially compressive strength, is an inadequate measure of concrete durability [e.g., 8-10] Because of the well-known complexity of the sulfate attack mechanisms [e.g., 11,12], this is valid also for sulfate attack. Thus, it is impossible at the present time to relate the probability of sulfate attack (or the "resistivity" to sulfates of a particular concrete composition) directly to a single variable such as the strength or volume stability of the concrete at a selected time. That is the reason why deterioration of concrete due to sulfate attack has been assessed in a number of ways [13], e.g., visual deterioration, wear rating, compressive and tensile strengths, volume expansion, and mass change.

Progress of the deterioration must be characterized both by the rate at which and the degree to which the sulfate-related reactions have already altered (and continue to alter) the concrete matrix, and that subsequently result in loss of mechanical properties such as permeability, volumetric stability, and strength. In line with others, Khatri and Sirivavatnanon [14] concluded recently that both permeability (a non-linear consequence of water-to-cementitious material ratio) and the type of binder (portland cement, blended cements, etc.) play an important role in sulfate attack. In reality, the situation is even more complex, because variables other than permeability to aggressive solutions and cement type are involved.

It is this complexity and non-linearity that make it so difficult to design all-encompassing test methods or standards, and why it is most important not only to deliver a quality concrete structure but also to maintain it properly. Note that despite these uncertainties in detail, we know enough to eliminate sulfate attack in virtually all cases.

Can Physical Sulfate Attack be Disassociated from Chemical Sulfate Attack?

In our view, the common separation of "chemical" and "physical" sulfate attack mechanisms, such as given in the present ACI "Guide to Durable Concrete" [15] and in numerous other publications, is an incorrect oversimplification. This is because the complex physico-chemical processes of "sulfate attack" are interdependent as is the resulting damage.

Physical sulfate attack, usually evidenced by efflorescence (the presence of sodium sulfates, Na_2SO_4 and/or $Na_2SO_4.10H_2O$) at exposed concrete surfaces, **is not only a *cosmetic* problem, but is the visible manifestation of possible problems *within* the concrete matrix**. We fully agree with Young [16] who states that: *"... physical processes don't occur without chemical changes"* and that often observed sulfate salt crystallization *"is the result of superposition of an upward flux of sulfate driven by capillary effects on regular sulfate attack."*

Both the chemical and physical phenomena observed are representations and consequences of a complex process referred to *in toto* as sulfate attack, and their separation is misleading: a popular opinion is not necessarily a correct opinion. We are looking forward to hearing the arguments of other speakers on the correctness or incorrectness of such separation.

Are "Sulfate-Resisting" Cements an Adequate Remedy Against Sulfate Attack? What is the Role of Fss (C_4AF)?

The use of ASTM Type V and other similar cements to limit the amount of ettringite formed may be helpful in most cases, but is no panacea in the absence of the design and production of effectively impermeable concrete. As a matter of fact, in some types of sulfate solutions (e.g., rich in sulfate **and** Mg^{2+} ions), no concrete is durable unless designed and cured to give a low-permeability, disconnected pore system. The same applies for concrete made with other cementitious materials, e.g., a combination of Type II cement and fly ash.

The role of C_3A: The dominant influence on the probability of sulfate attack of clinker C_3A content has been well known for over 70 years [e.g., 17-19]. At lower C_3A content, as reflected in standards, this probability decreases. In our view, lower C_3A content increases the sulfate resistance of otherwise well produced concrete (low permeability, adequate curing). The beneficial action of low C_3A content in **high w/(c+m), permeable** concrete is less obvious and there are only very limited data specifically addressing this issue. In such cases – and/or under complex conditions of concrete exposure -- the protective action of low C_3A cement may be minor or be entirely irrelevant.

The role of the Fss (C_4AF, ferrite solid solution): All of us are aware of the fact that Fss is not an unreactive clinker component under all sulfate attack conditions. Depending on the "reactivity" of Fss and the chemical environment during the sulfate attack, the aluminate and ferrite ions may be released at different rates, thus supplying "raw material" for formation of expansive ettringite. Results in support of this view have been presented by Riedel [20] and Matthews and Baker [21], amongst others. Aluminate ions from Fss will react with sulfate and other cement components by the same mechanism(s) as aluminate ions from any other source. So the issue is not the aluminate ion's origin, but its availability. High w/c+m concrete will be more predisposed to release aluminate ions than dense concrete. In a way similar to aluminate ions, ferrite ions may also participate in the sulfate reactions ... by similar, not entirely understood mechanisms [19].

We expect to hear more about the role of "sulfate-resisting" cements from other authors and in the discussions.

Is Low Water-to-Cement Ratio Really Important?

The importance of low w/c+m to prevent sulfate attack is undisputed by most. This importance was recognized years ago by ACI, and it is reflected in ACI 201's Guide to Durable Concrete [15] and in the Uniform Building Code [22].

Novokshchenov [23] reported on a case of severe sulfate attack in "high water/cement ratio concrete" in Saudi Arabia, and relates the attack, among others, to concrete permeability (related to w/c+m). His data also show clearly that, in addition to w/(c+m), cement composition and sulfate concentration, the rate of attack depends on the atmospheric conditions, the attack being more severe when concrete is exposed to cycling conditions (wetting-drying, low-high temperature) than when continuously exposed to sulfate containing water. Reporting on forms of acid attack, MacFarlane [24] states that *"concrete of high quality is markedly more resistant than poor concrete."* Our interpretation of the reported "high quality": proper curing and low permeability.

The effect of surface evaporation on sulfate attack is highlighted also by Lossing [25] who -- discussing deterioration of concrete pavements in Mississippi after only 7 to 8 years after construction -- notes that *"the original disruption occurs at the surface due to more rapid evaporation of moisture"* and believes the disintegration *"progresses downward with time, finally resulting in disruption of the entire depth of concrete."* He considers evaporation of water from the surface of the concrete to be one of the needed conditions for the progress of sulfate reactions. In our view, this is a contributing, but not a necessary condition for the progress of sulfate attack (i.e., concrete may deteriorate when totally submerged or when buried in sulfate soils) and is clearly caused by the higher rate of evaporation from high porosity (higher w/(c+m)) concrete surfaces.

The above quoted data are in support of our view that for concrete placed in potentially "very severe" sulfate environment, as defined by ACI's Guide to Durable Concrete [15] and the Uniform Building Code [22], it would be advisable to take the most conservative approach, such as to use w/(c+m) as low as 0.4 in addition to high-quality supplementary materials (silica fume, fly ash, slag). This is presently being considered by ACI 201. Such a move would eliminate the effect of soil sulfate concentration which is known often to be highly variable within very small areas.

There are several problems with determination of the degree of aggressivity of sulfates at a particular site. Among others, at issue is the best analytical technique to be used for sulfate determination but also, most important, the variability of the sulfate form and content. As an example, the soluble sulfate content of some soils varies from negligible to well above 2 percent not only on a global scale, but also on a local scale. In other words, the sulfate concentrations even within a particular development may show large variations over very short distances. Swenson & MacKenzie [26] reported, as an example, a 1922 case in Winnipeg, where on a plot measuring 50 by 50 feet (ca 16 by 16 metres) the ground water sulfate concentration varied from 3,200 ppm to 17,570 ppm.

As a consequence of the soil variability and local conditions, the sulfate concentrations in ground water can also vary within a wide range both globally and locally, from negligible to well above 10,000 ppm. Other than sulfates, the ground waters may contain additional anions such as chlorides, phosphates, and bicarbonates. The most prevalent cations are usually sodium, calcium and magnesium, sometimes potassium.

As one would expect, soils differ not only in their concentration of various chemical species, but also with respect to their plasticity, permeability, pH, etc. Sulfates in soils may also vary with respect to the distribution of relevant cations (usually between Na^+, K^+, Mg^{2+}, and Ca^{2+}), all of which may play a significant role in concrete deterioration {19]. All of these compounds are highly soluble; thus, in the presence of water (especially at high w/(c+m)), the soil composition may guarantee a continuous supply into concrete of soluble sulfates.

Because of the above, we believe that, in cases of highly variable local sulfate concentrations, the maximum w/(c+m) corresponding to the highest sulfate level found should be prescribed by the applicable standards for the whole site.

It is not only the w/(c+m), but also the cement content that influences sulfate resistance. For example, Harboe [18] in agreement with others states that *"an increase in cement content increases the sulfate resistance of concrete of all types of cements"*. Stark [27], summarizing a long-term study of concrete exposed to sulfate soils, states that *"... the cement factor, with attendant changes in water-cement ratio, was the most significant factor affecting the resistance of concrete to sulfate attack"*. On a closely related issue, Swenson & MacKenzie [26] report that *"lean concretes were found to disintegrate rapidly when surrounded by disintegrated concrete and pure water. This indicates the presence in the affected concrete of excessive amounts of available sulfates."* In other words, both a minimum, though not excessive, cement content **and** an adequately low w/(c+m) are required to produce a low-permeability, sulfate attack-resisting concrete.

It is also probable that low soil or ground water sulfate contents are not necessarily a guarantee of concrete durability. It is known that, under proper temperature-humidity conditions, sulfates from the ground water may penetrate through a highly-porous concrete, concentrate at and/or beneath the exposed surface, and consequently lead to physical and chemical deterioration. To prevent this from happening, a low water-to-cement ratio and a properly cured concrete mixture, capable of reaching a high degree of maturity, are recommended.

The effect of w/(c+m) on durability of concrete, considering the affects of additional variables in play, is most complex; this is clearly true also with respect to sulfate attack. Such complexity often leads to controversial interpretations of the available knowledge. It is now evident, however, that lowering the w/(c+m) decreases the sensitivity of any concrete to other possible variables, such as cement composition and type of admixtures; thus w/c is probably the most important variable to control. This fact is related to lower porosity and permeability of concrete at lower w/(c+m), leading to diminished flow of aggressive species such as sulfates through the matrix.

Water-to-Cementitious Material Ratio or Cement Type?

It should be clear from the above discussion that the use of sulfates-resisting cement, as defined by today's standards, is not by itself an adequate remedy against external sulfate attack.

This was known already to Kalousek et al. [28]: *"... limitations on C_3A and C_4AF contents are not the ultimate answer to the problem of sulfate attack."*

They state that *"Restrictions of cements to those meeting the present day specifications for Type V cement does not appear justified."* Similarly, Mehta [29] states: *"Permeability of concrete rather than the mineralogy of cement appears to be the most important factor in sulfate attack."*

We believe that a low-porosity, low-permeability concrete made with ASTM Type I or similar cement will, under otherwise constant conditions, perform in high-sulfate conditions better than a Type V concrete made at higher w/c leading to open, continuous porosity. Low permeability and adequate curing are the reasons why sulfate attack cases are rare in low w/(c+m) concrete structures, such as massive structures. On the other hand, sulfate attack is quite common in sulfate-exposed concrete structures made of concrete with inadequate content of Type V cement at high w/(c+m). The atmospheric conditions (e.g., evaporation due to temperature changes), coupled with inadequate curing, may add other dimensions to the problem, especially in the case of thin structures exposed partially to an underground source of sulfate.

Let us quote from the conclusions of Reading [30]: *"The performance of the 40-year old concrete on Ft. Peck Dam* ***is surprisingly good*** *considering the* ***severe*** [sulfate] ***exposure and only modest controls on the cement,*** *and emphasizes the* ***importance of a high quality mixture****"* [highlighted by the present authors].

We also believe that the intent of the developers of sulfate-resisting cements such as ASTM Type V **was not** simply to replace one quality (low w/c giving the required strength) by another quality (sulfate resistance). To the contrary, the intent was to add an **additional level of protection** for otherwise good quality concrete desired for structures built at that time. That is the reason why **both** w/c and cement type are prescribed by the Uniform Building Code [22].

Standards and Test Methods: Are they Accurate and Relevant?

Nobody would argue that the existing standards and tests are absolute or even close to perfection. These issues have been debated for years [12, 13, 28, 29, 31-33] and will continue to be controversial. As a matter of fact, relevant issues are presently discussed both by ASTM and by the ACI 201 committee on Concrete Durability. Meaningful progress has been made, but additional changes are required as our knowledge of the underlying phenomena grows.

The reasons for the inadequacies of the sulfate attack-related standards are due, among others, to the following facts:

- The sulfate attack phenomena are complex and non-linear
- The available tests are developed for mortars rather than concrete

- A complete lack of rapid test methods
- Reliance on an oversimplified single measure such as compressive strength or volumetric expansion
- Prevention often relies on the type of cement mineralogy, not always an adequate remedy by itself
- Difficulties with reliable and representative sampling of the sulfate content at the site
- Lack of enforcement

These issues will certainly be discussed by others. However, in our view, the presently applicable standards and tests have to be tightened, better enforced and, possibly, replaced by standards more accurately reflecting field conditions. Such judgement was very politely expressed years ago by Harboe [18]: *"... our present requirements for sulfate resistant concrete are not considered to be excessively conservative"* and somewhat more forcefully by Mehta and Gjorv [12]: *"The existing methods of testing relative sulfate resistance of cements are not satisfactory."* Their statements are still valid.

WHAT NEW TECHNICAL KNOWLEDGE IS DESIRED?

In our view, the existing knowledge on the mechanism(s) of sulfate attack is adequate to prevent damage in most cases [34]; however, this knowledge has to be disseminated, accepted as valid, and applied. A thorough discussion of research needs, particularly with respect to standardization of sulfate attack tests, was presented in 1991 by Cohen and Mather [3].

The purpose of this Seminar is to formulate and answer important external sulfate attack-related detailed technical questions and to recommend actions to minimize such damage. We do not pretend to be able to make exhaustive recommendations by ourselves. Allow us, however, to suggest a few technical items that we believe are important to include in the upcoming discussions and that need to be further explored.

- What is the relationship between the rate of sulfate attack and the multiple variables represented by the complex characteristics of modern cements? More specifically, how can we relate the rate of potential sulfate attack to the following and other "qualities" of cements:
- distribution of clinker compounds in a particular cement,
- chemical and mineralogical compositions of the clinker minerals, and their "reactivity" (how to best define?),
- form, reactivity, and amount of sulfates in clinker and cement,

- fineness of cement and its individual components,
- effect of minor components, including free lime and alkali sulfates.

- What are the most important parameters leading to sulfate resistance of a specific cement under changing conditions of water (better, pore solution) transport? What is the w/c+m level above which concrete cannot be protected by cement composition? How can the porosity, permeability, and diffusivity be engineered to give sulfate-resistant concrete?

- What are the chemical and environmental conditions under which the aluminate ions from the Fss contribute to expansiveness of the cement paste? What are the conditions under which the ferrite ions participate in formation of expansive ettringite?

- What are the basic processes and differences, under otherwise constant conditions, governing deterioration of concrete components by sulfates with different cations? What are the mechanisms of interactions of such cations with respect to the prevailing mechanism of deterioration? Under what conditions is it probable that deterioration will/will not occur as a result of volumetric expansion of the paste caused by ettringite (or gypsum?) formation? What is the effect of simultaneous chlorination, carbonation, etc.?

- What are the best methodologies for determination of sulfate content and form in clinkers, hydraulic cements, soils, aggregates, and ground water?

- What are the critical characteristics of fly ashes, slags, and other mineral admixtures that often, but not always, enable concrete sulfate resistance?

- Is strength a proper characterization of the progress of sulfate attack? Is tensile strength more sensitive to sulfate attack then compressive strength? Is volume change an adequate measure of sulfate-related damage? How best to characterize and measure sulfate-related deterioration of concrete when caused by mechanisms other than formation of expansive ettringite?

- What are the best failure criteria? What are the critical experimental parameters needed to enable prediction of the rate and time of failure of concrete exposed to and attacked by sulfates of various compositions? What additional information and methodologies are needed to improve modeling methodologies that would predict behavior and failure of concrete exposed to different sulfate-bearing environments?

- What **additional** external remedies (drainage, coatings, etc.) should be considered in severe sulfate environment? Under what conditions?

CONCLUDING REMARKS

Unfortunately, concrete is the "Cinderella" of man-made materials. Public support for concrete research is limited; industrial funds are minimal. Concrete mix design and processing are often abused. Why is this so? There are several reasons: concrete is a very forgiving material which is used in large quantities, it is an inexpensive commodity, it is considered to be a low-tech (unappealing) material and, most important, its internal complexity is much higher than that of most other materials, including ceramics, metals, and polymers.

The chemical and physical intricacies of concrete formation and degradation are intellectually most interesting and extremely challenging to study; the available know-how on sulfate degradation was generated by first-class chemists, mineralogists, and physicists. Unfortunately, while many of the "lab-crete" researchers have little appreciation for the "real-crete" requirements and techno-economic problems encountered in the real world, routine interpretations are often done by practicing civil engineers, having typically inadequate appreciation for the properties and durability of materials. Concrete can be studied and deconvoluted only by close cooperation of multidisciplinary groups consisting of chemists, physicists, materials scientists, engineers, geologists, mineralogists and, yes, management with vision.

We would like to share with you the following quote from Chris Hall [35]:

" ... chemical reactivity is so much an intrinsic feature of cement-based materials that we must place concrete chemistry at the head of the research agenda for rational concrete technology." Are we ready to accept this fact?

***

Concrete deterioration due to sulfates can be easily eliminated by application of existing knowledge and common sense based on experience. That includes:

- use of proper mix design, including appropriately low w/c and cement type selected for the expected environmental/atmospheric conditions of exposure; adequate cement content; design for long-term durability rather than compressive strength; **and**
- proper processing of concrete, including adequate mixing to obtain a homogeneous mixture, proper consolidation to enable formation of a

continuous and impermeable concrete matrix, and – most important - adequate curing to guarantee long-term chemical and physical stability; **and**

- minimizing access of ground water and other carriers of aggressive ions to the concrete structure; **and**
- timely maintenance and repair of the structure.

REFERENCES

[1] Gebhardt, R.F. "Survey of North American Portland Cements: 1994", Cement, Concrete, and Aggregates, CCAGDP, **17**, 2, 145 (1995),

[2] PCA, "Portland Cement: Past and Present Characteristics", Concrete Technology Today, July 1996, 1 (1996)

[3] Cohen, M.D. and Mather, B., "Sulfate Attack on Concrete – Research Needs", ACI Materials Journal, January-February, 62 (1991)

[4] Lawrence, C.D., "Delayed Ettringite Formation: An Issue?," in *Materials Science of Concrete* (J. Skalny & S. Mindess, Eds.), Volume IV, The American Ceramic Society, 1995, pp.113 (1995)

[5] ASTM Symposium on Internal Sulfate Attack, San Diego 1997 (Hooton D. and Roberts, R.L. (Chairmen); presentations to be published in Cement Concrete Aggregate, Vol.21, (1999)

[6] Papers in *Ettringite -- The Sometimes Host of Destruction* (Erlin, B.,. Skalny, J. and Hill, G., Eds.), American Concrete Institute SP-177, in press (1998)

[7] Wong, G. Sam and Poole, T., "Sulfate Resistance of Mortars Made Using Portland Cement and Blends of Portland Cement and Pozzolan or Slag," WES Technical Report SL-88-34 (1988)

[8] Mehta, P.K., "Durability – Critical Issues for the Future", Concrete International **19**, 7,27 (1997)

[9] Neville, A., "A 'New' Look at High-Alumina Cement", Concrete International **20**, 8, 51 (1998)

[10] Wang Kejin, Igusa, Takeru, and Shah, S.P., "Permeability of Concrete – Relationships to its Mix Proportions, Microstructure, and Microcracks", in *Materials Science of Concrete, Special Issue: The Sidney Diamond Symposium* (M.Cohen et al, Eds.), The American Ceramic Society, Westerville, OH, 45 (1998)

[11] Lee, F.M., *The Chemistry of Cement and Concrete*, Chemical Publishing Co., New York (1971)

[12] Mehta, P.K. and Gjorv, O.E., "*A New Test for Sulfate Resistance of Cements*", Journal of Testing and Evaluation, JTEVA **2**, 6, 510 (1974)

[13] Hobbs, D.W., *Minimum Requirements for Durable Concrete*, British Cement Association, U.K.(1998)

[14] Khatri, R. P. and Sirivivatnanon, V., "Role of Permeability in Sulfate Attack," Cem.Concr.Res.**27**, 8, 1179 (1997)

[15] Guide to Durable Concrete (1992), section 2.2, ACI 201.2R-92

[16] Young, F., Sulfate Attack (letter to the Editor), Concrete International **20**, 8, 7 (1998)

[17] Thorvaldson, T., Vigfusson, V. A. and Larmour R.K., "The action of Sulfates on the Components of Portland Cement", Trans. Royal. Soc. Canada, 3rd Series, **21**, Section III, 295 (1927)

[18] Harboe, E. M., "Longtime Studies and Field Experiences with Sulfate Attack," in *Sulfate Resistance of Concrete*, American Concrete Institute SP-77, 1 (1982)

[19] Taylor, H.F.W., *Cement Chemistry*, 2nd edition, Thomas Telford Services Ltd., London (1997)

[20] Riedel, W., "Corrosion Resistance of Cement Mortars in Solutions of Magnesium Salts," Zement-Kalk-Gips **26**, 286 (1973)

[21] Matthews, J. D. and Baker R. S., "An Investigation of the Comparative Sulfate Resistance of Ordinary and Sulfate-Resisting Portland Cements and their Blends with Fly Ash and Blastfurnace Slag," BRE Note N116/76, Garston, U.K. (1976)

[22] Uniform Building Code, International Conference of Building Officials, Whittier, CA (1994)

[23] Novokschenov, V., "Investigation of Concrete Deterioration Due to Sulfate Attack – A Case History," in *Concrete Durability* (J. M. Scanlon, Ed.), American Concrete Institute SP-100 (1983)

[24] MacFarlane, I. C., "Corrosion of Concrete by Bog Waters – A Literature Review," NRC Canada, November 1958

[25] Lossing, A., "Sulfate Attack on Concrete Pavements in Mississippi", presentation at the 44th Highway Research Board Meeting, Washington, DC, January 1965

[26] Swenson, E.G. and Mackenzie, C. J., "Contribution of Thorbergur Thorvaldson to Cement and Concrete Research: A Historical Review," in *Performance of Concrete: Resistance of Concrete to Sulfate and Other Environmental Conditions* (E. G. Swenson, Ed.), University of Toronto Press (1968)

[27] Stark, D., "Longtime Study of Concrete Durability in Sulfate Soils," in *Sulfate Resistance of Concrete*, American Concrete Institute SP-77, 21 (1982)

[28] Kalousek, G. L., Porter, L.C. and Benton, E.J., "Concrete for Long-term Service in Sulfate Environment," Cem.Concr.Res. **2**, 79 (1972)

[29] Mehta, P. K., "Sulfate Attack on Concrete – A Critical Review," in *Materials Science of Concrete III* (J. Skalny, Ed.), The American Ceramic Society, Westerville, OH, 105 (1992)

[30] Reading, T. J., "Physical Aspects of Sodium Sulfate Attack on Concrete," in *Sulfate Resistance of Concrete*, American Concrete Institute SP-77, 75 (1982)

[31] Mather, K., "Tests and Evaluation of Portland and Blended Cements for Resistance to Sulfate Attack," in *Cement Standards Evolution and Trends*, ASTM STP-663, 74 (1978)

[32] Dunstan, E. R., "A Spec Odyssey – Sulfate Resistant Concrete for the 1980's", in *Sulfate Resistance of Concrete*, American Concrete Institute SP-77, 41 (1982)

[33] Spellman L and Skalny J., "ASTM Standards and Materials Research," in *Materials Science of Concrete* IV (J. Skalny & S. Mindess, Eds.), The American Ceramic Society, Westerville, OH, 391 (1995)

[34] DePuy, G.W., "Chemical Resistance of Concrete", in *Concrete and Concrete-Making Materials"*, ASTM STP 169C, Philadelphia, PA, 263 (1994)

[35] Hall, C., "Research Needs and Opportunities", in *Penetration and Permeability of Concrete: Barriers to organic and contaminating liquids"*, RILEM Report 16, E&FN SPON, London, p.329 (1997)

SULPHATE ATTACK IN CONCRETE:
Urgent Needs for Renewal in Concrete Research

G. M. Idorn
International Consultant
Naerum, Denmark

ABSTRACT

Demographic changes in the world will require increasing investment in concrete buildings and construction in regions with hitherto unknown potentials for intense sulphate attacks. The conventional concepts and methodology of concrete research do not suffice to ensure durability of field concrete under such conditions. Recognition of the non-linear nature of the conversion of the chemical energy in the reactions of cementitious systems to mechanical work, such as cracking and volumetric expansion, suggest the desirability of applying fuzzy-logic and chaos theory to the analyses and interpretation of laboratory simulations and field investigations. The investments and mindshifts required for this call for cement industry leadership, especially in the current economic situation.

CONCRETE RESEARCH IN CONTEXT

The precious Wallace collection of antiques in London was founded by British Marquesses whose "spiritual affinities were with the ancient regime of 18th century Paris, when sensuous pleasure in life and art was not harassed by utilitarian scruples" (1). In the same century, Isaac Newton and other outstanding fellows of the Royal Society, who were leaders in Academia, industry and commerce, achieved formidable advances in the natural sciences and exerted equally determined efforts in support of the transfer of technological innovations (2).

The last international conference I recall which has concentrated on the deliberate transfer of new scientific discoveries to progressive industrial technology was in 1983; incidentally it was arranged by the Royal Society (3). Although there are still pleasures in working on the sciences behind cement and concrete technology without being "harassed by utilitarian scruples", we are now in a situation where such harassment has become increasingly common.

In the 19th century, collaboration between scientific researchers and cement industry and civil engineering leadership made decisive progress towards the resistance of concrete against sulphate reactions in marine exposure in Europe. They studied and renewed the ancient use of pozzolans, and early in the 20th century found that ground, granulated slag was effective in blends with Portland cement. The research was based primarily on experience with port and coastal concrete constructions and extensive field test plots, along with intensive international co-operation and close association with the scientific approaches by cement industry chemists; all with vigorous top-down support and participation (4, 5). From the 1920's, advances in cement chemistry in North America and Europe brought forward the low C_3A cements as *sulphate resistant cement.* They were developed on the basis of the conventional perception of the chemistry of sulphate reactions; their preventive effects were classified and documented primarily by laboratory mortar bar testing.

A NEW WORLD SITUATION – NEW REQUIREMENTS FOR CONCRETE

The title of the present seminar, *sulphate attack mechanisms*, refers to its defensive origin. It was an omen that the association of the infamous delayed ettringite formation (DEF) with excessively high curing temperatures in pre-cast concrete (6), was not unravelled as a milestone in research and development under industrial and scientific leaderships, but as a result of extensive litigation in North America and Europe. There is, however, now a new situation emerging which sustains the need for the change from a defensive to a progressive strategy for research on sulphate resistance of concrete under present day exposure conditions.

The third millennium's global socio-economic development must emphasise the population growth in the belt of the globe between the Tropic of the Cancer and the Tropic of the Capricorn. Concrete, the indispensable building material for development in this region, must be produced to be able to endure exposure conditions which can be more intense than any steam curing procedure in industrialised pre-casting. But there will also be tremendous construction investments in arctic regions, where expansive calcium carbonate hydration has been found vigorously taking place in sea water at close to 0°C (5).

There are also still large, undeveloped areas with aggressive sulphatic soils in some of the CIS countries, Mongolia and China, and presumably also in South America and several parts of Africa. In other words, the advance of scientific research concerning different types of hydration reactions, including those comprising sulphates (and ASR), is much more important for progressive technology development in the future than for assessing "blame" in past concrete failures.

NEEDED CHANGES TO THE RESEARCH PARADIGM

The DEF story re-emphasized the validity of the Arrhenius equation as governing the interdependence of the temperature regime of cement hydration and the build-up of the solid structure and strength of hardening cement paste. It also focused attention on the abrupt qualitative change to instability of ettringite in concrete at about +70°C. Hence, this situation has served to remind us that the traditional "room temperature hegemony" of conventional research and testing, which was appropriate in the past when sulphate attack was confined to marine concrete in the benign north temperate, climatic environments, is no longer sufficient. Both research and practice must now take into account the potential for failures with concrete made and used under the current and future energy intensive technology and performance requirements.

Along with our reliance on room temperature testing, the traditional use of "linear expansion" of pastes, mortars and concretes as an indicator of deleterious reactions, which has persisted for over a century, must be terminated. An analogy with hydraulic research is appropriate. If that was confined to dealing with laminar flow of liquids in laboratory modelling, hydraulic engineering would experience one catastrophic failure after another, because the energy conversion under natural and engineering circumstances causes flows of liquids to be a chaotic mix of laminarity and turbulence.

For analyses of the corresponding nature of energy conversion in deleterious chemical reactions in concrete, another analogy, namely to ASR, is useful, because early researchers appreciated the distinction between:

- the chemical reaction, i.e. the dissolution of susceptible silica in alkali hydroxide;
- the mechanical work exerted by the conversion of energy in the reaction.

Two ASTM test methods, C289 and C227, emerged from application of this dualistic concept; however, there was little appreciation of their inherent, fuzzy complexity, which was too complicated for laboratory and test convenience. It is noteworthy that P.E. Grattan-Bellow in 1996 in a review of many years' development of the ASR test methods (7), concluded that neither of the tests could disclose whether a susceptible aggregate would cause a deleterious or harmless reaction in field concrete. In other words, the research community had consistently chosen to consider ASR to be a linear dynamic conversion of energy to mechanical work, despite the obvious inconsistency with the observed course of the reaction in field concrete.

For sulphate reactions the situation is different in some respects. There does not now exist an ASTM method which explicitly measures the chemical reaction - including ingress, dissolution and re-crystallisation of sulfate-containing compounds. They are merely methods which measure mechanical effects such as linear expansion of cement paste, mortar bars or concrete specimens. Moreover, the reaction sites are not, as in ASR, macro aggregate particles with visible cracks (nuclei of volumetric expansions) radiating into the ambient cement paste, but sub-microscopic cement clinker and paste-derivative compounds which are difficult to localise and uncertain to identify as origins of sub-microscopic fracture and expansion.

The lesson from mineralogy that the process of ettringite formation changes abruptly to dissolution at about +70°C is an undeniable indication that chemical processes in concrete are generally non-linear. It is also true, therefore, that one-dimensional, i.e. linear expansion measures, may be misleading indicators of the mechanical effects of such processes. This includes situations in which the chosen laboratory procedures confine the energy conversion to occur at a constant temperature for reasons of convenience, thus violating the true kinetics of the reactions instead of modelling the reality of field concrete.

NEW CONCEPTUAL APPROACHES

Personal experience with several thousand field inspections and investigations of deleteriously affected concrete exposed to conditions from arctic to tropical, involving aggressive conditions lasting from two thousand to only a few years, justifies my view: *research must develop system protocols for scientific field concrete studies, and must implement them vigorously.* Detailed information on the concrete and the environmental exposure circumstances must be recorded. Important concrete structures exposed, *inter alia*, to sulphate reactions should be chosen for long-term programmes as in the past, but with the benefits of updated methodology. Emphasis should be on the heterogeneous (i.e. turbulent) course of deleterious reactions and their mechanical effects through their developments in field concrete, supported by thorough petrographic examinations of macro to sub-microscopic sites of reactivity. Such dissections should lead to laboratory studies, including simulation modelling, in which the simplistic retention of linear expansion testing at constant temperatures must be abandoned.

There are great promises in Gao Fa Liang's application of Fuzzy Logic (8) for analysis and interpretation of laboratory testing of concrete specimens, because the complex of chemical reactions and mechanical effects under realistic circumstances is indeed fuzzy (9). For the up-scaling from such modelling to field concrete it is once again worthwhile to refer to ASR due to the tremendous accumulation of experience with this reaction over close to 60 years (5). Numerous investigations of field concrete affected by harmful ASR have shown substantial, and often confusing, variability in the extents and features of surface map-cracking and of the rare volumetric expansion. This is consistent with the observation that thin section examinations of such concretes reveal astoundingly varying:

- numbers of sites of microcrack formation in reacting particles;
- intensities of radiating microcracks and amounts of exuded alkali-silica gel in such cracks and in voids;
- evidence of deterioration such as clusters of ettringite in available void space.

Systematic accumulation of such thin section observations compared with the visual symptoms of the reaction in field concrete are evidence of the dissipative, non-linear dynamic progress of the reaction and its mechanical effects. Hence, it ought not to be surprising that simulations and "linearised" experimental and test measurements can produce a confusing number of different histories of the course of the reaction, and therefore make predictions illusory .

Wang et al, (10) found that the features of surface cracking in ASR affected structures are fractal, and this is typically also the case on the micro-scale of evidence of the reaction such as in "reaction rims", solidified exuded gel, microcracking etc. This fractality along with other visual evidence of ASR in field concrete suggests convincingly that researchers ought to recognise the reactions as truly chaotic, and to see the modern theory of Chaos (11, 12) as a challenging door opener in the quest for improved understanding of ASR.

Sulphate reactions in field concrete represent, like ASR, a non-linear dynamic conversion of chemical energy to mechanical work. The visual features of the effects of these reactions - surface crumbling, appearance of crystalline secondary compounds, etc. - are truly fractal (crumbling is micro-mapcracking), and the "growth" of mechanical effects from the sub-microscopic nuclei to macro-scale is bewilderingly unpredictable. Although hardly anywhere systematically explored, the indices are irrefutable: the chaotic nature of sulphate reactions must be clarified in order to understand them and - why not - make them useful.

CONCRETE RESEARCH IN THE MARKET PHASE

Since any deleterious reaction in field concrete takes differing random courses towards final histories in accordance with different starting conditions and intermediate fractal "jumps", laboratory modelling cannot be planned to simulate one or other particular solution, but must be applied for examination of the various possible courses of the reality. This inevitable consequence of adopting fuzzy-logic and chaos-theory as basic principles in new research on concrete durability requires the introduction of comprehensive computer programming and operations, profound changes in mindset, substantial economic investments and determined, strategic leadership.

The basic laws of the liberal market economy which rules the world says that:

- cement and concrete production and use are indispensable for further socio-economic progress in the world;
- cement and concrete industries must have growth opportunities in order to acquire investor interest, and must have financial strength to invest in R&D as a prerequisite for growth.

REGION	1996		1997	
	Sales Bill. US$	Number of companies	Sales Bill. US$	Number of Companies
USA	2,127	2	3,760	1
Europe	25,923	8	73,692	10
Japan	120,625	13	28,030	3
Emerging 200	18,787	-	8,601	5
TOTAL	167,482		114,083	

TABLE 1. Sales in Billion US$ and number of companies in the industrial sectors ***Building Materials and Construction*** *entered among the GLOBAL 1000 (highest in market value, 1996 and 1997) in the industrial regions, and among 200 of highest market value in the "emerging markets" (i.e. developing countries). For comparison, USA has of all countries the highest number of entered companies - 480 in 1997, and the four leading IT-companies, all American, account for a total sales of 157,831 Bill. US$*

Table 1, based on 1996 and 1997 surveys (13) of the status of the 1000 commercial enterprises of the highest market value in the world, shows for the two categories Building Materials (and components), and Construction, which approximately represent cement and concrete construction and uses, that the financial power concentrations in the field are in Europe, albeit with Japan still second despite the later years' recession. (Industries in the People's Republic of China and the CIS countries do not appear in the survey).

As the stagnating 850 million population in the industrial world do not represent potentials for further growth of the homelands' enterprises, it is the increasing 4400 mill. population in the developing continents and countries which account for the future markets for the industries and professional civil engineering enterprises. As mentioned in the introduction, these new major regions for socio-economic development comprise the sub-tropical/tropical belt of the world with the heat as the energy intensity source; but also colder regions of China and the CIS countries possess little known geological and environmental circumstances where construction investments are bound to come. As that happens for the provision of energy, housing, industry and commodity production for the increasing populations, corresponding investments in R&D are required to keep the source materials' depletion and the building and construction expenses at a minimum. It is the planning and management of such research and its transfer which will require strong, persistent industrial commitments. This points especially to the leadership in the major concentrations of cement industries, where the financial power is, or, as implied in Table 1, to organisations such as Portland Cement Association, Verein Deutscher Zementwerke, Japan Cement Association, CEMBUREAU, and their leading cement industry corporation members.

CONCLUDING REMARKS

Research on sulphate attack mechanisms cannot enjoy progress without being an integral part of new approach to concrete research for progressive technology as urgently needed service to mankind. Some researchers who feel they live primarily for personal independence in their work may regret a new development with the focus on utilitarian requirements to their individual or institutional planning and ways of operation. But fewer would prefer to see the legal professions as the major fund-raiser for concrete research, with *ad hoc* emphasis on subjects changing from one backward oriented controversy to another.

Moreover, this perspective of the new development does not include merely the conventional and inexpensive automatic continuation of research which in reality is obsolete. No, the perspective now demands from the leadership commitments for profoundly new, expensive and difficult but promising attitudes to R&D, of benefit to both the producers and the users.

REFERENCES

1. Hughes, P., "The Founders of the Wallace Collection". 55 pp. The Trustees of the Wallace Collection, Manchester Square. London. 1992
2. Sobel, Dava: "Longitude. The True Story of a Lone Genius Who Solved the Greatest Scientific Problems of his Time". 184 pp. Fourth Estate Ltd., London 1995
3. Technology in the 1990's: "Developments in hydraulic cements". Proceedings a Royal Society discussion meeting held on 16 and 17 February, 1983. Eds. Sir Peter Hirsh, J.D. Birchall, D.D. Double, A. Kelly, G.K. Moir and C.D. Pomerc 207 pp. The Royal Society, London. 1983.
4. Poulsen, A: "Cement in Sea Water". the International Association for Testing Materials, 5th Congress. 59 pp. Also in French and German. Copenhagen, 1909
5. Idorn, G.M.: "Concrete Progress. From Antiquity to the third Millennium". 35 pp. Th. Telford Publishers, Ltd. London. 1997
6. In the United States District Court for the District of Maryland. MDL. Docket No. 827. "Final Judgement Order". 34 pp. 1995
7. Grattan-Bellew, P. E.: "A critical reciew of accelerated AAR Tests". Proceedi of the 10th ICAAR. Ed. A. Shayan. CSIRO Division of Building Construction and Engineering. Melbourne, Australia, 1996.
8. Gao Fa Liang: "A new way of predicting cement strength - Fuzzy Logic". Cement and Concrete Research, pp. 883-888, Vol. 27, No. 6. 1997.
9. "Fuzzy sets and systems". Official Publication of the International Fuzzy Syste Association. 4 pp. (Leaflet). Elsevier Science, New York.
10. Wang, T., Nishibayashi, S. and Nakano, K.: "Fractal analyses of cracked surfac in concrete". pp. 426-433. Proceedings, 10th ICAAR Melbourne 1996.
11. Glück, J.: "Chaos - Making a new Science". 314 pp. Viking Penguin Inc. New York, 1987.
12. Thompson, J.M.T.: "The Principia and contemporary mechanics: chaotic dynamics and the new unpredictability ". Notes and Records of the Royal Society. 4297-122. 1988.
13. Weber, J.: "The Global 1000". Business Week. pp. 42-79. 13 July 1997.

THE ROLE OF ETTRINGITE IN EXTERNAL SULFATE ATTACK

P.W. Brown, The Pennsylvania State University, University Park, PA, USA

H.F.W. Taylor, Maundry Bank, Lake Road, Coniston, Cumbria LA21 8EW, UK

ABSTRACT

The reactions and microstructural changes taking place when cement paste or concrete made with ordinary or sulfate resisting portland cements is attacked by solutions of sodium or magnesium sulfate are summarized and related to the observed physical changes. Damage in field concretes may take the form of loss of cohesion and strength rather than expansion, but these effects are related. The main cause of damage in some instances may be C-S-H destruction rather than ettringite formation, but the latter nevertheless plays an essential part in the process. Hypotheses to explain the mechanism of expansion from ettringite formation are reviewed. The simple view that expansion can be attributed solely to the increase in solid volume is untenable, but most of the numerous other hypotheses have strengths and weaknesses, and further experimental work is needed to determine which, if any, is correct. Such work should take into account the conditions existing in the field and be related to field observations.

INTRODUCTION

Ettringite formation in fresh concrete is regarded as beneficial in controlling the rate of set, but its formation in hardened concrete can be associated with distress. There is no consensus nomenclature for different types of sulfate attack. This paper deals only with external sulfate attack, which we will define as that in which the source of the SO_4^{2-} is external to the concrete. This distinguishes it from internal sulfate attack, in which the source of the SO_4^{2-} is inside the concrete, and which thus includes delayed ettringite formation (DEF) and expansion due to excessive sulfate content, whether from the cement or other constituents of the concrete. Although the mechanisms of external and internal sulfate attack are not fully established, the events causing microstructural damage are likely to have features in common. We shall also deal only with pastes, mortars or concretes made using portland cements without slags, pozzolans or other additions.

ETTRINGITE FORMATION IN CEMENT PASTE AND CONCRETE

Tricalcium aluminate (C_3A) may be present in portland cement in amounts up to at least 14%, as estimated using the Bogue calculation. In the absence of a retarder, the rapidity of its reaction can cause a premature stiffening, known as "flash set", which makes mixing, placement and finishing difficult. In the absence of a retarder, the hydration of C_3A in a cement paste yields the AFm phases, C_4AH_{13} and C_2AH_8, the hexagonal, platey crystals of which form an interlocking network. In the presence of calcium sulfate, different products are formed, partly as a layer over the C_3A surfaces which is less permeable. They are also of different morphology, and the C_3A thus reacts more slowly and gives products less able to cause stiffening. The most easily detectable of these products is ettringite, the formula of which is $C_6A\bar{S}_3H_{32}$, or, more meaningfully in terms of ionic constitution, $Ca_6[Al(OH)_6]_2(SO_4)_3\cdot 26H_2O$. Ettringite forms acicular, hexagonal crystals, some of which appear to grow outwards from the C_3A surfaces. The retardation of C_3A hydration has been attributed to the ettringite coating (1, 2), but it has been queried whether this layer is sufficiently impermeable to have this effect (3) and the retardation is perhaps more probably due to an underlying layer of hydrous alumina (4, 5).

The amount of gypsum needed to control setting is typically in the region of 2%, but gypsum also accelerates the hydration of alite, and larger amounts are commonly added to increase early strength. Normally, the proportion used is such as to maximize one-day compressive strength, and is typically in the region of 3–4%. At proportions above the optimum, the early strength decreases, and at still higher proportions deleterious expansion can occur. This last effect is associated with ettringite formation. The cause of the decrease in early strength at contents nearer to but still above the optimum is not clearly understood, but could be similar.

It is widely stated that, beginning some 24 hours after mixing, the content of ettringite decreases in quantity and finally disappears. This is attributed to the fact that the calcium sulfate has been consumed, whereas C_3A continues to react, leading to decomposition of ettringite and formation of monosulfate ($C_4A\bar{S}H_{12}$). This is, however, only one of a number of possible reaction sequences. Its extent must depend on the amount of excess C_3A (6), and is probably substantial only with Type I cements relatively high in that constituent. As long ago as 1960, Copeland et al. (7) observed the presence of ettringite in several 10-year old cement pastes. Present day cements are considerably higher in SO_3 than those made at that time; this probably increases the tendency of ettringite to persist. The reaction sequence probably depends on the relative availabilities of $Al(OH)_4^-$, SO_4^{2-}, OH^- and CO_3^{2-} at each stage of reaction, and ettringite, monosulfate, and carbonate-containing AFm phases are all possible consituents of mature cement pastes.

Because small crystals are unstable relative to larger ones, ettringite tends to recrystallize in any available cavities, assuming that water is present for transport. The presence of ettringite, as detected in a distressed concrete by petrography, SEM or XRD therefore does not necessarily mean that the ettringite has caused the distress. Even where ettringite formation has caused the distress, it may not be that which is most obviously detectable, but that present as much smaller crystals within the paste. Erlin (8) noted that the presence of ettringite or other phases at a paste-aggregate interface shows only that something has caused the paste to expand, and that in order to establish the most probable cause or causes of the distress, more detailed consideration of various relevant factors was needed.

REACTIONS IN EXTERNAL SULFATE ATTACK

External sulfate attack on a paste, mortar or concrete proceeds through the advance of a reaction front into the material (9, 10, 11). With a paste of low w/c ratio, immersed in a 0.25 molar Na_2SO_4 or $MgSO_4$ solution, the effects can be undetectable at depths greater than 0.5–1 mm after 6 months (9). With a poor quality field concrete of high w/c ratio, it can affect all parts of a substantial mass of material (11). The initial reactions in the sequence of attack tend to be those observed at the greatest depths.

Attack by Na_2SO_4 Solutions

In the first stage of reaction, ettringite is formed. Equation 1 shows this, assuming that monosulfate is the source of aluminate ions, but these could also be supplied by other phases, such as unreacted C_3A. Ettringite formation requires a supply of Ca^{2+}, which is initially obtained by dissolution of $Ca(OH)_2$ (Equation 2). When no more $Ca(OH)_2$ is readily available, decalcification of C–S–H begins

(Equation 3). As the surface is approached, the SO_4^{2-} concentration in the pore solution increases, and gypsum is formed (Equation 4). This further increases the demand for Ca^{2+}. Close to the surface, decalcification is extreme, and ettringite is probably the only crystalline phase that remains (provided the pH is the range where ettringite remains stable).

$$Ca_4Al_2(OH)_{12}{\cdot}SO_4{\cdot}6H_2O + 2Ca^{2+} + 2SO_4^{2-} + 20H_2O \Rightarrow Ca_6Al_2(OH)_{12}(SO_4)_3{\cdot}26H_2O \quad (1)$$

$$Ca(OH)_2 \Rightarrow Ca^{2+} + 2OH^- \quad (2)$$

$$1.7CaO{\cdot}SiO_2{\cdot}aq. \Rightarrow xCa^{2+} + 2xOH^- + (1.7-x)CaO{\cdot}SiO_2{\cdot}aq. \quad (3)$$

$$Ca^{2+} + SO_4^{2-} + 2H_2O \Rightarrow CaSO_4{\cdot}2H_2O \quad (4)$$

Decalcification of C–S–H has been considered to be insignificant with Na_2SO_4 solutions (12) but has been detected using X-ray microanalysis (9). Charge balance in the pore liquid could be maintained by outward migration of OH^- or inward diffusion of Na^+ or both. In the former case, the alkali cation plays no part in the reaction. In the latter, it is involved only in that its presence allows the OH^- concentration in the pore solution to increase. It has been suggested that this can cause concomitant alkali silica reaction (13). However, the pH values of solutions in which test specimens are immersed increase, showing that the OH^- is at least partly lost to the solution. In field concrete, sodium carbonates may form.

Attack by $MgSO_4$ Solutions

The reactions are modified by effects resulting from the low solubilities of brucite ($Mg(OH)_2$) and magnesium silicate hydrates. Because of these low solubilities, penetration of Mg^{2+} ion into the material is normally very restricted, but can be substantial in porous concretes of high w/c ratio (11). A composite layer of brucite and gypsum is typically formed at the surface. Magnesium silicate hydrate, identified by XRD and X-ray microanalysis as poorly crystalline serpentine ($M_3S_2H_2$; 9, 14) can also be formed. The precipitation of these phases creates a very strong demand for OH^- and the resulting fall in pH can be sufficient in the regions strongly attacked to decompose all the hydrated calcium silicate and aluminate phases, including ettringite, the Ca^{2+} that is released being precipitated as gypsum. Measurements of pH show that, with $MgSO_4$, relatively little OH^- is lost to the external solution. In regions sufficiently distant from those in which the Mg-containing phases are precipitated, the demand for OH^- is more moderate and the reactions are the same as with alkali sulfate solutions.

The reaction sequences described above, for attack by either Na_2SO_4 or $MgSO_4$, are those observed with ordinary (Types I or II) portland cements. Sulfate-resisting (Type V) portland cements behave similarly, but all the effects are less marked (14).

Combined Attack by Sulfate and Carbonate

If CO_2 or CO_3^{2-} is present in addition to SO_4^{2-}, destruction of C–S–H can be extensive and even complete, causing severe softening or cracking of the concrete or mortar (15, 16). This effect is due to the formation of thaumasite (Equation 5):

$$3Ca^{2+} + \text{“}SiO_3^{2-}\text{”} + CO_3^{2-} + SO_4^{2-} + 15H_2O + \Rightarrow Ca_3[Si(OH)_6]CO_3.SO_4.12H_2O \qquad (5)$$

Thaumasite formation can be accompanied by normal sulfate attack or carbonation or both.

The crystal structure of thaumasite is closely similar to that of ettringite, with $Si(OH)_6^{2-}$ in place of $Al(OH)_6^{3-}$ and ($2CO_3^{2-} + 2SO_4^{2-}$) in place of ($3SO_4^{2-}$ + $2H_2O$). Its morphology is similar to that of ettringite, with which it is easily confused using light microscopy or SEM, but its XRD powder pattern, though similar, is distinct and it is readily distinguished by X-ray microanalysis. If unsubstituted and unmixed with other phases, it contains no Al_2O_3. Although it has been shown that ettringite is capable of forming solid solutions (17) recent SEM/EDX studies of field concrete have shown ettringite and thaumasite to form intimate mixtures but suggest the absence of ettringite-thaumasite solid solutions (18).

As may be seen from Equation 5, the requirements for thaumasite formation include silicate ions, represented conventionally in that equation as

SiO_3^{2-}. These can only be obtained from unreacted clinker phases or C–S–H. The extent of reaction is limited only by the availability of Ca^{2+} or silicate and can proceed until the C–S–H is completely destroyed; no significant quantity of aluminate ion is required, and the use of a sulfate-resisting portland cement does not confer protection. It can form very rapidly in concrete containing finely divided calcium carbonate. Thaumasite is formed most readily at temperatures below 5°C; factors favouring its formation in mortar or concrete include constantly high humidity, temperatures around 4°C, a little available aluminate ion and an adequate supply of sulfate and carbonate (19). The German literature (VDZ reports for 1981-84) suggests that failure in heat treated concrete in which thaumasite or mixed crystals of ettringite and thaumasite were observed is related to thaumasite formation. Failed railroad ties and panels were observed to contain large amounts of thaumasite and mixed crystals. Recently, thaumasite formation was observed in deteriorating foundations of homes in the southwestern USA where low temperatures are unlikely (18).

Sulfate is taken up by C–S–H when cement is cured at 80°C (20). Cement pastes hydrated for 5 hours, ground to fine powder, and stored at 5°C in H_2O suspension, and carbonated with CO_2 formed thaumasite; ettringite converted to AFm. The preparation of thaumasite seems to require also the presence of a small supply of aluminate ions, probably because the initial formation of a trace of ettringite is necessary to nucleate it (21).

PHYSICAL CHANGES IN EXTERNAL SULFATE ATTACK

Most laboratory tests to determine the relative susceptibilities of concretes or mortars made with different cements have been based on expansion measurements, though some have been based on ones of strength (22-24). A laboratory study on mortars immersed in Na_2SO_4 solutions (Figures 1 and 2) showed the two quantities to be related. The initial increase in strength and the delayed beginning of significant expansion were attributed to to the filling of pores by the products of sulfate attack, and the subsequent reversal of these effects to disruption of the material when this was no longer possible. The results also confirmed earlier observations (25, 26) that the rate of attack is highly pH dependent. Khatri et al (27) made similar observations on the effect of pH on the expansion of concretes, and also confirmed numerous earlier observations that expansion depends both on the nature of the cement and the permeability of the concrete. Immersion in solutions of controlled and relatively low pH thus accelerates the changes, and may offer a more realistic approximation to field conditions.

Mehta (24) concluded from a number of observations on field concretes that the principal manifestations of external sulfate attack were loss of adhesion and strength; long-term exposure of permeable concrete could transform the material into a puttylike mass with no strength. The observations described above show that expansion data may nevertheless give some indication of probable field performance. The loss of strength and cohesion could be at least partly explained by microcracking, but, especially if Mg^{2+} is present, destruction of C–S–H probably also has an important effect.

THE IMPORTANCE OF ETTRINGITE FORMATION IN EXTERNAL SULFATE ATTACK

Various workers (e.g. 28-30) have reported the formation of ettringite and gypsum, but there has been disagreement as to their relative amounts and whether or not the degree of expansion is related to the quantity of ettringite formed. Probably a majority of studies have shown that there is no linear relationship between the quantity of ettringite formed and expansion or other physical properties; ettringite formation can probably be extensive before damage is observed. Probably any hypothesis in which expansion is attributed directly or indirectly to ettringite formation can account for these observations, since some ettringite can be formed in places that do not result in stress.

In an XRD study of phase composition at different depths, Wang (31) found relatively little ettringite. Bonen and Cohen (32, 33) and Bonen (34) found only minor amounts of ettringite in pastes or mortars attacked by $MgSO_4$ solutions. Using X-ray microanalysis of polished sections, Gollop and Taylor (9,14) could not find any single-phase regions of ettringite in pastes of ordinary or sulfate resisting portland cements attacked by Na_2SO_4 or $MgSO_4$ solutions, though many analyses were compatible with mixtures of ettringite with C–S–H at or below a micrometre level. The XRD data obtained by Wang (31) on comparable specimens showed, however, that ettringite was present in the regions in which these mixtures occurred.

Gollop and Taylor (9) also observed that cracking occurred nearer the surface than the regions in which ettringite was believed to be present, and concluded that expansion was probably an indirect or delayed result of ettringite formation. Using XRD and X-ray microanalysis, Yang et al. (35) studied the interfacial zones of mortars attacked by Na_2SO_4 solution. They reported the presence only of gypsum, and obtained evidence that the sulfate penetrated mainly through the interfacial zones. These observations are generally consistent with those carried out on field concretes. Carbonated and decalcified near-surface zones, which contain gypsum and may show significant structural damage, overlaid regions where the pH remained high enough for ettringite to remain stable (11).

Some of these data cause one to wonder whether the damage from sulfate attack is caused by ettringite formation. Wang (31) concluded that damage was due mainly to formation of gypsum. In materials attacked by $MgSO_4$ solutions, the most obvious and direct source of damage is the destruction of C–S–H. The formation of ettringite nevertheless appears to play a vital role. This is shown by the abundant evidence that external sulfate attack, including that by $MgSO_4$, is most serious with cements high in C_3A, and much reduced with sulfate resisting cements, which are low in Al_2O_3. In blends containing moderate proportions of slag, damage is lessened if the slag is low in Al_2O_3 but increased if it is high in Al_2O_3 (22). The data suggest that ettringite formation causes the initial disruption of the material, and that this facilitates further reactions which can in some cases, at least, be more damaging.

The absence of agreement on matters involving quantitative contents of ettringite is probably due, at least in part, to problems inherent in its determination, both in general and more specifically in materials that have undergone external sulfate attack. There is no consensus as to the reliability of either XRD or thermal methods. XRD determination is subject to error arising from differences in crystallinity between sample and reference material (29). Both XRD and thermal methods are affected if partial dehydration has occurred, whether as a result of processes occurring before sampling or through localized heating when the sample is extracted or ground. Such dehydration occurs very easily (36). In materials that have undergone external sulfate attack, the ettringite is not uniformly distributed but is present, as shown above, in a zone within which there are probably gradients of ettringite concentration. Analyses of bulk material are therefore of very limited significance, and even if smaller samples are analyzed, estimation of the total amount of ettringite present is clearly far from straghtforward.

The data on pastes obtained using SEM with X-ray microanalysis show that the ettringite is formed in crystals of or below micrometer dimensions that are intimately mixed with C–S–H, and also that cracking occurs in pastes as well as in mortars or concretes. The damaging processes therefore take place, at least in part, within the paste, which is caused to expand. This leaves it open whether larger ettringite crystals observed in mortars or concretes are purely a result of recrystallization, or whether their formation and growth contributes further to the damage. The SEM and microanalyis data also do not show the mechanism by which paste expansion occurs. These questions are addressed in the remaining part of this paper.

HYPOTHESES OF ETTRINGITE RELATED EXPANSION

Various hypotheses have been advanced to explain expansion associated with ettringite formation. They can be grouped into several major categories.

1 Expansion Occurs because the Formation of Ettringite is Accompanied by a Large Increase in Solid Volume

This hypothesis has been widely espoused. However, the increase in solid volume when ettringite is formed from any likely starting materials is about the same as that occurring when C–S–H and $Ca(OH)_2$ are formed from tricalcium silicate (37), and in many other reactions in which liquid water is consumed and a solid product or products formed. Any reaction of this type proceeds only to the extent that either sufficient space is available, or the expansive forces are sufficient to disrupt the surrounding matrix. If neither condition is satisfied, reaction and expansion stop. This situation occurs in the hydration of cement pastes of low w/c ratios. In the hydration of normal cement pastes, even if saturated conditions are maintained, the extent of expansion is very small. An increase in solid volume may be a necessary condition for expansion, but it is not a sufficient one.

2 Ettringite Formation Causes Expansion if the Ettringite is Formed by Topochemical Reaction

This point of view has been variously expressed. Lafuma (38), who appears to have been the first to propose it, considered that ettringite formation caused expansion if it was formed in situ from C_3A, without any passage of ions through the solution. However, the crystal structures of C_3A and ettringite are totally unrelated, and it is inherently difficult to see how such a process could occur without the existence of a thin intervening layer, at least, of solution. Apart from this, ettringite formation from sources other than C_3A can cause expansion; in DEF, for example, hydrated compounds having true solubilities are interconverted, and expansion occurs. More directly relevant, the microanalytical evidence shows that in the case of external sulfate attack, the ettringite is not formed in space previously occupied by C_3A, or even at the surfaces of any unreacted C_3A residues that may remain, but in intimate admixture with C–S–H. The hypothesis thus appears unlikely.

A second significant shortcoming of the topochemical hypothesis is that ettringite exhibits a true solubility. The $CaO–Al_2O_3–CaSO_4–H_2O$ phase diagram at room temperature (Figure 3) illustrates the solubility range of ettringite (39). Consideration of this range in relation to the dissolution of C_3A shows that the topochemical model cannot apply. Rather, ettringite forms by a through solution mechanism in which C_3A dissolves and ettringite precipitates. This quaternary system is a model system for the formation of ettringite in a cement paste. The solubilities of the relevant phases are affected by the presence of alkali, but the overall mechanism of the solubilization of an aluminate source and the precipitation of ettringite still applies.

3 The Origin of Ettringite Expansion is Crystal Growth Pressure

Two mechanisms of expansion caused by pressure from growth of ettringite crystals might be considered. Debate on this matter has centered on the case of DEF, but is probably applicable also to external sulfate attack. The most obvious microstructural feature of DEF is the formation of cracks, typically some 20 μm wide and wholly or partly filled with ettringite, which are observed at aggregate interfaces and elsewhere. According to one point of view, expressed, for example, by Diamond (40) these cracks result from the growth of the ettringite crystals within them. Alternatively, it has been held that the formation of ettringite as microcrystals within the paste causes the latter to expand, thereby producing the cracks under discussion, and that the ettringite within them results from recrystallization, which could occur either simultaneously or subsequently. This view has been expressed by Johansen et al. (41), Erlin (8) and others. In support of paste expansion, Johansen et al. cited observations that the width of the bands of ettringite surrounding aggregate particles was proportional to the size of the particle, but the generality of this observation has been disputed (40).

Intermediate possibilities exist; Marusin (42) considered that the process began with expansion of the paste and that damage increased progressively as ettringite was deposited in cracks in the paste and at aggregate interfaces. The growth of ettringite crystals must be governed by local supersaturation, which could occur both in the paste and at paste-aggegate interfaces, so that neither hypothesis is inherently untenable. It has, however, been questioned whether the growth of ettringite crystals formed by recrystallization could exert sufficient pressure to cause distress (37).

Observations by Diamond and Ong (43) on limestone mortars suggest that in DEF both ettringite formation and expansion are either limited or strongly retarded by the strength of the surrounding material. This is consistent with the hypothesis that expansion results from crystal growth within the paste.

Several investigators have shown that expanion is not simply related to the amount of ettringite formed, but depends on where it is formed (44-46). If $Ca(OH)_2$ is present, the ettringite forms close to the surfaces of the aluminate particles, and expansion can be large. If $Ca(OH)_2$ is absent, it forms throughout the paste or in voids and expansion is minimal. The alumiante source may be anhydrous (e.g. C_3A) or hydrated (e.g. C_4AH_{13} or $C_4A\overline{S}H_{12}$). In the presence of $Ca(OH)_2$, supersaturation is high at the aluminate surfaces, and $Al(OH)_4^-$ ions react locally; in its absence, migration of $Al(OH)_4^-$ occurs more readily. Deng Min and Tang Mingshu (46) considered that crystal growth and swelling pressure from absorption of water both contributed to expansion, but that crystal growth pressure was a much more important effect. These considerations also imply that the loci of expansion are locations where the degree of suprsaturation with respect to ettringite are greatest, e.g. locations where C_4AH_{13} or $C_4A\overline{S}H_{12}$ have formed.

Factors affecting crystal growth pressure: Thermodynamic theory indicates that the maximum pressure that can be exerted by a growing crystal is given by the expression

$$P = \frac{RT}{V} \ln \frac{K}{Ks} \tag{6}$$

where P is the pressure in MPa, R is the gas constant (8.3 $J \cdot K^{-1} \cdot mol^{-1}$), T is the temperature in K, V is the molar volume of the substance ($m^3 \cdot mol^{-1}$) and K/Ks is the degree of supersaturation, defined as the activity product of the substance, K, divided by its value at saturation. Applied to ettringite at $K/Ks = 2.4$, this gives $P \approx 3$ MPa (37). However, this is the *maximum* pressure that can be exerted. Scherer (47) noted that the actual pressure is less, and depends on other factors, including the energy of the interface between the crystal and the pore wall and, for an acicular crystal such as ettringite growing across a pore, the yield or buckling strength of the crystal. Another factor concerns the size of the regions through which the growing crystals propagate, which must be sufficiently large that the stress field can interact with the large flaws that control the strength. Even if the crystallization pressure is large, the stress within a single pore cannot cause failure, because it acts on too small a volume. The theory suggests that compact crystals growing in small pores generate more swelling pressure than the long needles seen in open cracks. The further significance of these considerations for external sulfate attack has yet to be examined.

4 Expansion is Due to Bridging

Expansion could be associated with growth of ettringite crystals in another way, if growth occurs in a gap that undergoes repeated and reversible changes in width. Such changes could result from various causes, such as variations in temperature, stress or humidity. Thus, aggregate, paste and ettringite all have different coefficients of thermal expansion, and cement paste is subject to humidity-driven dimensional variations unlikely to affect ettringite or aggregate, and which affect both its external size and that of pores within it. If ettringite growth can be sustained during periods when the gap is wide, the crystal will exert a growth pressure when the gap dimensions decrease.

As with crystals growing in a gap of otherwise fixed size, the ability of ettringite to behave in the manner described above would depend on the strength of its crystals. Abo-El-Enein et al. (48) studied the effects of compaction on products of the C_3A-gypsum-water reaction. At ~50 MPa, cracking occurred, sometimes breaking ettringite crystals; at 5–20 MPa the crystals were not broken, the fractures passing between them. The authors tentatively concluded that the cohesive forces within the crystals were not much stronger than those holding one crystal to the next. Compressive strengths in this range are higher than the tensile strength of C–S–H, and the process could therefore eventually cause the latter to develop microcracks in the vicinity where an ettringite crystal had grown. Thus, the bridging effect thus provides an alternative possibility for expansion from crystal growth pressure.

For any theory based on crystal growth pressure to be valid, it must contrast the expansive behaviour of ettringite with the very much smaller expansive effects produced by other solids which form in hardened cement pastes, including especially C–S–H and $Ca(OH)_2$. The morphology of ettringite is dominated by its needle-like habit, indicating that a single growth direction is strongly preferred. In contrast, C–S–H is essentially non-crystalline and is capable of accommodating its formation to the space available. $Ca(OH)_2$ is crystalline, and when formed from free lime can cause expansion, but lacks a single direction of preferred growth and has little or no expansive effect when formed from the silicate phases in a cement paste.

5 Expansion is Due to a Change in the Gel Properties of C–S–H; the Formation of Ettringite is a Side Issue

Thorvaldson (49) considered that "the volume changes in the mortars are controlled by osmotic forces concerned with the swelling and shrinkage of gels, [and] that the chemical changes condition the gel system and destroy cementing substances while the formation of crystalline material is incidental to these reactions." In support, he quoted various observations showing that there is no relation between amounts of either ettringite or gypsum formed and expansion. Probably a majority of more recent studies support this. However, these changes could occur in the absence of ettringite.

DEF is accompanied by changes in the chemistry of C–S–H gel. Scrivener and co-workers (50, 51) found that in mortars 1 day after heat treatment, the S/Al ratio of the C–S–H gel was ~1.5 for mortars that subsequently expanded and ~0.5 for ones that did not. No ettringite was detected at this stage. When expansion occurred, the S/Al ratio dropped, apparently due to crystallization. On the other hand, Wang's evidence, mentioned earlier (31), showed that in pastes attacked by Na_2SO_4 solution, material having a S/Al cement of ~1.5 could also be a mixture of C–S–H with ettringite. A S/Al ratio of ~1.5 in C–S–H gel observed using X-ray microanalysis can thus arise from admixture of C–S–H with ettringite, but does not necessarily do so. One possibility, discussed later, is that it results from admixture of C–S–H with an amorphous, dehydration product of ettringite. The aluminate and sulfate ions might also be present as substituents or as sorbates in the C–S–H itself.

The addition of gypsum increases the rate of hydration in pastes of C_3S or cement, but the intrinsic strength of the hydration products decreases (52-54). Bentur's work (52, 53) showed that the pore structure was altered. He considered that the Ca/Si ratio of the C–S–H was increased; this could be accounted for by sorption of Ca^{2+} and SO_4^{2-}. Rasheeduzafar et al. (30) considered that, in sulfate attack on cement mortars, ettringite was initially formed, followed by gypsum. The environment around the C–S–H and ettringite, dominated by $Ca(OH)_2$ and OH^-, was thus replaced by one dominated by gypsum and SO_4^{2-}. This optimized the conditions for entry of SO_4^{2-}ions into the C–S–H, causing loss of strength and stiffness. This hypothesis could explain the observation (9) that expansion occurs later than the formation of ettringite. However, pastes of C_2S or C_3S are only attacked very slowly by Na_2SO_4 solutions (49). In addition, the hypothesis does not explain why aluminate ions have any effect.

6 Expansion is Caused by Sorption or Imbibition of Water.

Mehta (55) suggested that ettringite crystals of colloidal size imbibed water and thereby caused expansion. Generally, imbibition of water and resultant swelling involve gel-like materials of indefinite composition, which have the flexibility to expand without breaking. Ettringite does not appear to exhibit these necessary features. Three possibilities might be considered, viz

(a) separate microcrystals of ettringite attract water to their surfaces;

(b) aggregates of such crystals similarly attract (in this case, imbibe) water; and

(c) ettringite crystals take up water internally.

None of these hypotheses can be sustained as it stands. It is not obvious why ettringite crystals, whether isolated or in aggregates, should attract water to their surfaces more strongly than does C–S–H, which surely has a much higher surface area. As to (c), there is no room for more water in the crystal structure of ettringite. A higher hydrate with $36H_2O$ is reported to exist at high relative humidities (56), but no XRD data were quoted for it. The XRD pattern could not be identical with that of the 32-hydrate, and the patterns obtained for cement pastes, either before or after sulfate attack, give no indication of the presence of such a phase. These patterns include ones obtained from pastes studied in a saturated condition (7).

7 Expansion is Caused by a Reversal of Local Desiccation

A modified Mehta hypothesis may be possible. Formation of ettringite entails the incorporation of a substantial proportion of water into a solid product. If this water cannot be supplied rapidly enough from an external source, the following two hypotheses might be considered.

The ettringite is formed initially in a partly dehydrated condition; expansion occurs when it subsequently takes up water from an external source or elsewhere in the paste: This hypothesis is identical with variant (c) of that proposed by Mehta, if a lower hydrate is substituted for ettringite, and does not suffer from the objection raised to the latter. Ettringite is easily dehydrated from 32 to $12H_2O$

with shrinkage of the hexagonal lattice dimensions *a* and *b*, which are normal to the needle axis, from 1.12 nm to 0.84 nm, though the crystallinity normally deteriorates greatly. Abo-el-Enein et al. (57-59) reported some interesting results on dehydration and rehydration isotherms and isobars. They concluded that a step corresponding to loss of the most easily lost water occurred at a loss of 16 moles, but their data seem to fit 20, which agrees with evidence from analogous minerals (e.g. despujolsite, $Ca_3[Mn(OH)_6]\cdot(SO_4)_2\cdot 3H_2O$, 60).

The existence of a very poorly crystalline dehydration product of ettringite might explain the many claims to have found "amorphous ettringite". The latter is clearly a contradiction in terms, as ettringite is a mineral name that implies a particular *crystalline* structure and composition, but an amorphous dehydration product is a reasonable concept.

The ettringite is formed fully hydrated as a result of taking some of its water from the surrounding C–S–H; expansion occurs when the latter regains its water: This could explain observations that expansion occurs only after the ettringite has been formed and that damage can increase where significant increase in ettringite content is not detected. It is less clear whether this could be true also of the previous one. Assuming that one or other of these hypotheses is correct, the choice between them would depend on the relative affinities of partly dehydrated ettringite and partly dehydrated C–S–H for water.

With any hypothesis that assumes the initial formation of a partly dehydrated product, there is clearly a question as to whether the postulated local shortage of water could arise in a saturated material.

CONCLUSIONS

The views that have been proposed to explain the processes causing damage from external sulfate attack are manifold. We have attempted to summarize some possible mechanisms of expansion and to point out the strengths and weaknesses of each. Implicit in a diversity of views regarding the mechanism is the implication that none is completely valid and that further experimental work will be required to fully elucidate the mechanism of expansion. Particularly relevant in this respect is the observation that the expansion observed in the laboratory need not be matched in field concrete. Rather, loss of cohesion may occur. Further experimental work should take into account the conditions existing in the field and be related to field observations.

In addition to accounting for this observation, any theory of ettringite expansion must account for the facts that ettringite forms by a through-solution mechanism, exhibits a true solubility and must occupy more space than that which was initially available to it. Finally, the absence of a linear relationship between the amount of ettringite produced and the expansion observed must also be recognized.

ACKNOWLEDGMENT

PWB gratefully acknowledges the support of NSF grant CTS 93-09528.

REFERENCES

1 H.E. Schwiete, U. Ludwig and P. Jäger, "Investigations in the System $3CaO \cdot Al_2O_3$-$CaSO_4$-H_2O"; pp. 353–367 in *Symposium on Structure of Cement Paste and Concrete*. Special Report 90, Highway Research Board, Washington, DC, 1966.

2 M. Collepardi, G. Baldini and M. Pauri, "Tricalcium Aluminate Hydration in the Presence of Lime, Gypsum or Sodium Sulfate," *Cem. Concr. Res.*, **8** [5] 571–580 (1978).

3 P.K. Mehta, "Scanning Electron Micrographic Studies of Ettringite Formation," *Cem. Concr. Res.*, **6** [2] 169–182 (1976).

4 W.A. Corstanje, H.N. Stein and J.M. Stevels, "Hydration Reactions in Pastes C_3S+C_3A+$CaSO_4 \cdot 2aq$+H_2O at 25°C. I," *Cem. Concr. Res.*, **3** [6] 791–806 (1973).

5 P.W. Brown, "Phase Equilibria and Cement Hydration"; pp. 73–93 in *Materials Science of Concrete I*. Edited by J.P. Skalny. American Ceramic Society, Westerville, OH, 1989.

6 W.A. Klemm and L.D. Adams, "An Investigation of the Formation of Carboaluminates"; pp. 60–71 in *Carbonate Additions to Cement* (STP 1064). Edited by P. Klieger and R.D. Hooton. American Society for Testing and Materials, Philadelphia, PA (1990).

7 L.E. Copeland, D.L. Kantro and G. Verbeck, "Chemistry of Hydration of Portland Cement"; pp. 429–465 in *Chemistry of Cement. Proceedings of the 4th International Symposium, Washington 1960*, Volume 1. National Bureau of Standards Monograph 43, US Department of Commerce, 1962.

8 B. Erlin, "Ettringite – Whatever You May Think it Is"; pp. 380–381 in *Proceedings of the 18th International Conference on Cement Microscopy*. Edited by L. Jany, A. Nisperos and J. Bayles. International Cement Microscopy Association, Duncanville, TX, USA, 1996.

9 R.S. Gollop and H.F.W. Taylor, "Microstructural and Microchemical Studies of Sulfate Attack. I. Ordinary Portland Cement Paste," *Cem. Concr. Res.*, **22** [6] 1027–1038 (1992).

10 J.G. Wang, "Sulfate Attack on Hardened Cement Paste," *Cem. Concr. Res.*, **24** [4] 735–742 (1994).

11 P.W. Brown and S. Badger, "The Distributions of Bound Sulfates and Chlorides in Concrete Subjected to Mixed NaCl, $MgSO_4$, Na_2SO_4 Attack," *Cem. Concr. Res.*, accepted.

12 F.M. Lea, pp. 345 et seq. in *The Chemistry of Cement and Concrete*, 3rd. ed. Arnold, London, 1970.

13 K. Pettifer and P.J. Nixon, "Alkali Metal Sulphate – A Factor Common to both Alkali Aggregate Reaction and Sulphare Attack on Concrete," *Cem. Concr. Res.*, **10** [2] 173–181 (1980).

14 R.S. Gollop and H.F.W. Taylor, "Microstructural and Microchemical Studies of Sulfate Attack. III. Sulfate-Resisting Portland Cement: Reactions with Sodium and Magnesium Sulfate Solutions," *Cem. Concr. Res.*, **25** [7] 1581–1590 (1993).

15 J.H.P Van Aardt and S. Visser, "Thaumasite Formation: A Cause of Deterioration of Portland Cement and Related Substances in the Presence of Sulphates," *Cem. Concr. Res.* **5** 225-232 (1975)

16 N.J.Crammond, "Thaumasite in Failed Cement Mortars and Renders from Exposed Brickwork," *Cem. Concr. Res.* **15**, 1039-1050 (1985).

17. H. Poellmann, H.-J. Kuzel and R. Wenda, "Solid Solution of Ettringite, Part I: Incorporation of OH^- and CO_3^{2-} in $3CaO{\cdot}Al_2O_3{\cdot}3CaSO_4{\cdot}32H_2O$," Cem. Concr. Res. **20**, 941-947 (1990)

18 P.W. Brown, unpublished results.

19 N.J. Crammond and P.J. Nixon, *in 6th International Conference on the Durability of Building Materials and Components*, S. Nagataki, T. Nireki and F. Tomosawa, Eds., p. 295, E&FN Spon, London (1993).

20 H.-M. Sylla "Reactions in Cement Stone Due to Heat Treatment," *Beton* **3** [11], 449-456 (1988).

21 H.F.W. Taylor, *Cement Chemistry*, 2nd Ed., pp. 373-374 Thomas Telford London (1997)

22 F.W. Locher, "Zur Frage des Sulfatwiderstands von Hüttenzementen" [On the Question of Sulfate Resistance of Slag Cements]. *Zem.-Kalk-Gips*, **19** [9] 395–401 (1966).

23 P.W. Brown, "An Evaluation of the Sulfate Resistance of Cements in a Controlled Environment," *Cem. Concr. Res.*, **11** [5/6] 719–727 (1981).

24 P.K. Mehta, "Sulfate Attack on Concrete – A Critical Review"; pp. 105–130 *in Materials Science of Concrete III*. Edited by J. Skalny. American Ceramic Society, Westerville, OH, 1992.

25 T. Merriman, in Fort Peck Dam Specifications, 1933; quoted by R.H. Bogue on p. 717 of "The Chemistry of Portland Cement" (2nd ed.). Reinhold, New York (1955).

26 P.K. Mehta and O.E. Gjørv, "A New Test for Sulfate Resistance of Cements," *J. Test. Eval.* **2** [6] 510–514 (1974).

27 R.P. Khatri, V. Sirivivatnanon and J.L. Yang, "Role of Permeability in Sulfate Attack," *Cem. Concr. Res.,* **27** [8] 1179–1189 (1997).

28 L.Heller and M.Ben-Yair, "Effect of Sulphate Solutions on Normal and Sulphate-Resisting Portland Cement," *J. Appl. Chem.,* 20–30 (1964).

29 N.J. Crammond, "Quantitative X-ray Diffraction Analysis of Ettringite, Thaumasite and Gypsum in Concretes and Mortars," *Cem. Concr. Res.*, **15** [3] 431–441 (1985).

30 Rasheeduzzafar, F.H. Dakhill, A.S. Al-Gahtani, S.S. Al-Saadoun and M.A. Bader, "Influence of Cement Composition on the Corrosion of Reinforcement and Sulfate Resistance of Concrete," *ACI Mater. J.*, **87** [2] 114–122 (1990).

31 J.G. Wang, "Sulfate Attack on Hardened Cement Paste," *Cem. Concr. Res.,* **24** [4] 735–742 (1994).

32 D. Bonen and M.D. Cohen, "Magnesium Sulfate Attack on Portland Cement Paste. I. Microstructural Analysis," *Cem. Concr. Res.,* **22** (1) 169–180 (1992).

33 D. Bonen and M.D. Cohen, "Magnesium Sulfate Attack on Portland Cement Paste. II. Chemical and Mineralogical Analysis," *Cem. Concr. Res.,* **22** (4) 707–718 (1992).

34 D. Bonen, "A Microstructural Study of the Effect Produced by Magnesium Sulfate on Plain and Silica Fume-Bearing Portland Cement Mortars," *Cem. Concr. Res.*, **23** 541–553 (1993).

35 S. Yang, X. Zhongyi and T. Mingshu, "The Process of Sulfate Attack on Cement Mortars," *Advn. Cem. Bas. Mater.*, **4** [1] 1–5 (1996)

36 J. Millet, A. Bernard, R. Hommey and A. Poindefert, "Sur la dosage de l'ettringite dans les pâtes de ciment et les mortiers" [On the determination of ettringite in cement pastes and mortars], *Bull Liaison Lab. Ponts Chaussées*, [109] 91–95 (1980).

37 H.F.W. Taylor, "Sulfate Reactions in Concrete – Microstructural and Chemical Aspects"; pp. 61–78 in *Cement Technology (Ceramic Transactions, Vol. 40).* Edited by E.M. Gartner and H. Uchikawa. American Ceramic Society, Westerville, OH, 1989.

38 H. Lafuma, "Théorie de l'expansion des liants hydrauliques" [Theory of the expansion of hydraulic binders], *Rev. Matér. Constr. Trav. Publ.* (243) 441–444 (1929).

39 P.W. Brown, "The Implications of Phase Equilibria on Hydration in the Tricalcium Silicate-Water and the Tricalcium Aluminate-Gypsum-Water Systems"; pp. 231-238 in *8th International Congress on the Chemistry of Cement, Vol. 3*. Abla Gráfica é Editora Ltda, Rio de Janeiro, 1986.

40 S. Diamond, "Delayed Ettringite Formation – Processes and Problems," *Cem. Concr. Comp.*, **18** [3] 205–215 (1996).

41 V. Johansen, N. Thaulow and J. Skalny, "Simultaneous Presence of Alkali-Silica Gel and Ettringite in Concrete," *Advn. Cem. Res.*, **5** (17) 23–29 (1993).

42 S.L. Marusin, "SEM Studies of DEF in Hardened Concrete," pp. 289–299 in *Proceedings of the 15th International Conference on Cement Microscopy*. Edited by G.R. Gouda, A. Nisperos and J. Bayles. International Cement Microscopy Association, Duncanville, TX, USA, 1993.

43 S. Diamond and S. Ong. "Combined Effects of Alkali Silica Reaction and Secondary Ettringite Deposition in Steam-Cured Mortars; pp. 79–90 in *Cement Technology (Ceramic Transactions, Vol. 40).* Edited by E.M. Gartner and H. Uchikawa. American Ceramic Society, Westerville, OH, 1989.

44 T. Nakamura, G. Sudoh and S. Akaiwa, "Mineralogical Composition of Expansive Cement Clinker Rich in SiO_2 and Expansibility," pp. 351-365 in *Proceedings of the 5th Internatinal Symposium on the Chemistry of Cement, Tokyo, 1968*, Vol. 4. Cement Association of Japan, Tokyo, 1969.

45. M. Okushima, R. Kondo, H. Muguruma and Y. ono, "Development of Expansive Cemettn with Calcium Sulphoaluminate Cement Clinker," pp. 419-38 in *Proceedings of the 5th Internatinal Symposium on the Chemistry*

of Cement, Tokyo, 1968, Vol. 4. Cement Association of Japan, Tokyo, 1969.

46 Deng Min and Tang Mingshu, "Formation and Expansion of Ettringite Crystals," *Cem. Concr. Res.* **24** [1] 119-126 (1994) and other references therein.

47 G. Scherer, "Crystallization Pressure from Salts in Small Pores," *Cem. Concr. Res.,* accepted.

48 S.A. Abo-El-Enein, T.M. Salem and E.E. Hekel, "Thermal and Physico Chemical Studies on Ettringite. I. Hydration Kinetics and Microstructure," *Cemento*, **85** (1) 47–58 (1988).

49 T. Thorvaldson, "Chemical Aspects of the Durability of Cement Products"; pp. 436–466 in *Proceedings of the 3rd International Symposium on the Chemistry of Cement.* Cement and Concrete Association, London, 1954.

50 K.L. Scrivener, "Delayed Ettringite Formation and Concrete Railroad Ties"; pp. 375–377 in *Proceedings of the 18th International Conference on Cement Microscopy.* Edited by G.R. Gouda, A. Nisperos and J. Bayles. International Cement Microscopy Association, Duncanville, TX, USA, 1993.

51 M.C. Lewis, K.L. Scrivener and S. Kelham, "Heat Curing and Delayed Ettringite Formation"; pp. 67–76 in *Materials Research Society Symposia Proceedings,* Vol. 370. Edited by S. Diamond, S. Mindess, F.P. Glasser, L.R. Roberts, J.P. Skalny and L.D. Wakeley. Materials Research Society, Pittsburgh, PA, 1995.

52 A. Bentur, "Intrinsic Strength and Microstructure of Hydrated C_3S," *Cem. Concr. Res.*, **6** [4] 583–590 (1976).

53 A. Bentur, "Effect of Gypsum on C_3S Pastes," *J. Am. Ceram. Soc.*, **59** [5/6] 210–213 (1976).

54 P.K. Mehta, D. Pirtz and M. Polivka, "Properties of Alite Cements," *Cem. Concr. Res.*, **9** [4] 439–450 (1979).

55 P.K. Mehta, Mechanism of Expansion Associated with Ettringite Formation," *Cem. Concr. Res.*, **3** [1] 1–6 (1973).

56 H. Pöllmann, H.-J. Kuzel and R. Wenda, "Compounds with Ettringite Structure," *Neues Jahrb. Mineral. Abhandl.*, **160** (2) 133–158 (1989).

57 S.A. Abo-El-Enein, S. Hanafi and E.E. Hekel, "Thermal and Physico Chemical Studies on Ettringite. II. Dehydration and Thermal Stability. Hydration Kinetics and Microstructure," *Cemento*, **85** (1) 121–132 (1988).

58 S.A. Abo-El-Enein, S. Hanafi, T.M. Salem and E.E. Hekel, "Thermal and Physico Chemical Studies on Ettringite. III. Dehydration-Rehydration under Reduced Water Vapour Pressures," *Cemento*, **86** (4) 239–250 (1989).

59 S.A. Abo-El-Enein, T.M. Salem and E.E. Hekel, "Thermal and Physico Chemical Studies on Ettringite. IV. Modes of Interaction with Water and Associated Energetics," *Cemento*, **89** (1) 31–42 (1992).

60 C. Gaudefroy, M.-M. Granger, F. Permingeat and J. Protas, "La despujolsite, une nouvelle espèce minérale" [Despujolsite, a New Mineral Species], *Bull. Soc. fr. Minéral. Cryst.*, **91**, 43–50 (1968).

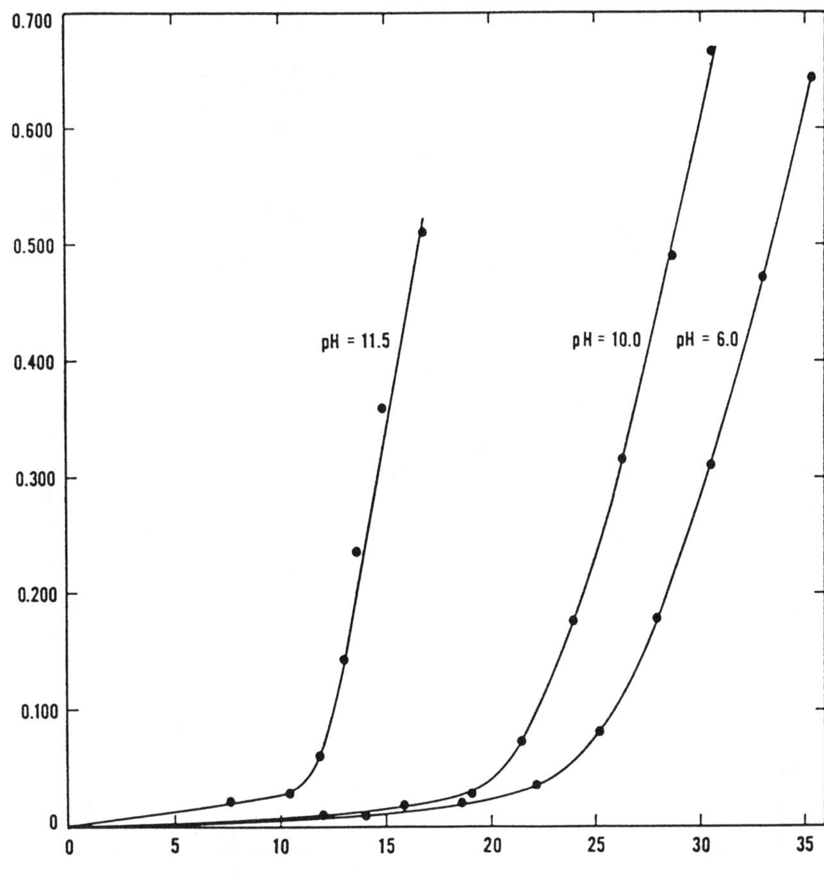

Figure 1. The relationship between sulfate uptake per unit area and the expansion of mortar bars immersed in Na_2SO_4 solutions maintained at constant pH values (after *Ref.* 23).

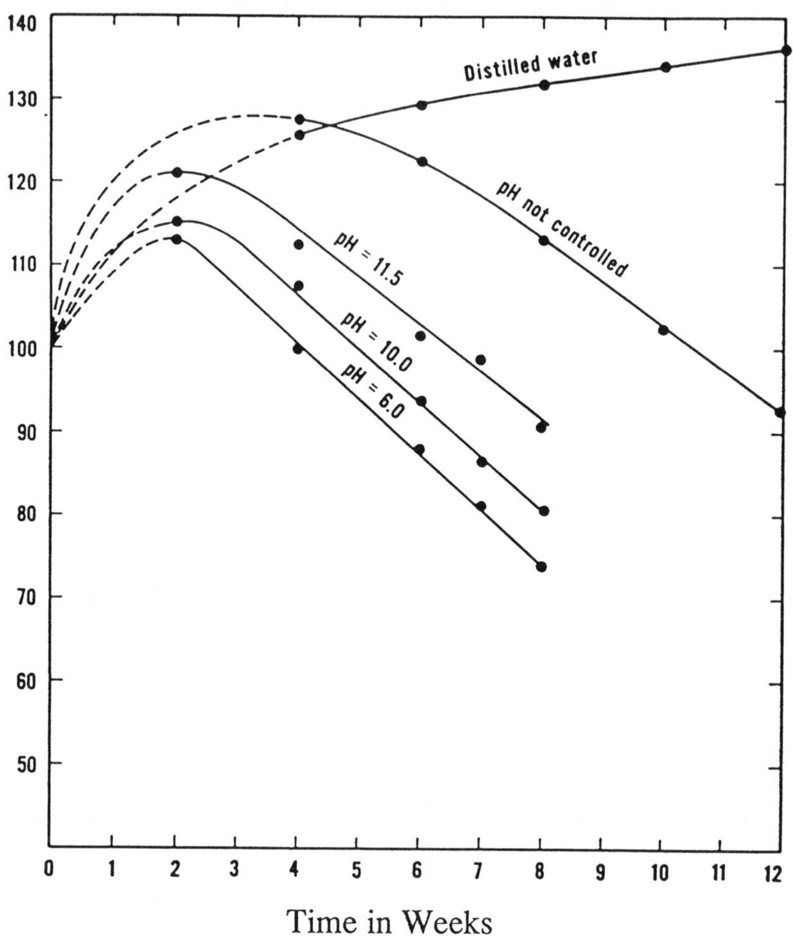

Figure 2. The effect of immersion in Na_2SO_4 solutions on mortar cube strength (after *Ref.* 23). Compared to control cubes immersed in water, those immersed in sulfate solutions eventually undergo significant reductions in strength.

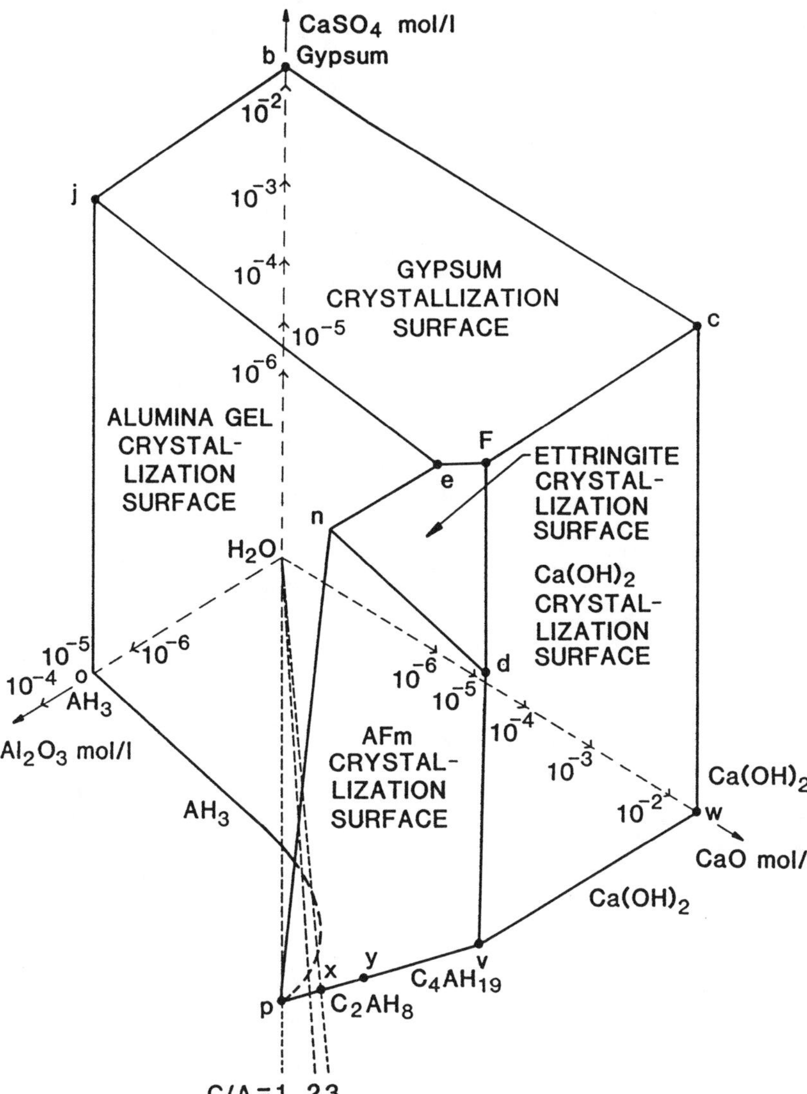

Figure 3. The metastable quaternary CaO-Al_2O_3-$CaSO_4$-H_2O diagram (after *Ref.* 39). The stability range of ettringite is shown.

REACTIONS BETWEEN CEMENT PASTE COMPONENTS AND SULFATE IONS

F.P.Glasser
Department of Chemistry
University of Aberdeen
Old Aberdeen
AB24 3UE
Scotland

ABSTRACT

Sulfate attack on Portland cement concretes takes many forms: the reactions could involve internal as well as external sulfate. Isochemical sulfate reactions are related to sulfate availability rather than amount: several simple calculations are used to illustrate differences. Conditions for external, non isochemical attack are reviewed. It is shown that the mineralogy and internal chemistry reflect a series of local equilibria calculation of thermodynamic equilibria have been modified to include metastable but persistent states and to reflect diffusion gradients can be applied; predictions agree well with results of laboratory experiment and field exposure, although at present it is not possible to predict reaction kinetics. The modification of cement specification is considered, as is and the possibility of using blended cements to mitigate attack.

INTRODUCTION

The writer is grateful to Brown and Taylor for making available at a preliminary stage a draft of their contribution. Their paper reviews and puts in perspective much of the literature. I draw on the same sources, except where indicated, and reading their contribution is an essential preliminary to my contribution. Because Brown and Taylor have presented a balanced view of the literature, I have felt able to discuss and speculate about the implications of current research and on the future development of sulfate-resistant cements.

MANUFACTURE AND COMPOSITION OF PORTLAND CEMENT

Portland cement clinkers have been produced for *ca* 150 years. During that time, their compositions - chemical as well as mineralogical - pyroprocessing and clinker finishing have evolved. Blezard [1] gives a historical account of the development of Portland cement from a technological perspective.

Typical meso-Portland cements characteristic of the mid - 19th century, relative to current production, were richer in belite but poorer in alite and were more coarsely ground. As a consequence set and strength gain occurred more slowly than in modern product. The intrinsic rapid set of modern clinker requires that a set retarder be added. Sufficient sulfate is usually interground with clinker to achieve the required set characteristics. This sulfate is normally supplied by adding gypsum but other rapid release calcium sulfates may be used (hemi-hydrate, active anhydrite). Of course the clinker will also contain sulfate, some of which is readily soluble and which helps control the duration of the initial period of workability . In any event, the readily soluble sulfate does not much affect the rheological aspects of hydration beyond the first few hours or days.

Broadly, the impact of clinker sulfate on hydration may be envisaged by dividing this sulfate into two sources: rapidly soluble and sparingly soluble. Rapidly soluble sulfate forms include $(K,Na)_2SO_4$ solid solutions, calcium langbeinite, syngenite etc. The rapid-release forms also include interground gypsum which is the main source of readily soluble-sulfate in many cements. Some slow release sulfate will also be present, mainly in solid solution in belite, but also contained in potentially soluble phases which are mechanically occluded within unhydrated clinker particles. Various selective extraction procedures have been proposed to determine separately the quick release and slow release contributions but these methods are not very satisfactory and involve arbitrary distinctions between types [2]. Nevertheless, the hydration products obtained in a maturing paste are very sensitive to the total quantity of sulfate as well as the amount available for reaction at any point in time. These balances, together with the tendency to increasing overall sulfate contents, are factors which could potentially affect the long-term performance.

Changes in Portland cement and their consequences may be envisaged in two ways. The first way is familiar: in terms of changes which occur to the bulk composition of the paste - mineralogical, chemical or both. But not all components of the cement hydrate and react at the same rate and because these intermediate stages of reaction persist for several months, years, or even longer it is convenient to define a second type of composition, which is the reactive fraction of substance. If all phases in the system were to react at the same rate, the second type of composition would be an equal fraction of the first but if their reaction rates are unequal, a new balance has to be sought by multiplying the total of each phase by α_1, α_2, α_3, ... where α is the fraction reacted of phases 1, 2, 3, at a particular time. As has been shown in respect of sulfate which is held in rapid release and slow release forms, sulfate reactivity is unlikely to occur at the same rate as aluminate reactivity particularly at early (0-28d) stages of the hydration process. These concepts will be employed subsequently and examples given of their impact on hydration.

REACTION OF ALUMINATE AND SULFATE

When Portland cement is first formulated, liquid water is plentiful. As a consequence of the abundance of water and the solubility of cement components, the mix water becomes strongly alkaline: within seconds of mixing its pH exceeds 12. Much of the early hydration occurs through solution; solids dissolve forming an initially supersaturated solution which subsequently discharges its supersaturation by precipitation. Many of the early hydration reactions are conditioned by the composition of the aqueous phase which in turn, is a function of the ability of solids to continue to dissolve and maintain aqueous species concentrations and the resulting balances between dissolution and precipitation processes. I focus on changes in the aluminates although, as hydration proceeds towards completion, aluminates collectively may comprise only ~ 10% of the paste volume.

Ettringite, $Ca_3Al_2O_6.3CaSO_4.32H_2O$, is readily identified amongst the first hydrated phases to be formed. Its formation is largely conditioned as a consequence of the high pH, the high availability of sulfate during early stages of hydration and the rapid reaction of solid aluminates with sulfate. As hydration progresses, readily - available sulfate is consumed, with the result that ettringite may cease to be formed and may be partially resorbed; sulfate AF_m, $Ca_3Al_2O_6.CaSO_4.\sim 13H_2O$ increasingly forms. Comparison of the two formulae, written in the same style to facilitate comparison, shows that only $^1/_3$ as much $CaSO_4$ per mole of "Al_2O_3" is required to form AF_m as is required to form ettringite. Thus AFm formation is favoured in sulfate-deficient (relative to ettringite) environments.

The sulfate ratios used in these calculations are approximations as the phases may not have their ideal formulae. In particular, sulfate required to form AF_m may be conserved and extended by further substitution of $2OH^-$ for each SO_4^{2-} [3]. However, ideal formulae are adequate to depict and explain the observed features, although they may not necessarily be appropriate for quantitative calculations of the hydrate phase content.

VOLUME CHANGES ASSOCIATED WITH ETTRINGITE FORMATION

Fresh, plastic cements can tolerate some volume change without serious consequence and for that reason ettringite formation in fresh paste is not regarded with concern. Volume changes occurring after final set may be less acceptable. In fact, as will be shown, ettringite formation is not necessarily expansive. On account of its low physical density, *ca* 1.754g/cm^3, ettringite has often been viewed as "expansive" because other cement phases have higher densities; typically in the range 2.1-2.4 g/cm^3. But simple comparison of densities is potentially misleading. It can readily be shown both by calculation and example that ettringite formation is not necessarily expansive. For example, many proprietary cement systems developed for mine support and rock anchoring systems are based on rapid-hardening mixes which develop mainly ettringite, but which are nevertheless dimensionally stable. While these formulations develop much ettringite before final set occurs, much is also developed subsequently. It may be argued that these formulations typically have high w:s ratios and hence have much pore space in which ettringite can crystallise non - expansively. However sulfoaluminate cements have been in use in China for 25 years where some 10^6 tonnes/yr of clinker are manufactured. This clinker is normally blended with additional gypsum and, upon hydration at typical w:s ratios, 0.3-0.7, ettringite is the major paste constituent. Much of this ettringite is formed after final set, yet mortars and concretes formulated with this cement, even at low w:c ratios, normally shrink slightly, albeit less than Portland cements [4]. It is true that expansive versions of these sulfoaluminate cements are also manufactured, and these perhaps hold lessons about why ettringite is expansive under some circumstances but not in others. Nevertheless, the formation of ettringite is not synonymous with expansion in closed systems with fixed total sulfate contents. This important distinction, between ettringite formation in 'open' and 'closed' systems, will be discussed subsequently.

Table I. Volume Changes Attending Ettringite Formulation

Ettringite Formation Reaction	ΔV	
	cm^3	relative*
$C_3A + 3[CaSO_4.2H_2O] + 26H_2O$	-68.2	91
$C_3A + 3CaSO_4 + 32H_2O$	-87.5	89
$C_3AH_6 + 3[CaSO_4.2H_2O] + 20H_2O$	-21	97
$C_3AH_6 + 3CaSO_4 + 26H_2O$	-40.4	95

Notes: * Relative to the molar volume of ettringite, 715.3 cm^3, which is taken as 100. Other relevant data used in the calculations are given in the first column and in Table 2.

Table II. Data for Calculations of Specific Volume Changes

Phase Designation	Formula Weight	Density, g/cm^3	molar volume, cm^3
C_3A	270.2	3.03	89.17
$CaSO_4$	136.14	2.96	45.99
$CaSO_4.2H_2O$	172.17	2.32	74.21
C_3AH_6	378.29	2.52	150.00
Ettringite	1255.13	1.754	715.6
$Ca(OH)_2$	74.1	2.241	33.07
C_4AH_{13}	560.48	~ 2.00	280.2
H_2O	18.00	1.00	18.00

Notes. Density values from H.F.W. Taylor (ed.) "The Chemistry of Cements" Vol 2, Academic Press, London (1964) except for ettringite, which is taken from "Dana's System of Mineralogy (Gaines, et.al, editors) Wiley, New York (1996).

Calculation also supports the contention that ettringite formation is not necessarily expansive. Table 1 calculates specific volume changes relative to one mole of ettringite for its formation from various solid precursors and water: supporting data used in the calculation are given in Table 2. The first reaction in Table 1, for C_3A, gypsum and water, was also calculated by Taylor [5] with results identical to those given here. He noted "There is no general relation between the changes in the solid volume in a cementitious material and the change in volume of the material as a whole". I confirm this view; even using physically dense reactants such as hydrogarnet a diminution in specific volume occurs upon ettringite formation. That this should be so seems at first sight contrary to common sense. However I attribute the diminution in specific volume to the packing of water within the crystal structure of ettringite, which is more efficient than the packing of H_2O molecules in liquid water with the result that, overall, ettringite utilises space more effectively than the reactants from which it forms. Water is an essential component, and its specific volume must be included in calculations of volume change. These considerations are relevant to fresh pastes and explain why ettringite formation at this stage is not expansive: water, aluminates, calcium and sulfate react with diminution in specific volume.

These considerations are relevant to a closed system, i.e., one in which no chemical migration occurs in or out of the system in the course of hydration (a more formal definition will be offered subsequently). Unless we have a clear conceptual understanding of phase development and reaction mechanisms in closed systems we will not be able to treat systems which are open to sulfate transport. To explore other conditions relevant to open systems we need to introduce and develop more background material.

EQUILIBRIUM AND NON-EQUILIBRIUM IN CEMENT HYDRATION

General Features

The term "equilibrium" is used in its thermodynamic sense to embrace the reactions between several solids and an aqueous phase: each phase has a free energy, and equilibrium is attained when the overall free energy of the system is minimised. Since surfaces also have an associated energy term, equilibrium concepts may be applied to the development of microstructures.

Hydration of cement occurs in response to a large thermodynamic driving force. Thus the overall Gibbs free energy change for the hydration reactions can be simplified as:

Cement clinker phases + H_2O = cement hydrates

The free energy change is strongly negative, i.e., reaction proceeds spontaneously and much of the driving force is attributed to bulk free energy changes in the state and condition of reactants and products. However, details are important and when examined in detail, not all the reactions and reaction products occurring in a cement paste necessarily represent the equilibrium state. The occurrence of C-S-H gel-like material is an example. This fact has often been used in support of the assertion that equilibrium thermodynamics are inapplicable to cement hydration. I reject this view. To reject application of one of the most generally useful and powerful tools known to materials science is wrong. Moreover, many other branches of materials processing and utilisation are characterised by complex processes which may involve the prolonged existence of metastable states; metallurgy and ceramics furnish numerous examples of how departures from equilibrium can be treated without loss of rigor. Thus the utility of thermodynamics is accepted, but it is also routine to modify strictly equilibrium thermodynamics to accept into assessment long-lived metastable states. I return to this theme subsequently.

Table III. Metastable Features of Cement Hydration

Feature or Process	Comment
Occurrence of C-S-H (amorphous) to the exclusion of crystalline phases	Gel does not much affect sulfate balances. Some sorption of sulfate may occur. Gel may physically occlude other phases, e.g. AF_m. However, as a first approximation C-S-H gel is not chemically involved in the reactions of sulfate with paste constituents.
Persistence of unhydrated clinker	Potential for future release to hydrates, particularly of Al and Fe held in ferrite. As release occurs it could alter the balance of cement hydrate mineralogy.
Formation of metastable aluminate hydrates	SO_4-AFm is unstable with respect to other phases, notably ettringite, at $T < 40°C$. OH-AF_m is unstable with respect to hydrogarnet, C_3AH_6, and AH_3 at all temperatures.

Table 3 lists some of the principal metastable features of cement hydration which are potentially relevant to the present discussion. Several of these features are explained by involving the principle of the reactive fraction, described previously. Thus the persistence of belite and especially of unhydrated ferrite, which withholds much Al and Fe from the reactive fraction of the paste until relatively late in the hydration process, coupled with the high availability of sulfate at early stages of hydration, helps explain why AF_t, so conspicuous amongst the early hydration products, reduces in amount with continued hydration: in a closed system, the ratio of reactive (Al_2O_3+Fe_2O_3) to reactive SO_3 tends to increase in the post-set phase. This encourages conversion of previously-formed AF_t to AF_m.

Other metastable features must simply be accepted, such as the formation and persistence of AF_m relative to more stable phases (Table 3) and especially of the occurrence and persistence of C-S-H gel. However, thermodynamic data for ettringite disclose that it is a stable phase over a wide range of alkaline pH's and temperatures typical of the internal environment of concrete [6-9].

Until now, we have considered ettringite formation in a closed system. By "closed" I mean closed to transport of matter but open to energy fluxes. This approach is broadly satisfactory to explain the course of cement hydration if we (i) further subdivide the mass of the system into two fractions, one consisting of reactive components and the other of unreactive components and, moreover, allow the subdivision of components to change progressively according to the time frame being considered and (ii) admit into the mineralogical and mass balance calculations solids know to be metastable but persistent. The conceptual framework thus developed will enable kinetics to be interfaced with 'steady-state' processes: steady-state is chosen in preference to 'equilibrium', because we wish to admit into assessment thermodynamically metastable phases such as OH-AF_m.

Let us revisit the question of potential expansivity by ettringite formation using some of the principles thus far enunciated. In the previous examples selected for calculation, shown in Table 1 and text, we concluded that ettringite formation from a wide range of precursors was not expansive; indeed, that a net diminution in specific volume will occur. However, if we alter the definition of the system, in this case by changing the system definition and boundaries, by transferring matter from the 'inactive' fraction to the 'active' fraction of mass, or some combination thereof, it may be necessary to perform fresh calculations. Two illustrative cases will be considered; both involve a core of solids which, it is envisioned, is surrounded by a semi-permeable membrane. The "system" is now redefined so as to include a number of inhomogenities; it is in effect divided into a number of subsystems which can only react with each other as the time frame progressively increases. Fig 1 shows the two types of subsystems and their essential links with the rest of the system. The first case is included, not because it necessarily corresponds to reality, but because it relates to the first example presented in Table 1. In this example, since all the components necessary to form ettringite are included within the envelope, the specific volume will diminish as ettringite is formed. But by redefining the system boundaries and content so that water from an external source becomes essential to completion of the reaction for ettringite formation, the nature of the calculation changes. Reaction to achieve equilibrium between the subsystems now proceeds according to the equation:

$$[C_3A + 3(CaSO_4.2H_2O)] + 26H_2O = \text{Ettringite}$$

Where square brackets indicate the solids which define the closed system, i.e., the core region in Fig 1. The volume of the solids reactants per mole of potential ettringite is 315.7 cm^3; once the necessary water, 468 cm^3, has diffused into the system the volume of ettringite thus formed is 783.8 cm^3. Obviously with these restraints the process has now become very expansive. It may be argued that it is unreal to suppose that the reactants will be totally dry, except of course for the structural water of gypsum. If we instead assume that the packing of grains of solid reactants in the core is 60%, an arbitrary but not unreasonable value, and the remaining pore space is water-filled, the isolated system volume increases to 526.2 cm^3, of which 210.5 cm^3 is water, i.e., 11.7 moles. Under these conditions less water has to diffuse in, (26-11.7), or 14.3 moles, so the reaction equation becomes:

$$[C_3A + 3(CaSO_4.2H_2O) + 11.7H_2O] + 14.3H_2O = \text{Ettringite}$$

The volume increase is less than previously calculated, but is still very large, increasing from 526.2 cm^3 to 783.8 cm^3. In this case, however, significant expansion is likely to be delayed until the space formerly occupied by liquid water becomes filled with solid ettringite. Further calculations, to show that the packing density of the reactants in the subsystem would have to be unrealistically low so as to make the process non-expansive, are left to the reader. It is also worth repeating that in both examples selected for calculation, as well as a wide range of other examples and conditions, ettringite forms because it is the thermodynamically - stable phase under these conditions.

One other specimen calculation which is believed to be somewhat more representative of "real life" situations is presented, in which the solid core consists of C_4AH_{13} and $Ca(OH)_2$ in stoichiometric proportions of Ca and Al to form ettringite. In this case both sulfate and water must diffuse inwards forming ettringite according to the reaction:

$$[C_4AH_{13} + 2Ca(OH)_2] + 3SO_4^{2-} + 17H_2O = \text{Ettringite}$$

The volume of reactant solids per mole of potential ettringite, 346 cm^3, is now somewhat greater than in a previous case but, since the volume of ettringite formed is 783.8 cm^3, the overall process is still very expansive.

The calculations are purely illustrative; the writer does not believe that membranes form in cements which are permeable only to SO_4^{2-} ions and water but not to other species. It is also very doubtful if regions consisting of only C_4AH_{13} and $Ca(OH)_2$ occur in a mature paste and, even if they were to occur, it would be fortuitous if the reactants were present in the precise stoichiometric ratio to form ettringite. But the examples usefully highlight that (i) the nature of calculations, together with their limiting assumptions and system definitions, biases the results and conclusions and (ii) many of the theories in the literature (see Brown and Taylor) invoke diffusion of water alone, or of water and sulfate together, to explain the origin of physical expansion in mature pastes. These theories are plausible and although different in format are often not mutually exclusive.

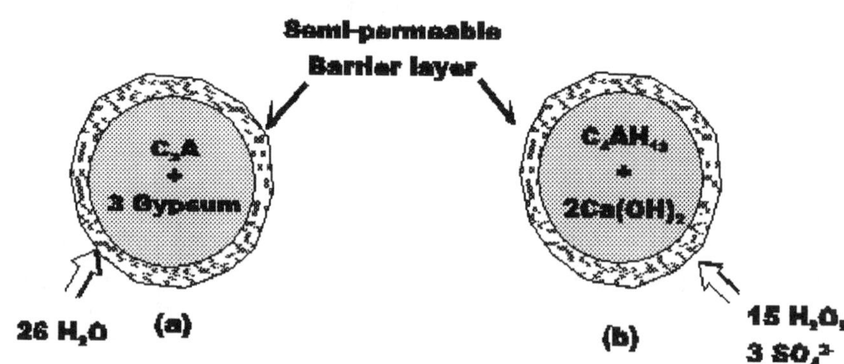

Figure 1 -Schematic, showing some types of inhomogenity leading to delayed ettringite formation. See text for numerical calculations of the expansive potential.

The mechanisms depicted in Figure 1 are frequently used to convert dimensionally stable (or nearly so) cements to expansive formulations. The occurrence of excess free lime in clinker is a chemically simple example: free lime often becomes "dead burnt" during clinkering and much does not hydrate until after it has become covered by a paste membrane formed from the more reactive components. However, as water diffuses slowly through the hardened paste, the free lime hydrates to $Ca(OH)_2$ and this process, known as "lime unsoundness", is expansive because of the increase in volume of the solid phase. Thus the essentials of an expansive system involve (i) the possibility of a thermodynamically favourable reaction occurring after final set (ii) introduction of chemical inhomogenities into the system such that mass transport must occur to approach equilibrium and (iii) localised formation of a thermodynamically favoured phase such that a net volume increase occurs within one set of inhomogenities. It is not difficult to achieve the first condition but controlling the others has proven more difficult, which helps explain why historically it has proven difficult to formulate cements which achieve a controlled degree of expansion.

SUBDIVISION OF CLOSED SYSTEMS
COMPONENTS/PRECURSORS OF ETTRINGITE BECOME AVAILABLE FOR REACTION AT APPROXIMATELY EQUAL RATES
INNOCUOUS
ONE OR MORE KEY PRECURSORS OF ETTRINGITE IS/ARE PREVENTED FROM REACTING UNTIL LATE IN THE HYDRATION PROCESS
POTENTIALLY EXPANSIVE

Figure 2 Schematic diagram showing a classification of inhomogenities within a cement paste: ettringite formation in a closed system.

Figure 2 summarises differences between the two types of regime. Referring to some of the theories described by Brown and Taylor, the distinctions made here between expansive and non-expansive regimes do not require the presence of “amorphous ettringite or of “ettringite gel”, although they do not specifically forbid the existence of such substances. However, since water is the most readily diffusible constituent in paste, it may be preferable in some circumstances to think of regions within the paste which contain potential ettringite-forming solids but which are water deficient: these, it is envisaged, are linked by slow diffusion to water-surplus regions. Depending on the composition of the regions, other species may also need to diffuse.

These considerations can be applied to role of gypsum (or similar sulfate sources) in Portland cement clinker. One school of thought argues that the tendency towards increased sulfate content in modern cements also increases their potential for long-term expansion. On the face of it, the argument seems tenuous: modern Portland cements do not generally contain sufficient sulfate to convert all their alumina content (and perhaps some of the iron oxide) to ettringite. That is, the cement is undersaturated in sulfate with respect to the total quantity of sulfate required to convert all aluminate to ettringite. So nothing is changed critically by marginal increases in sulfate content. Circumstances which favour expansion could only arise if sulfate-rich and aluminate-rich regions were to survive into the mature paste, and if these regions were subsequently to react expansively by one or more of the mechanisms depicted previously. Thus rather special conditions are needed to achieve expansion and these are not necessarily directly related to clinker sulfate content.

I am thus unable to conclude whether the modern trend of increase in cement sulfate contents is, in general, acceptable: much depends on its distribution and reaction kinetics. Indeed, clinker microstructures and the distribution of sulfate between phases may vary even between different clinkers of nominally equivalent bulk chemistry. What is certain is that if all the sulfate, or almost all, is rapidly reactive the potential for subsequent closed-system expansion must be low. The need for additional knowledge concerning the existence and nature of diffusion barriers and the fraction α of sulfate and aluminate reacted within real pastes is inescapable. But on present knowledge, arguments that enhanced sulfate contents in clinker will lead to expansion must be regarded as lacking any tangible basis. Open systems, however, completely alter the mechanisms of expansion and the basis for calculation.

CONCRETE IN ITS SERVICE ENVIRONMENT

The processes whereby sulfate penetrates concrete and reacts with its constituents are shown diagramatically in Figure 3. The figure envisages the service environment to be sea water or brackish water and with a water/air interface, although of course it is applicable with appropriate modification to other environments such as clay or rock permeated by sulfate-containing groundwaters perhaps with evaporation from some portions. I return subsequently to some of the individual processes which occur. Sea water (Table 4) is a difficult service environment to analyse because it contains several constituents, of which Na, Mg, Cl, SO_4 and CO_2 are the most important components in terms of their potential reactivity with the paste constituents of concrete.

Thus cement and concrete systems do not remain as closed systems in their service environment. For example, a concrete pile intended for service in a marine environment might be precast and cured prior to being incorporated into a structure, during which time its behaviour closely approximates to a closed system. But once in service it reacts with its local environment and becomes part of an open system.

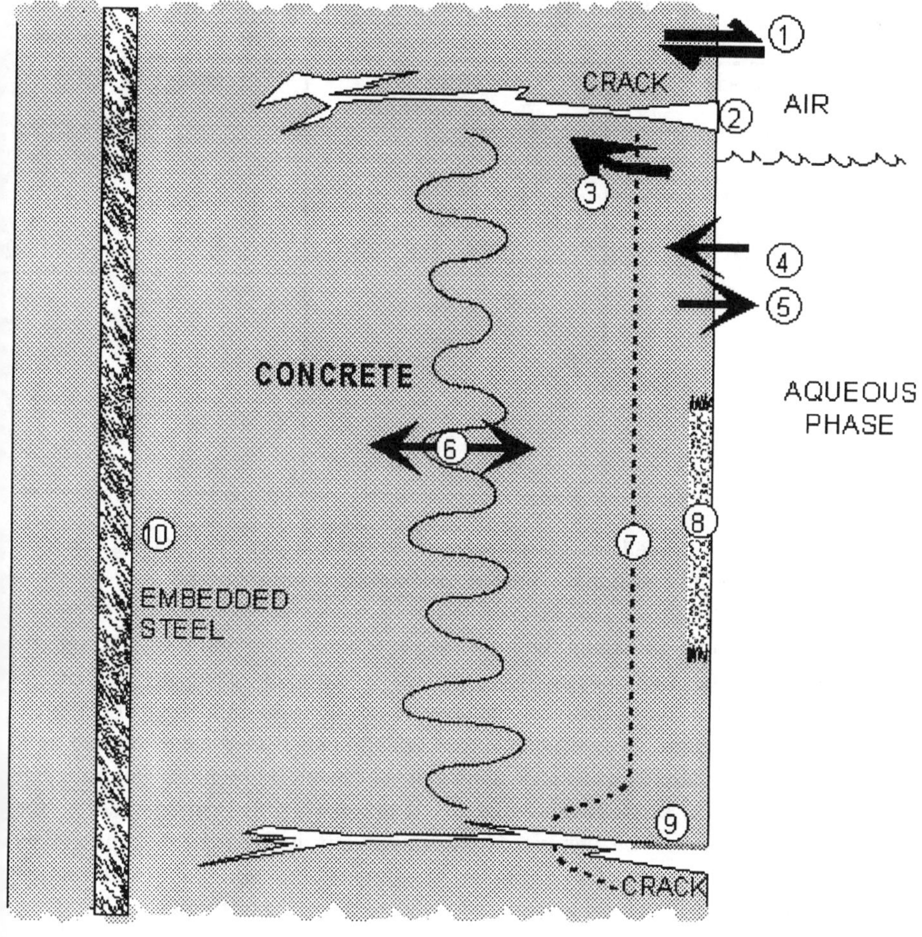

Figure 3 Diagram representing some of the chemical exchanges and processes occurring in an open system when a cement concrete is exposed to sea water.

The resulting chemical and mineralogical changes are important to most aspects of durability but, as discussion has foreshadowed, the consequences of these changes are much more difficult to evaluate owing to the chemical complexity of the system and the superposition of diffusion gradients.

Table IV. Major Dissolved Constituents in Sea Water

Ion/Species	Concentration, Parts per 1000	Percentage of Total %
Cl	19.00	55.0
Br	0.065	0.2
SO_4	2.65	7.7
HCO_3	0.14	0.4
F	0.001	< 0.1
H_3BO_3	0.03	~ 0.1
Mg	1.27	3.7
Ca	0.40	1.2
Sr	0.01	< 0.1
K	0.38	1.1
Na	10.56	30.6

Source: H.V. Svedrop "The Oceans" Prentice Hall, N.Y. (1942). p166. The concentration of chloride is adjusted to 19.00 parts per thousand.

The principle governing the in-service reactions is the desire of a system to reduce its internal free energy in response to changing chemistry. But a complication exists; these chemical changes are not uniform, but occur in response to chemical gradients. The driving forces which applied to cement, treated as a closed system, are altered as a consequence of the chemical potentials and potential gradients in the new service environment. If the new service environment is, for example, sea water, it is characterised by having relatively fixed chemical potentials and temperature; sea water is similar in composition worldwide and service environments typically have a narrow temperature range, *ca* 0° to 40°C.

We thus have three useful tools with which to approach external sulfate attack. These are (i) databases or compilation of thermodynamic properties of cement substances together with iterative programs for minimisation of free energy in chemically complex systems (ii) kinetic data on relative rates of reaction and relative diffusibility of species and (iii) practical experience, often of rather uncertain applicability, of the performance of concrete in sulfate-rich environments. The latter body of experience does, however, serve as an extremely valuable but rather qualitative reality check on mechanisms and kinetics.

Given the number of components involved, it is apparent that numerical evaluation of thermodynamic calculations relevant to sulfate durability could become a Herculean task. Portland cement contains six components, CaO-Al_2O_3-Fe_2O_3-SiO_2-SO_3-H_2O, and a sea water environment (for example) adds at least two more, NaCl and $MgSO_4$. Nor is it possible to omit some components. "Sulfate attack" in sea water is complex and the reaction of other saline components with cement paste components cannot be neglected. However, calculation of phase equilibrium in complex aqueous systems has made great advances in recent decades. Stimulated by geochemists, who needed to elucidate very complex natural systems, computer codes have been developed for this purpose. Very briefly, a "system" intended for calculation has to be user-defined (composition, temperature and pressure). Subsequent iterative calculations are performed automatically to minimise the free energy of the defined system. The output takes the form of listings of phase stability, mass balances, composition of the aqueous phase, phase percentages etc. Some 20 packages are available with which to perform the necessary calculations. They differ slightly in methodologies, for example in which option is used to correct for species activities, in their ease of coupling of equilibrium with transport and kinetic codes, etc. A unified database for cement substances has been developed by the writer and colleagues and is freely accessible as part of HATCHES, a much larger and maintained database [10]. Thus the sulfate concentrations necessary to induce mineralogical changes in the paste, and the nature of the change, even in the presence of chemically - complex aqueous solutions such as sea water, can be calculated.

While equilibrium and kinetics would at first sight appear not to be compatible, sequential and coupled reactions accompanied by changes to solids can often be successfully treated using purely equilibrium codes by the method of successive approximation. For example, in treating sulfate permeation from an external source, we might allow aqueous sulfate concentrations to rise in selected increments and determine the resulting sequence of mineralogical changes, assuming an unlimited flux of sulfate, albeit at fixed concentration, from an external source. Such a model is of course too simplistic to give numerical solutions to mass balances; ions other than sulfate also diffuse and dissolution of cement substance into the aqueous phase will also occur. Nevertheless, the impact of the constituents is not a simple function of their concentrations. In support of this consider the attack of a single salt, $MgSO_4$, on cement paste. Equilibrium calculations performed by the author and colleagues show that base exchange, $Ca(OH)_2 + Mg^{2+} \rightarrow Mg(OH)_2 + Ca^{2+}$, occurs under a wide range of salt concentrations at Mg concentrations $> 10^{-8}$M. Thus for all practical purposes, Mg is virtually insoluble at elevated pH values, typically ~ 10, conditioned by $Mg(OH)_2$

solubility. Since the kinetics of the base exchange reaction are also rapid, at least in the initial stages of reaction, Mg is very effectively removed. Sulfate may also react with Ca forming gypsum or anhydrite but both are much more soluble than the Mg product by several orders of magnitude. How much more soluble depends on pH, the presence of alkali and whether the precipitating phase is gypsum or ettringite, or both. But the large differences in solubility between Mg and Ca result in a zonal structure which is often characterised by a sharp separation between zones of Mg precipitation and those of sulfate precipitation. Because of the complex nature of the problem, only results of simplified calculations are presented in the references cited but even these, appropriately used, give insights into the nature of the processes occurring.

More complex codes which treat equilibria in diffusion gradients are available, so it is not necessary to use the method of successive approximations: see the paper by Marchand, *et al.* However, more sophisticated calculations also require a significant increase in user understanding of appropriate input data.

Up to this point, the narrative seems mainly to have concentrated on introducing additional complications, into quantifying sulfate attack, for example:

- The complex nature of the relationships between bulk composition and mineralogy.
- The virtual certainty that metastable solid cement phases will form and persist.
- Additional complications resulting from internal transport within a cement system, as well as those involving external transport, as occur when cement systems react with their service environment.
- The apparently contradictory conclusions resulting from calculations which, depending on input, give different specific volume changes attending ettringite formation from precursors.
- The need to evaluate the role of all the saline components in the system, not just sulfate.
- The occurrence of cracks and of boundaries, e.g. cement - sea water and cement - air, which complicate diffusion, paths and rates.

These complications are very real and their importance should not be minimized. On the other hand, we have also introduced some possibilities of managing these difficulties. For example:

- Introducing useful concepts for example which enable a distinction between closed and open systems, with different types of treatment applicable to each.
- Calculation of equilibria in chemically complex systems. Many of the systems relevant to ettringite formation and stability are well-known. Calculation also frees one from the tyranny of two-dimensional schemes for the representation of complex equilibria: problem - specific restraints can be imposed to obtain tailored computations relevant to specific environments.
- Equilibrium calculations can readily be modified to cater for the persistence of long-lived metastable phases or states, should these prove relevant.
- More than a century of empirical experience, of the behaviour of concrete in sea water and of laboratory exposure to sulfates, provide a reality check on calculations as well as giving limited data on reaction mechanisms and kinetics. I note that both calculation and experience conclude that ettringite formation in a reasonably homogeneous closed system is, in general, not expansive. Only when inhomogenities are introduced such that ettringite is induced to form within inhomogeneous domains is expansion likely to occur. In open systems, precursor domains to ettringite formation are much more likely to develop spontaneously.

It is concluded that calculations and experience give some insight into the possible mechanisms involved in sulfate expansion but do not uniquely support one of the mechanisms postulated in the literature to the exclusion of others. Indeed, several of the mechanisms in the literature could operate simultaneously to a greater or lesser extent, depending on circumstances. But many of these explanations concentrate on the behaviour of one component to the exclusion of others and, as such, probably rely on physically incomplete models.

Figure 3 focuses on a real environment and shows how complex physical mechanisms affect diffusion rates of sulfate into concrete. Exchanges of matter are shown by arrows. In region 1, atmospheric CO_2 can react with cement. However, splash and spray may also result in wetting by sea water and rain so some salt ingress occurs. If wet-dry cycling occurs, capillarity may draw an aqueous phase into the concrete. Cracks, region 2, may permit direct access of air and moisture to greater depths. At the air-water interface, 3, mass transport is often accelerated by sharp chemical potential gradients both horizontally and vertically. In sections of concrete which are fully immersed, mass transport occurs across a nearly stationary interface. Taylor and Brown describe microstructural details of sulfate migration, process 4, across interfaces, 3, such that a zonal structure develops spontaneously. In sea water a number of transport processes occur simultaneously with the result that Mg, NaCl and CO_2 migrate to different depths; this, and the reaction of these species with paste, give rise to the deeper zones. Of course some dissolution of cement paste into sea water also occurs, process 5. The net resultant of these inward and outward migrations controls the near-surface zonal structure. The depth of each zone and the rate at which it thickens cannot at present be predicted accurately. We do, however, have some empirical guides. For example chloride often migrates to considerable depth, 6. The main solubility control on its movement is the formation of Friedel's salt, which begins at chloride concentrations > 5mM and is complete at ~ 15 mM [11].Ettringite and gypsum, on the other hand, tend to form at shallower depths at or near the physical solid-aqueous interface. Moreover a layer, 8, consisting of $CaCO_3$, Mg-containing phases and an amorphous alumina-silica gel develops close to the solid-aqueous interface. Water-saturated cracks, 9, complicate the relative accessibility of cement paste to sea water components. The net permeation achieved depends upon the intrinsic cracking present in the matrix as well as cracking due to expansion of the paste in a zone near 7, and Fickian diffusion modified by the strength of the interaction between the diffusing species and paste. Cracking, feature 7, is often associated with formation of a zone rich in ettringite and gypsum. Eventually chloride and dissolved oxygen reach embedded steel, thereby accelerating corrosion and promoting additional cracking by another mechanisms, namely increase in volume of the corrosion product of steel.

I make three points concerning these complex reactions. Firstly, the sequence of reactions is, as a first approximation, controlled by the apparent diffusibility of cations and anions. Thus Na and Cl, which are only weakly bound to cement solids, are relatively much more diffusible than are either Mg or CO_3. The latter two species react much more strongly with paste, resulting in formation of insoluble products, which retards their diffusion. The second point concerns dissolution, which is also important. This is often the "invisible" part of experiments; investigators have tended to concentrate on mass gains and mineralogical and microstructural changes within the matrix but neglect mass losses. In any set of mass balance calculations it is essential to keep track of both gains and losses unless of course, one or more terms are shown to be negligible. Furthermore, mass transfer may occur *within* the matrix; some zones could be experiencing densification while at the same time, others are undergoing porosification. Densification is of special concern because it creates expansive pressures, as discussed previously. However, because of the intense alteration of the microstructure in these zones, the writer is unsure of the value of calculations which purport to calculate numerically a "swelling pressure". The concept may be applicable to an isolated pore or possibly a crack, but so many processes occur simultaneously, although separated in space, that it is difficult to evaluate a swelling pressure. The final point concerns the direction of reaction; taking into account the changing component concentrations with depth, each successive zone tends to approach the equilibrium phase distribution appropriate to its local chemistry. Thus the zones reflect a local equilibrium, or series of equilibria, developed in response to an overall driving force to reach equilibrium. Viewed in this way, the need for a unified description of the macro - and microstructural zoning becomes more comprehensible because, as noted previously, we have the ability to predict mineral stability within each zone. Alternatively, from knowledge of mineralogical zoning, we can predict possible ranges of pore fluid chemistry and pH within a zone. Less - readily predictable features and processes are (i) the relative geometric thicknesses of the zones (ii) the time dependence of the rate at which they migrate into the concrete (iii) the balance between precipitation and dissolution within each zone (iv) the morphology of product phases and the microstructure of the zones, although in principle it can be predicted, and (v) whether the concept of swelling pressures can usefully be applied.

DISCUSSION

It is perhaps helpful to begin by recapitulating what we know, what we suspect and what we do not know about sulfate attack. Physical chemistry has laid an adequate foundation to understand and explain the mineralogical and chemical changes which occur upon exposure of cements to different environments. While phase diagrams are a rather rigid format for representing and interpreting changes in complex systems of four or more components, they remain useful as tools for thinking. But the real utility of thermodynamics (application of which underlies construction of phase diagrams) is to enable data on stability and solubility to be interfaced with computer - based routines which permit user-defined conditions to be calculated. The phases included in the scope of calculations can, if desired, include those which are not strictly speaking equilibrium phases, provided they are sufficiently persistent to enable their thermodynamic properties to be measured for inclusion in a database. Much of the relevant database for calculations is for 25°C; but, since most cement deterioration occurs in the range 10-40°C, the limited temperature range of the presently-available database should not present a major impediment to calculations.

Defining a problem such as sulfate attack in such a way as to be amenable for calculation is a more difficult task. Considerable experience of defining reactions in multi-component systems is almost essential if calculation is to generate meaningful results. The writer is unable to offer other than general words of wisdom; he is still feeling his own way to solutions and commonly finds that it takes longer to devise packages suitable for stepwise calculation, and subsequently integrate and explain the results, than to actually perform calculations. However, this has an advantage not foreseen at the outset but which, with hindsight, could have been foreseen; that the computer forces a rigid logic and discipline upon the user. Incorrect logic and missing key data either results in failure to converge on a numerical solution or, more usually, no solution. While this may be frustrating, it does in the long run highlight deficiencies both in underlying data relevant to sulfate attack and in problem definition and solution. These are being addressed, albeit slowly. For example, the cation(s) associated with sulfate must be included in calculations, so in sea water it is as important to determine the reaction between paste and magnesium etc., as it is with sulfate. Another perhaps unforeseen dividend is that the computational routines often suggest logical ways of breaking down very complex processes into steps which interface with each other in parallel, or in series, or both. As this interfacing is being done, one often finds that empirical knowledge of sulfate attack reveals clues as to the steps and processes involved: several simple examples are developed in the text.

However, these developments lie at the present frontiers of knowledge. The writer is aware that engineers also need practical guidance: how can sulfate attack be avoided or mitigated? The guidance I offer can be based on three sets of assumptions, namely that:

I. Portland cement, as we know it today, will continue to be used in the future, or:
II. Portland cement as we know it today will be modified in mineralogy and composition, so as to make it more suitable for service in sulfate rich environments. Also,
III. new types of non-Portland cement will be formulated having enhanced sulfate resistance.

The option of using presently - available cements offers only limited prospects for improvement. Research may, it is true, contribute to modest but worthwhile improvements in its performance in sulfate - rich environments. I would highlight briefly several areas of improvement - briefly, because I believe they are well-known. We need to ensure optimum concrete formulation: cement fineness, cement content and water:cement ratios as well as optimising aggregate grading so as to reduce matrix permeability. Secondly, we need to reduce cracking which facilitates diffusion and thereby lowers the integrity of low-permeability concretes. Improving the quality of concrete delays the onset of deterioration by slowing reactions and also allowing more time for stress relief.

It would, however, be a pity if efforts to achieve better performance from Portland cement were to become beclouded by overly simplistic arguments concerning their bulk chemistry. Concerns over the increased level of sulfate in modern cements are a case in point. Most of this sulfate is in quick-release forms and normally contributes to ettringite formation by one of the non-expansive modes discussed in the text. To generate an expansive reaction in a closed system, part of the system must become physically isolated from the rest of the system such that ettringite can develop in response to equilibrium, but within physically semi-isolated regions, such that diffusion leads to a net increase in solid volume. As noted, this is most likely to be achieved if some unreacted sulfate sources persist until comparatively late in the hydration while at the same time, aluminate rich but sulfate poor regions persist elsewhere in the matrix. Any discussion of permissible levels of sulfate in OPC must recognise that potential for expansion rests on the *distribution* of potential ettringite-forming reactants as well as their *amounts*.

As the author understands the specifications for sulfate - resistant Portland cement, two kinds of requirement are imposed: one set of specifications addresses the problem by limiting the C_3A content, while another set limits chemical Al_2O_3. Given that cement concretes are generally unable to gain Al_2O_3 from their service environment but may often be able to gain sulfate; also that it is desirable to limit potential for ettringite formation in these circumstances, only the second type of specification makes physicochemical sense. The ferrite phase of Portland cement, which hosts significant Al_2O_3, may not be reactive in the short term but, as expectations of performance lifetimes of concrete constructions increase, ferrite must increasingly be counted as part of the reactive fraction of the clinker. Hence only specifications which limit chemical Al_2O_3 make sense if the intention of the specification is to limit the long-term potential for ettringite formation by control of clinker chemistry.

Sulfate attack could be mitigated by the judicious use of blending agents. It may be argued that slag and fly ash contain more Al_2O_3 than OPC, and that the resulting increase in bulk chemical Al_2O_3 brought about by adding these materials is potentially harmful in an open system. Indeed this might appear to follow from the logic of the preceding paragraph, but this is not necessarily so. Blending agents alter the mineralogy of the set cement, as is well known. But what is less well appreciated is that they also develop phases which chemically combine Al_2O such that it is not available, or is less available, for ettringite formation. An example of a phase formed preferentially to ettringite in Portland slag blends is the hydrotalcite-like phase M_4AH_{10} (but with some sulfate replacing OH) and C_2ASH (again with sulfate substitution). Further research is required to determine the aluminium balances between these phases and with ettringite and to determine the potential for ettringite formation in blended cements: direct comparison of system chemistries of Portland cement with those of blended cements is too simplistic. But this approach offers distinct possibilities that Portland cement, as made today can be modified so as to reduce its potential for expansive ettringite formation.

The third approach is to change cement types. In the past, use was made of so-called "supersulfated cements" [12]. These consisted of ternary blends of slag, cement and gypsum (or active anhydrite). The intention was to saturate the system in sulfate at the outset, so that potential for subsequent ettringite formation in service conditions was suppressed. I consider this principal to be soundly based. Unfortunately, these blends had other technical problems, such as false set and on account of this and other problems have not found general acceptance. Perhaps with modern technology these problems could be overcome. However other approaches are possible.

Calcium sulfoaluminate cements, based on mixtures of belite (C_2S), ferrite and calcium sulfoaluminate $4CaO{\cdot}3Al_2O_3{\cdot}SO_3$ with added gypsum are now being made commercially; in China; the technology is well developed, although awareness of these technological developments outside of China is poor. These cements are handled like Portland cements and gain strength rapidly. The iron-rich versions show outstanding resistance and dimensional stability upon prolonged exposure to sea-water.

The writer has had the opportunity to examine ettringite-rich cement concretes made using an iron - rich calcium sulfoaluminate-belite clinker and after exposure to sea water for 14 to 16 years; while this examination is not yet complete, it is apparent from the lack of mineralogical cracking and of alteration that these concretes are crack-free and mineralogically stable. The paste fraction is mainly ettringite. The specific example was formulated by centrifugal casting a fine concrete formulated with a clinker containing calcium sulfoaluminate-belite-iron rich ferrite, interground with ~ 30% gypsum. After 14-16 years exposure to sea water the matrix had altered to a maximum depth of 600-800 μm and the main impact of alteration was leaching.

Thus I conclude that we have a number of technological options to mitigate sulfate attack which could be exploited. The specification of cement intended for service in sulfate-rich environments needs to be based on physicochemical principles: short-term empirical laboratory testing is an inadequate foundation to determine cement specification.

SUMMARY

Some of the concepts underlying ettringite formation in cements and its connection to expansion are presented. The distinction between 'closed' and 'open' systems is also developed. Closed systems are the easiest to analyse. However, it is shown that increasing the sulfate content of a Portland cement can only make it expansive if late ettringite formation, occurring well after set, can be induced to occur. The technology of expansive cements can be applied to delineating at least some of the preconditions for expansion. These preconditions are most easily attained in open systems in which sulfate migrates into cement paste from an external source.

The complex mineralogical changes occur in response to the drive to achieve equilibrium and the methods of equilibrium thermodynamics can be used to assess and quantify these changes.

Open systems present challenges. Laboratory and field trials tend to disclose only the net balances between inward diffusion and reaction. Losses due to leaching and some internal migration, e.g., of calcium, are not readily elucidated and considerable care is required to determine mass balances with a view to calculating potentials for expansion. Nevertheless, much progress has been made to couple equilibrium thermodynamics with transport models with a reality check from field experience thus probably represents the way forward.

Some suggestions are made for improving the resistance of concrete to sea water. Improvement in concrete quality, to retard sulfate ingress, is well - recognised. It is recommend that the specifications for sulfate-resistant cement be reexamined to reflect physicochemical realities rather than short form performance, which is dominated by reaction kinetics; that new cement types with high intrinsic sulfate contents be evaluated more systematically especially for service in severe environments.

REFERENCES

1. R.G. Blezard. "The History of Calcareous Cements" pp 1-24 in Lea's Chemistry of Cement and Concrete. Edited by P.C. Hewlitt. Arnold (1998).
2. E.M. Gartner and F.J. Tang. "Formation and Properties of High Sulfate Portland Cement Clinkers" Cemento, **84** 141-164 (1987). (English and Italian).
3. F.P. Glasser and S.S. Stronach. "Stability and Solubility Relationships in AF_m phases I. Chloride, Sulfate and Hydroxide" Cement and Concr. Res. (Submitted).
4. Liang Zhang, China Building Materials Academy (personal communication) (1998).
5. H.F.W Taylor. "Sulfate Reactions in Concrete - Microstructural and Chemical Aspects" pp 61-78 in *Cement technology*. Edited by E.M. Gartner and H. Uchikawa (Ceramic Transactions, Vol 40). Amer. Ceram. Soc. OH, (1990).
6. D.Damidot and F.P. Glasser. "Thermodynamic Investigation of The $CaO-Al_2O_3-CaSO_4-K_2O-H_2O$ System at 25°C". Cement Concr. Res. **23**, 1196-1204 (1993).
7. D.Damidot and F.P. Glasser. "Thermodynamic Investigation of the $CaO-Al_2O_3-CaSO_4-CaCO_3-H_2O$ System at 25°C and the Influence of Na_2O". Advances in Cement Res. No.27, 129-134 (1995).
8. F.P. Glasser. "The Role of Sulfate Mineralogy and Cure Temperature in Delayed Ettringite Formation". Cement and Concr. Res. **18**, 183-193 (1996).
9. D.Damidot and F.P. Glasser "Thermodynamic Investigation of the $CaO-Al_2O_3-CaSO_4-CaCl_2-H_2O$ System at 25°C and the Influence of Na_2O" in *Proc. of the X International Congress on The Chemistry of Cement*" (Editor, H. Justness). Vol. 2, paper 3ii 091 (8pp) (1997).
10. HATCHES (Harwell-Nirex Thermodynamic Data Base for Chemical Equilibrium Studies. http:/www.nea.fr/abs/html/nea-1210.html. Note: cements data base complied by M. Paul, S. Stronach and F.P. Glasser will be available in the October, 1998 release.
11. U.A. Birnin-Yauri and F.P. Glasser. "Friedel's Salt, $Ca_2Al(OH)_6(Cl, OH)\cdot 2H_2O$: its Solid Solutions and Their Role in Chloride Binding" Cement and Concr. Res. (in press).
12. M. Moranville-Regourd. "Cements Made From Blastfurnace Slag", pp633-674 in "*Lea's Chemistry of Cement and Concrete*". Edited by P.C. Hewlitt, Arnold (1998).

MICROSTRUCTURAL ALTERATIONS ASSOCIATED WITH SULFATE ATTACK IN PERMEABLE CONCRETES

Sidney Diamond*and R. J. Lee**

* Purdue University, West Lafayette, IN 47907, U.S.A.

** R. J. Lee Group Inc., Monroeville, PA 15146, U.S.A.

ABSTRACT

Permeable concretes undergoing sulfate attack as a result of prolonged exposure to sulfate bearing ground water solutions have been extensively examined using backscatter mode SEM and EDXA. During the exposure sulfate bearing solutions have passed entirely through the concretes in question, leaving deposits of crystalline sulfates on their upper surfaces. The concretes display a progressive and eventually profound alteration of the internal microstructure. Some carbonation is identified as accompanying the sulfate attack, and is considered to be an unavoidable feature with sulfate attack in dry climates, where exposure to the attacking solution is typically discontinuous. Depositions of ettringite and gypsum occur, but deposition of these two phases forms only a small part of the many microstructural changes observed. Other substances formed include thaumasite, Friedel's salt, monosulfate, brucite, and magnesium silicate hydrate. Calcium hydroxide, normally a significant component of hardened cement paste, is removed by dissolution. Various alterations occur to the all-important C-S-H gel component, including local decalcification leading to the generation of low calcium C-S-H gel in some areas, of silica gel in others, and of magnesium silicate hydrate in still others. Some of the calcium appears to accumulate in areas of the paste groundmass that seem to be composed entirely very fine microcrystalline calcium carbonate. Previously unhydrated calcium silicates (alite and belite) in residual cement grains undergo dissolution in some areas, with the spaces created becoming filled in with secondary deposits of various kinds. In other areas selective removal of calcium from the calcium silicates occurs, leaving the space filled with either silica gel or sometimes magnesium silicate hydrate. The effects of local expansion and cracking in ettringite-rich zones are illustrated, but in these reactions expansion and cracking appears to be less important in sulfate attack than softening of the hardened cement paste associated with the other changes taking place.

INTRODUCTION

It is well understood that most of the properties of portland cement concrete reflect the influence of the hardened cement paste component. Hardened cement pastes in ordinary concrete display a set of characteristic features that vary slightly from place to place but are constitute an assemblage of the same set of microstructural components. Within a given concrete sample, details of these components, their mutual arrangements in space, and their chemical compositions are best examined by backscatter mode scanning electron microscopy (SEM) and the associated energy-dispersive X-ray analysis (EDXA) system.

All of the illustrations in this paper were acquired by this method. A few were obtained at Purdue University, but most were secured at the R. J. Lee Group, Inc laboratory. Most of the Lee Group micrographs are arranged in a characteristic pattern consisting of (1) an area imaged at lower magnification, (2) an enlarged image displayed to the right of the first image, of a sub-area outlined by an illuminated rectangle in the first image, and (3) an EDXA spectrum taken at a point in the higher magnification image marked by a small illuminated square.

The writers have had cause to examine separately and jointly specimens representing a very large number of concrete members which have been exposed to sulfate-bearing ground waters over periods of up to 10 years. The concretes in question have been placed at high water:cement ratios (typically 0.6 or more) and were apparently not subjected to rigorous curing procedures. Accordingly they are extensively porous and highly permeable They represent residential concrete exposed (directly or through exposed foundations) to sulfate bearing ground waters containing varying concentrations of sulfate, chloride and bicarbonate anions balanced by sodium, magnesium, and calcium cations. In many of the concretes ground water has been shown to have penetrated entirely through the mass of the material, as indicated by extensive deposits of ground-water derived salts (sodium sulfate, magnesium sulfate, etc.) on surfaces opposite those in contact with the soil.

Prolonged exposure to these sulfate bearing soil solutions has resulted in sulfate attack which has induced a number of changes in mechanical properties, the details of which have been studied by others. These changes stem from and reflect what we interpret as fundamental and irreversible alterations in the microstructure of the hardened cement paste. Such irreversible alterations constitute a major consequence of sulfate attack

In the present paper we provide several illustrations of the normal microstructural assemblage that constitutes the hardened cement paste in concrete not exposed to sulfate attack, and then illustrate many of the alterations in microstructure found to be induced by the sulfate attack processes.

ILLUSTRATIONS OF THE MICROSTRUCTURAL FEATURES OF CONCRETES NOT EXPOSED TO SULFATE ATTACK

Obviously, concretes prepared from various cements at different water:cement ratios and hydrated to varying degrees of effectiveness do not display exactly the same microstructure. Indeed, concrete microstructure is a "patchy" affair, with local microstructural details varying from place to place on a scale of a few hundred μm within a given concrete (1). Nevertheless, plain portland cement concretes not exposed to sulfate attack *always* display characteristic features including (a) areas within large cement grains that ordinarily remain unhydrated, regardless of the age of the concrete, (b) recognizable shells of dense 'inner product' or 'phenograin' C-S-H gel formed around all or most of these unhydrated cores, and similar C-S-H gel fully occupying the space formerly occupied by smaller cement grains; now completely hydrated (c) occasional hollow shell or "Hadley" grains, with a thin layer of C-S-H gel surrounding hollow (or mostly hollow) cores, (d) extensive regions of much more porous 'groundmass' or 'outer product' C-S-H gel, and (e) deposits of calcium hydroxide directly on many sand and coarse aggregate grain surfaces and locally in places within the groundmass. Occasional small (1-5 μm) crystals of ettringite or monosulfate can be found within the groundmass, but they are rarely seen in backscatter SEM examinations of normal concrete. All forms of C-S-H gel show almost identical chemical compositions, with the Ca:Si ratios being reasonably close to 1.7, and incorporating small contents of Al and S.

Within the residual unhydrated cement grains, unaltered alite, belite, and interstitial phases (C_3A and ferrite) may be separately distinguished with careful control of the backscatter electron image. EDXA ordinarily confirms that these cement components have their normal chemical compositions and are not at all hydrated.

Figure 1 illustrates the appearance of some of these features in a normal concrete. The dark area to the left is a sand grain. The roughly 50 μm long bright grain marked 'A' is a partly unhydrated cement grain, surrounded by a thin shell of 'phenograin' C-S-H gel ('B'). The much smaller grain marked 'C' is dense C-S-H gel marking a fully hydrated cement grain. There are several small hollow shell hydration grains, one of them incompletely hollowed out, in the immediate area surrounding the location marked 'D'. "E' marks one area of C-S-H gel deposit in a porous groundmass region. An extensive deposit of CH marked 'F' almost surrounds a 50 μm chip of the crushed sand used in this concrete. A smaller deposit of CH of the same brightness, marked 'G' can be seen in the groundmass at the lower right portion of the micrograph , and others are visible as well.

In mature concrete there is a certain blending together of some of these features and they may become more difficult to separate clearly.

However, the least subtle and at the same time most characteristic feature of the microstructure of normal concrete is the occurrence of peripheral deposits of calcium hydroxide around sand grains. A very clear (although larger than usual) example of such a deposit is shown in Figure 2. The existence of such deposits serves as an easy marker for concrete that has not been significantly altered by sulfate attack.

Chemically speaking, CH acts as a 'preservative' for the C-S-H gel in concrete undergoing sulfate attack. Its presence in a given area indicates the local existence of pore solution of pH in excess of 12.5, which is the normal condition. The pH levels of most concrete pore solutions range from about 13.2 to sometimes as much as 14, depending mostly on the alkali content of the cement used. In sulfate attack the pH drops progressively. As it attempts to drop below 12.5, the CH acts as a buffer, but is itself dissolved in the process. When the CH in an area is depleted, the local C-S-H itself is progressively de-calcified and a host of microstructural changes, as documented later in this paper, may take place.

GENERAL FEATURES OF THE ALTERATION PATTERN IN SULFATE ATTACK

Previously it has been mentioned that, in the many sulfate attack exposures studied here, penetration of sulfate - bearing ground waters entirely through concrete slabs and other structural units is indicated by the deposition of salts (sulfates and sometimes chlorides) on surfaces opposite to the surface of contact, as for example upper surfaces of garage slabs. Figure 3 illustrates the appearance of one such deposit, which here is entirely sodium sulfate (thenardite). The precipitation of such salts necessarily implies that a high concentration of the ions involved (sodium and sulfate) exists in the uppermost portion of the concrete.

In such circumstances , the general pattern of alteration of the microstructure of the concrete is bi-modal. In addition to the progressive alteration upward from the surfaces of slabs exposed to the ground water, there is progressive alteration in the opposite direction, starting from the opposite surface and extending downward into the concrete. To the degree that the sulfate attack is incomplete, a central zone may often be found that is often relatively unchanged, i.e. that may still contain CH and a largely unaltered microstructure.

In this paper we have attempted to illustrate the general characteristics of the alterations observed in the microstructure of the altered zones, without attempting to provide details of the zonal pattern found within any particular concrete sample.

FEATURES OF THE ALTERED MICROSTRUCTURE

General remarks

The features comprising the altered microstructure developed in these concretes are numerous, complex, and not easy to describe in unambiguous fashion. They involve (i) substances normally present but removed from the altered zone, (ii) substances not entirely removed but chemically and microstructurally altered, and (iii) new substances deposited within the altered zone of the concrete.

In traditional accounts of sulfate attack, for example as reviewed by Thorvaldsen (2) many years ago and by Mehta (3) more recently, it is generally held that sulfate attack is primarily a matter of formation and deposition of ettringite and to some extent gypsum within the confines of the concrete. In the present work we provide evidence that sulfate attack involves far more drastic and complex microstructural changes to cement paste than merely the deposition of gypsum and ettringite.

Laboratory studies of sulfate attack on cement pastes and mortars, for example by Gollop and Taylor (4), Bonen and Cohen (5,6) and Bonen (7), reported very substantial changes in affected cement paste, including dissolution of the calcium hydroxide, decalcification of affected C-S-H gel, and when the attacking solution is rich in magnesium sulfate, conversion of some of the C-S-H gel to magnesium silicate hydrate. These studies are helpful in interpreting the microstructural changes taking place in field concretes undergoing sulfate attack by naturally-occurring ground water solutions. However, the latter show far more complex alterations.

In laboratory experiments, small samples are usually exposed to concentrated sulfate solutions continuously for prolonged periods. In real sulfate attack on concrete, sulfate bearing ground water solutions do not flow to and through the concrete uniformly over time. Rather periods of dry weather produce intermittent drying of the concrete and interrupt the attack. Under such circumstances a degree of carbonation may be expected to accompany sulfate attack. In the present concretes, there are definite indications that some carbonation occurred not only at the surfaces exposed to air, but also at the underside of slabs in contact with the soil. This may be a consequence of the presence of bicarbonate ions in the soil water. In any event, some of the features of the sulfate attack - altered microstructure may be influenced by carbonation, which complicates the interpretation of the sulfate attack phenomena.

Actually most sulfate attack occurs in moderately dry climates (the Western U.S., the Prairie States of Canada, South Africa, etc.). In consistently wetter climates sulfate deposits have been leached out of soils over geological time by an excess of precipitation over evaporation, and sulfate attack is uncommon. In moderately dry climates where sulfate attack typically occurs, extensive intermittent dry periods are likely between rains and in consequence some carbonation may always be associated with sulfate attack. Laboratory studies in which specimens are continuously immersed in sulfate solutions thus may not fully reflect the complications experienced by concretes in the field.

Absence of Calcium Hydroxide in the Altered Zone

As mentioned previously, a clear signal that the paste is being altered in the sulfate attack processes is the loss of the normal deposits of calcium hydroxide around the sand and coarse aggregate grains, and within the paste. It is not easy to depict the *absence* of a phase in a broad region of the concrete in a micrograph, but such absence is a common and characteristic feature of altered zones.

Removal of CH locally does not necessarily leave a gap or space, since redeposition of other components (C-S-H gel) or deposition of new substances may occur within the spaces created.

Figure 4 provides an indication of an instance where the latter appears to have occurred. Instead of a characteristic CH deposit around the sand grain in the upper right corner, a deposit of gypsum occupying the same space is found. There is no evidence of local expansion or cracking in the vicinity, as might be expected if the deposit had been 'inserted' without prior dissolution of the previous phase.

In general, zones of sulfate attack extend well beyond carbonated zones in these concretes. Carbonation would also result in the removal of some CH, but the CH is completely absent throughout the zone where recognizable evidences of sulfate attack are found.

Altered C-S-H Gel

Many of the changes within the paste reflect alteration rather than complete removal or replacement of the pre-existing components. This is particularly true with C-S-H gel.

In some areas that have undergone carbonation as well as sulfate attack this alteration is marked by the development of a "salt and pepper" morphology, as exemplified in Figure 5. In this paste irregular patches of bright material alternate with patches of darker appearance. The darker portions represent C-S-H gel regions moderately or heavily depleted of calcium, but showing significantly higher contents of magnesium, aluminum, and sulfur than normal for C-S-H gel. In contrast, the bright areas have retained most of their calcium. The removal of calcium from C-S-H gel is clearly not a uniform process over a region of hardened cement paste, but reflects local variations. Indeed, in some areas the brighter zones may reflect local accumulation of Ca over what might have been present originally.

In some areas, often those beyond the influence of carbonation, there is a "smearing out" of the texture of the C-S-H gel; this is accompanied by heavy deposition of ettringite. Such an area is depicted in Figure 6, where ettringite has not only filled air voids but is heavily deposited in small pockets with the C-S-H mass, apparently leading to local expansion and cracking.

Altered Residual Cement Grains

A major kind of alteration induced by sulfate attack involve residual unhydrated cement grains. The locations of residual cement grains can be recognized easily in most of the concretes investigated here, since most were placed using ASTM Type V cement. The C_4AF phase found within large unhydrated grains in such cements is resistant to hydration, even under ongoing sulfate attack, and is easily recognizable by its brightness and its elongated morphology.

Figure 7 shows an early stage of alteration in a very large grain, about 80 μm long. The bright thin features within it are C_4AF. The dark outer zone of the residual cement grain is heavily hydrated. There appears to be a ragged border delimiting this outer hydrated zone from the inner portion, which consists mostly of three rounded grains which are obviously composed of C_2S. However, the EDXA analysis indicates a much reduced calcium content compared to that of unaltered belite, presumably reflecting selective leaching out of calcium.

Another large former cement grain is shown in Figure 8. This grain seems to have undergone much further dissolution, and no unhydrated calcium silicate remains. The interior of the grain is partly empty and partly occupied by hydration product of low Ca:Si ratio. The chloride content of the internal hydration product is significant; evidently the ground water in this case contained a significant concentration of chloride as well as of sulfate. A number of similar instances have been found that show identical morphology and overall composition but lack the chloride, suggesting that the chloride is not necessary for this type of alteration.

Another mode of alteration of residual unhydrated cement grains apparently involve selective removal of *all* of the calcium seemingly without any dissolution of the original calcium silicate component. Figure 9 shows an instance of this. The relics of the altered C_2S appear solid and fills the space of the original C_2S within the cement grain. It is hydrous, as indicated by the gray level, and it contains nothing but silica, retaining no calcium whatever. It appears that the attacking solution has gently removed all of the calcium and converted the C_2S *in situ* to a silica gel.

Selective removal of calcium from former cement grains does not necessarily occur without incorporation of another cation. Figure 10 shows an area where the similar *in situ* alteration product contains a significant content of magnesium, obviously abstracted from a magnesium-rich attacking solution. Indeed, as indicated in Figure 11, the magnesium content may be high enough to suggest the formation of serpentine ($6MgO.4SiO_2.4H_2O$). Gollop and Tayler (4), considered that "cryptocrystalline serpentine" might be the magnesium silicate hydrate alteration product they found in cement paste exposed to strong magnesium sulfate solutions. On the other hand, Bonen and Cohen (6) indicated that the "MSH" formed in their laboratory study was $2MgO \cdot SiO_2.xH_2O$, which is rather richer in MgO than serpentine, and Cole and Heuber (8) reported that an even more MgO-rich phase having the composition $4MgO \cdot SiO_2 \cdot xH_2O$ was formed in concrete attacked by sea water.

In many areas former cement grains appear to have been subjected to complete dissolution of its calcium silicates (leaving only the C_4AF as a marker) and then to occupation of the space created by of entirely foreign species. The types of material deposited reflect the full gamut of sulfate attack deposits.

Figure 12 shows a large former cement grain in which the material deposited in the hollowed spaces in the left half of the grain is monosulfate. Deposition of monosulfate seems particularly favored in cements containing fly ash, an example of which is shown in Figure 13. Ettringite deposits in such places is probably more common in plain concretes, an example of which is shown in Figure 14. However, Friedel's salt (monochloroaluminate) may be found where the attacking solution is rich in chloride, as shown in Figure 15.

This does not exhaust the list of kinds of material that can be found in the spaces created by dissolution of the calcium silicates in former cement grains. Even gypsum and brucite ($Mg(OH)_2$) have been found in such spaces.

Deposition of New Phases Within the Cement Paste Generally

The illustrations above of new phases deposited within former cement grains leads naturally to a consideration of the deposition of sulfate attack reaction products within the confines of the cement paste in general.

Indeed, the usual account of sulfate attack found in textbooks considers sulfate attack as being *defined by* the formation of ettringite and gypsum in paste, and considers only the purported consequences of such formation. For example, Neville (9, p. 76) describes sulfate attack thusly. "In hardened cement, calcium aluminate hydrate can react with a sulfate salt from outside the concrete in a similar manner: the product of addition is calcium sulfoaluminate, forming within the framework of the hydrated cement paste. Because the increase in volume of the solid phase is 227 percent, gradual disintegration of the concrete results. A second type of reaction is that of base exchange between the calcium hydroxide and the sulfates, resulting in the formation of gypsum with an increase in the volume of solid phase of 124 percent. *These reactions are known as sulfate attack*" (italics supplied by the present authors).

It is obvious from the examinations detailed previously in this paper that sulfate attack is much more complicated than a simple formation of ettringite and gypsum within the framework of the paste, and of course the mechanisms of formation cited by Neville are not correct. There is no calcium aluminate hydrate in ordinary concrete, and calcium hydroxide does not convert in situ to gypsum by base exchange. Nevertheless, deposition of ettringite and gypsum (and a number of other phases) within the framework of the cement paste are prominent features of sulfate attack, and illustrations are provided below.

Sometimes the most easily seen feature of sulfate attack is the generation of layered deposits of gypsum more or less parallel to the entry surface of the sulfate-bearing solutions. A typical example is shown as Figure 16. Light-colored parallel bands of gypsum, typically 40 to 80 μm thick are separated from each other by ca. 150 - 200 μm intervals, and dominate a zone several mm thick adjacent to the surface of contact with the soil solution below. The gypsum invades the groundmass of the hydrated cement paste, the contact between cement paste and aggregate, and even cracks within the individual aggregates. The occurrence of such deposits have been reported by various workers in reports of laboratory studies of sulfate attack (DEF).

Figure 17 shows finer detail of such a gypsum deposit in another concrete. The thick gypsum layers depicted here are clearly within a zone formerly occupied by hydrated cement paste. Nevertheless there is no trace in the accompanying EDXA analysis of any residue of the C-S-H gel formerly present. Thus the gypsum has not *infiltrated* into the paste; either it has mechanically forced the C-S-H gel above and below it apart to make room for it, or else the C-S-H gel has been dissolved locally leaving a space for the gypsum to deposit. It appears that the former is much more likely. Figure 18 shows a similar area where a cracked aggregate seems to have been forced apart by gypsum crystallization pressure. Note the euhedral character of some of the gypsum crystal grain boundaries in the space created.

Another phase commonly precipitated at or near surfaces of contact with sulfate solutions is brucite, $Mg(OH)_2$. Brucite deposits are commonly found at the surface zone in laboratory studies of magnesium sulfate attack (4,6,7) and in sea water attack (10) and may constitute a continuous outer layer. They are much less prominent in the present concretes, since the magnesium content of the ground water is typically much less than in sea water or magnesium sulfate solutions used in sulfate attack studies.

The other component whose deposition is classically associated with sulfate attack is ettringite. Ettringite does not occur (as gypsum does) in massive layers near the surface in contact with the ground water. This may be due to in part to its lack of stability within carbonated cement paste. As indicated earlier, somewhat carbonated zones tend to occur at the bottom surface, as well as at the upper surface of the concretes.

The instability of ettringite in such zones apparently does not extend to thaumasite ($C_3S \cdot CS \cdot H_{27}$) the calcium sulfate carbonate silicate analog of ettringite. A massive deposit of thaumasite at the bottom edge of a concrete is shown in Figure 19. Thaumasite has usually been associated with low temperature exposure conditions, but evidence is accumulating that it can be stable at normal and even high ambient temperatures as well. As indicated in Figure 20, thaumasite may be deposited within the concrete, as well as at the contact with external solution. In some places, such as seen in Figure 21, analyses indicate that mixed thaumasite and ettringite, (or more likely, a solid solution between them) occurs. Thaumasite is usually considered to be especially deleterious in concrete, and its presence as a consequence of sulfate attack in warm climates is not a favorable sign.

Ettringite itself then is ordinarily found only beyond the zone of carbonation in sulfate attack, which confines it to the restricted spaces of the interior of the concrete. However, it is extremely ubiquitous in occupying such spaces, and may dominate a local field by presence in every conceivable void or gap. Figure 22 shows the usual appearance of an ettringite-filled air void; Figure 23 shows ettringite in rims or 'gaps' surrounding aggregate grains; Figure 24 shows ettringite dominating an area of cement paste by local insertion in elongated zones and smaller voids; and Figure 25 shows an ettringite deposit within a hollow cenosphere in a fly ash-bearing concrete.

The same forces that lead to the deposition of ettringite in sulfate attack may induce the deposition of monosulfate in similar areas if the local concentration of sulfate is insufficient to support ettringite deposition. Such conditions often apply in fly ash bearing concretes, but may occur in any concrete. Figure 26 shows a void-filling monosulfate deposit entirely analogous to the ettringite deposit in Figure 22. While a small amount of monosulfate is present in concrete not subject to sulfate attack, such massive void-filling deposits are not found in concretes that have not been subjected to sulfate attack. Figure 27 shows a monosulfate deposit dominating a paste area in plain concrete, and Figure 28 shows locally ubiquitous monosulfate deposits in a fly ash concrete.

The deposition of phases such gypsum, ettringite, and monosulfate provide clear, and sometimes spectacular evidences of sulfate attack. The introduction of other species within the framework of the hardened cement paste may be less spectacular, but equally important. For example, Figure 29 shows an area where the general paste groundmass has undergone the removal of nearly all of its calcium, but instead of leaving a silica gel, sufficient magnesium has been introduced locally to convert the entire gel locally to a magnesium silicate. The fact that such a response may cover quite a broad area of cement paste is further indicated in Figure 30. The fate of the calcium removed from the original C-S-H gel is unclear. In some instances it seems to be accumulated in adjacent 'paste' areas that yield only calcium EDXA signals, perhaps indicating that such areas consist of extremely fine microcrystalline calcium carbonate. An example is shown in Figure 31.

Magnesium-Bearing Rims Formed Around Voids

Another unspectacular feature of sulfate attack as seen in these concretes is the development of so-called 'magnesium rich rims' around voids in the altered paste. Such rims are unimportant in terms of the volume of paste affected, but are consistent and repeatable markers for sulfate attack. The name is misleading; since the quite consistent compositions indicated by EDXA contain calcium, silica, and aluminum as well as magnesium. As seen in Figure 32, the usual peak height ratio are about 2:1:0.5:0.5 for Ca, Si, Al, and Mg, respectively. Rims of such compositions are found around empty voids, as in Figure 32; voids containing ettringite deposits, as in Figure 33; or voids containing monosulfate, etc.

EXPANSION AND THE DEVELOPMENT OF CRACK NETWORKS

While not a feature of altered microstructure *per se*, local expansion and development of a network of cracks are common feature in concretes subjected to sulfate attack. Such microcracking is most commonly seen in areas of extensive deposition of ettringite, and presumably arise from the expansion induced locally by such deposition. Illustrations are provided in Figures 34 and 35.

As seen in Figure 35, particularly in the upper left corner, the crack network is seen to include some debonding and the formation of open rims around portions of the sand grains. In our experience such debonding and partial rim formation is not as extensive as it often is in cases of delayed ettringite formation, but it does occur in external sulfate attack as well. An illustration of an extensive open rim around a coarse aggregate grain is provided in Figure 36.

DISCUSSION

The sulfate attack features documented here provide only a rough overview of the many complex changes that occur in the hardened cement paste in concrete as a result of external sulfate attack. The responses are locally variable, and of course reflect variations in age, water:cement ratio, location within a structure, and of course the varying compositions of the sulfate-bearing ground waters responsible for the attack.

In research and in the general profession sulfate attack has traditionally been associated with expansion associated with ettringite formation. Type V or "sulfate resistant" portland cement is formulated with a limitation on the C_3A content for the sole purpose of limiting ettringite formation. Field observers, however, have stressed the importance of softening and exfoliation of the concrete rather than of expansion. For example, St. John et al. (11, p. 253) indicated that "The visible effects of sulfate attack are exfoliation of the outer layers of the concrete, corrosion, and reduction of the hardened paste to what is commonly described as a 'soft pug' or a mush, and the deposition of salts on surfaces and in exfoliation cracks."

Some of the causes for the conversion of the hardened paste to a 'soft pug' or a mush begin to be apparent in the various alterations of the hardened cement paste described in this report.

It should be noted that the specific concretes studied here are not at the final stages of reaction, and do not necessarily display the full effects of changes that may eventually be brought about.

CONCLUSIONS

Microstructural changes observed in high water:cement ratio concretes exposed to sulfate attack from external sulfate bearing solutions containing varying concentrations of sodium, magnesium, calcium, sulfate, and chloride are not confined to the deposition of ettringite and gypsum. Rather they involve fundamental alterations of the hardened cement paste. The picture is complicated by the simultaneous development of carbonated zones, not only near surfaces exposed to the air but to a lesser extent, near surfaces in contact with the sulfate-bearing ground waters. The changes observed within the paste include dissolution of calcium hydroxide, decalcification of some of the C-S-H gel, and the effects of reactions with the previously unhydrated calcium silicates in large residual cement grains. It was found that massive deposits, not only of gypsum, but also of of brucite, and occasionally of thaumasite may occur near the external surface of contact of the concrete with the soil. Thaumasite, ettringite, monosulfate, and Friedel's salt may deposit freely in available pore space within the paste structure Some of the partly decalcified C-S-H gel may be converted to silica gel or to magnesium silicate hydrate, with some or all of the lost calcium accumulating in adjacent areas. Calcium silicates within the former cement grains may dissolve, leaving mostly empty space; they may decalcify leaving silica gel; or they may decalcify and incorporate magnesium, generating magnesium silicate hydrate. If emptied, the spaces left may subsequently be filled with ettringite, monosulfate, Friedel's salt, or other secondary deposits. Finally, narrow rims of a magnesium- and aluminum – bearing calcium silicate of almost constant composition are found around many voids, and may be a good marker for sulfate-attack induced changes.

ACKNOWELDGMENTS

The assistance of Sadananda Sahu of the R. J. Lee Group Inc. is is gratefully acknowledged.

REFERENCES.

(1) S. Diamond and Jingdong Huang, "The Interfacial Transition Zone: Reality or Myth?" Proc. of the RILEM International Conf. On the Interfacial Transition Zone, Haifa, March 1998., A. Bentur, ed. in press (1998).

(2) T. Thorvaldson, "Chemical Aspects of Durability of Cement Products," Proc. 3rd Intl. Symp. On the Chemistry of Cements, London (1952).

(3) P. K. Mehta, ""Sulfate Attack on Concrete – A Critical Revue," pp. 104-130 in Materials Science of Concrete, III J. P. Skalny, ed., Amer. Ceramic Soc., Westerville, OH (1992).

(4) R.S. Gollup and H. F. W. Taylor, "Microstructural and Microanalytical Studies of Sulfate Attack. I. Ordinary PortlandCement Paste," Cem. Concr. Res. 22, 1027-1038 (1992).

(5) D. Bonen and M. D. Cohen, "Magnesium Sulfate Attack on Portland Cement Paste. I Microstructural Analysis," Cem. Concr. Res. 22 159-168 (1992).

(6) D. Bonen and M. D. Cohen, "Magnesium Sulfate Attack on Portland Cement Paste. II.Chemical and MineralogicalAnalysis," Cem. concr. Res. 22 707-718 (1992).

(7) D. Bonen, "A Microstructural Study of the Effect Produced by Magnesium Sulfate on Plain and Silica Fume Bearing Portland Cement Mortars," Cem. Concr. Res. 23 541-553 (1993).

(8) W. F. Cole and H. V. Hueber, Nature, 171 354 (1953).

(9) A. M. Neville, "Properties of Concrete," Fourth edition, 844 pp., John Wiley and Sons, Inc. (1996).

(10) N. R. Buenfeld and J. B. Newman, "The Development and Stability of Surface Layers on Concrete Exposed to Sea Water,"Cem. Concr. Res. 16 721-732 (1986).

(11) D. A. St. John, A. W. Poole, and I. Sims, "Concrete Petrography," Arnold Publishers, London, 474 pp. (1998)

LEGENDS FOR FIGURES

1. Appearance of normal features of concretes not subject to sulfate attack, including (A) unhydrated cement, (B) dense inner product or phenograin C-S-H gel surrounding unhydrated cement, (C)inner product or phenograin C-S-H gel constituting fully hydrated cement grain, (D) region of small hollow shell hydration grains, (E) groundmass or outer product C-S-H gel, (F) calcium hydroxide surrounding a sand grain chip, (G) deposit of calcium hydroxide within the groundmass
2. Large deposit of calcium hydroxide on the surface of a sand grain.
3. Deposit of sodium sulfate removed from the upper surface of a concrete slab.
4. Gypsum deposit apparently replacing a former calcium hydroxide deposit around a sand grain.
5. 'Salt and pepper' morphology developed from C-S-H in affected zones.
6. 'Smeared out' texture of C-S-H gel in some areas of heavy ettringite deposition.
7. Early stage of alteration of a large former cement grain.
8. Former cement grain where no unhydrated calcium silicate material is left.
9. Former cement grain where calcium has been selectively leached out, leaving a solid residue of silica gel.
10. Former cement grain in which calcium has been leached out, but partly replaced by magnesium.
11. Former cement grain incorporating a higher content of magnesium.
12. Former cement grain with some of the C_2S dissolved and monosulfate deposited in the space so created.
13. Monosulfate deposit in a former cement grain in a fly ash-bearing concrete.
14. Ettringite deposit in former cement grain in a plain concrete.
15. Friedel's salt deposit in former cement grain in a plain concrete.
16. Layered deposits of gypsum parallel to the surface of contact with sulfate-bearing solution. The bottom of the concrete can just be seen at the bottom of the micrograph.
17. Layered but nearly massive deposits of gypsum near the surface in contact with sulfate-bearing solution.

18. Gypsum penetrating a cracked sand grain and seemingly forcing the several parts further apart.
19. Massive deposit of thaumasite at the bottom surface of concrete in contact with sulfate-bearing solution.
20. Deposit of thaumasite within a concrete. Note the lower part of the deposit apparently forming an occupied rim around the sand grain to the left of it.
21. A layered deposit of mixed thaumasite and ettringite (or a solid solution between them) within a sulfate-affected concrete.
22. A classical view of ettringite invading and filling a large air void within the concrete. The large cracks within the ettringite mass are shrinkage cracks induced by evacuation and drying of the sample.
23. Ettringite occupying a rim around a sand grain.
24. Ettringite dominating the local paste morphology by insertion into the pre-existing C-S-H gel mass.
25. Ettringite deposited in and filling the open space in a hollow fly ash cenosphere in a fly ash-bearing concrete.
26. A deposit of monosulfate filling an air void; such deposits are not found in concrete not exposed to sulfate attack.
27. Extensive monosulfate deposit mimicking the texture of "smeared out" C-S-H gel.
28. Area of locally ubiquitous monosulfate depsoits in a fly ash-bearing concrete.
29. Area of C-S-H gel decalcified and converted to magnesium silicate hydrate.
30. Broader area of C-S-H gel converted to magnesium silicate hydrate.
31. Area in which only Ca is recorded by EDXA, presumably comtaining cryptocrystalline caclcium carbonate.
32. Magnesium-bearing rim around an empty void.
33. Magnesium-bearing rim around an ettringite-containing void.
34. Area of microcrack network formation associated with extensive local ettringite deposition.
35. Another area of extensive network cracking associated with local ettringite deposition. Note the partial debonding and open rim formation around some of the sand grains.
36. An area illustrating extensive open rim formation around a coarse aggregate grain.

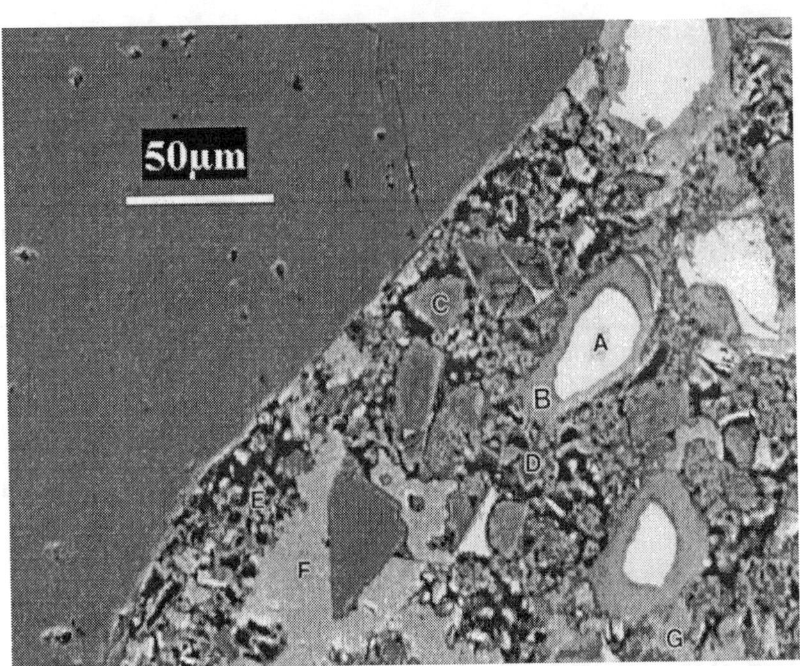

Fig. 1. Appearance of normal features of concretes not subject to sulfate attack, including (A) unhydrated cement, (B) dense inner product or phenograin C-S-H gel surrounding unhydrated cement, (C)inner product or phenograin C-S-H gel constituting fully hydrated cement grain, (D) region of small hollow shell hydration grains, (E) groundmass or outer product C-S-H gel, (F) calcium hydroxide surrounding a sand grain chip, (G) deposit of calcium hydroxide within the groundmass

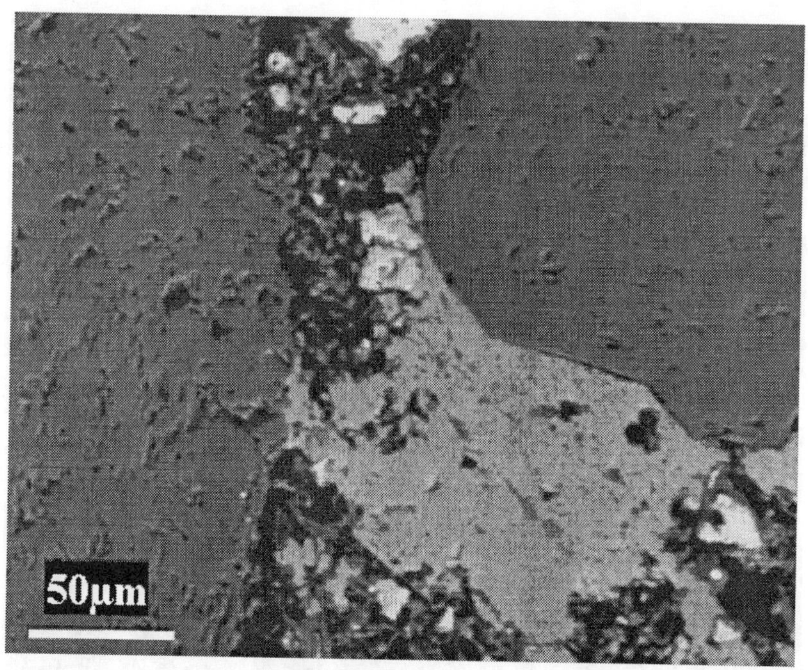

Fig. 2. Large deposit of calcium hydroxide on the surface of a sand grain.

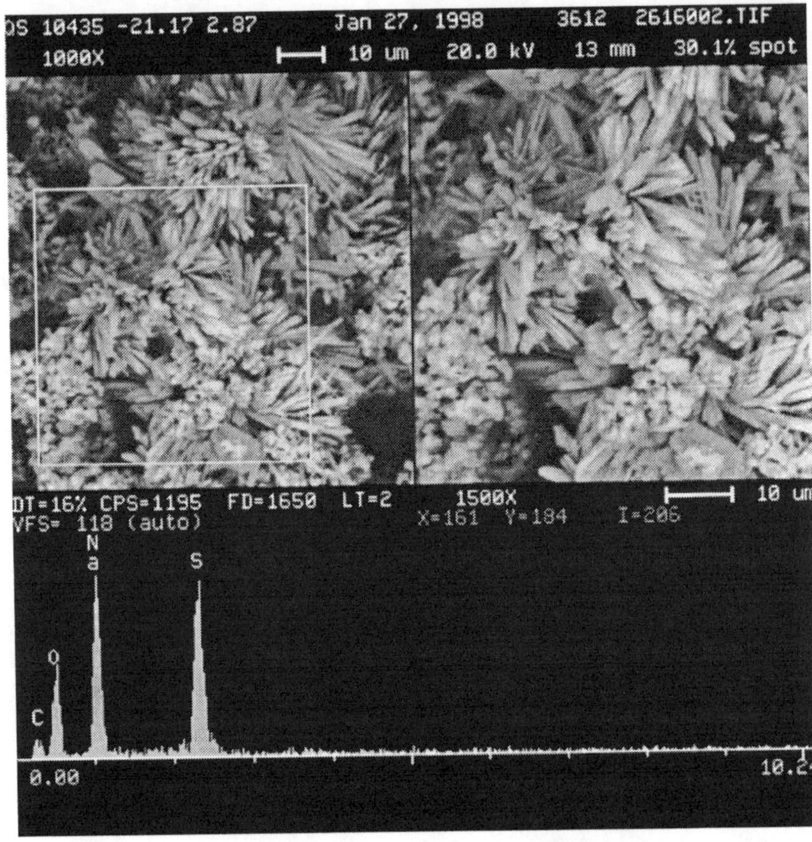

Fig. 3 Deposit of sodium sulfate removed from the upper surface of a concrete slab.

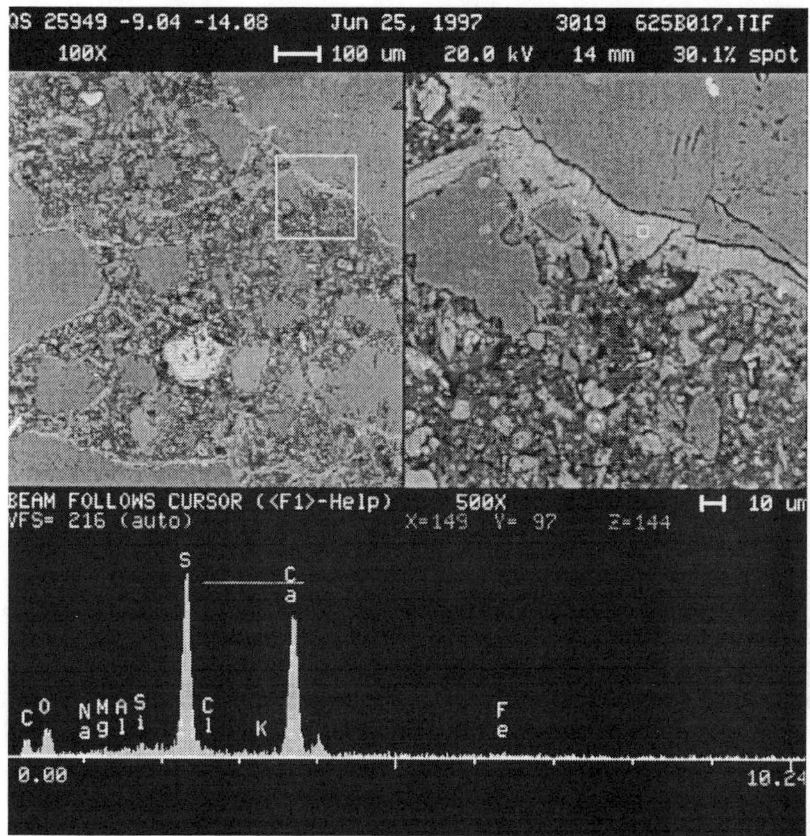

Fig. 4. Gypsum deposit apparently replacing a former calcium hydroxide deposit around a sand grain.

Fig. 5. ‘Salt and pepper’ morphology developed from C-S-H in affected zones.

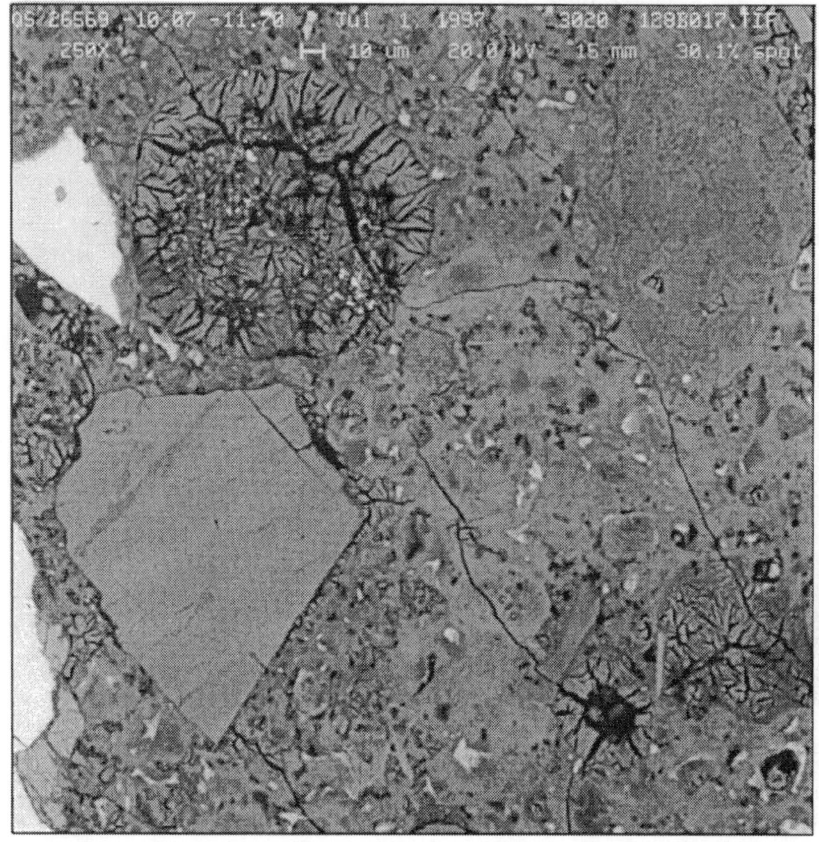

Fig. 6. 'Smeared out' texture of C-S-H gel in some areas of heavy ettringite deposition.

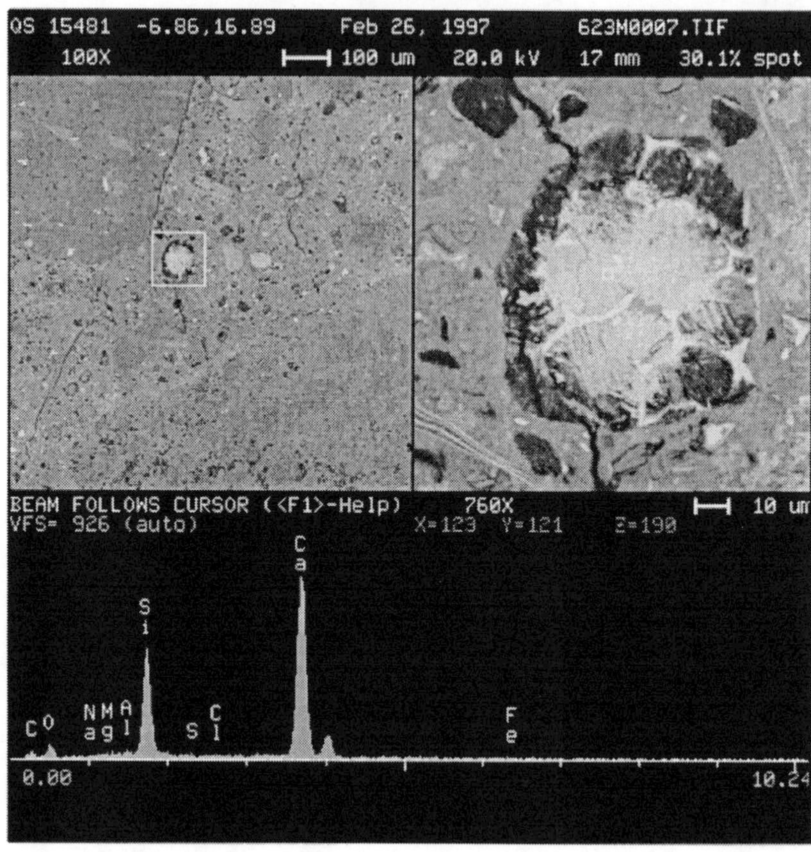

Fig. 7 Early stage of alteration of a large former cement grain.

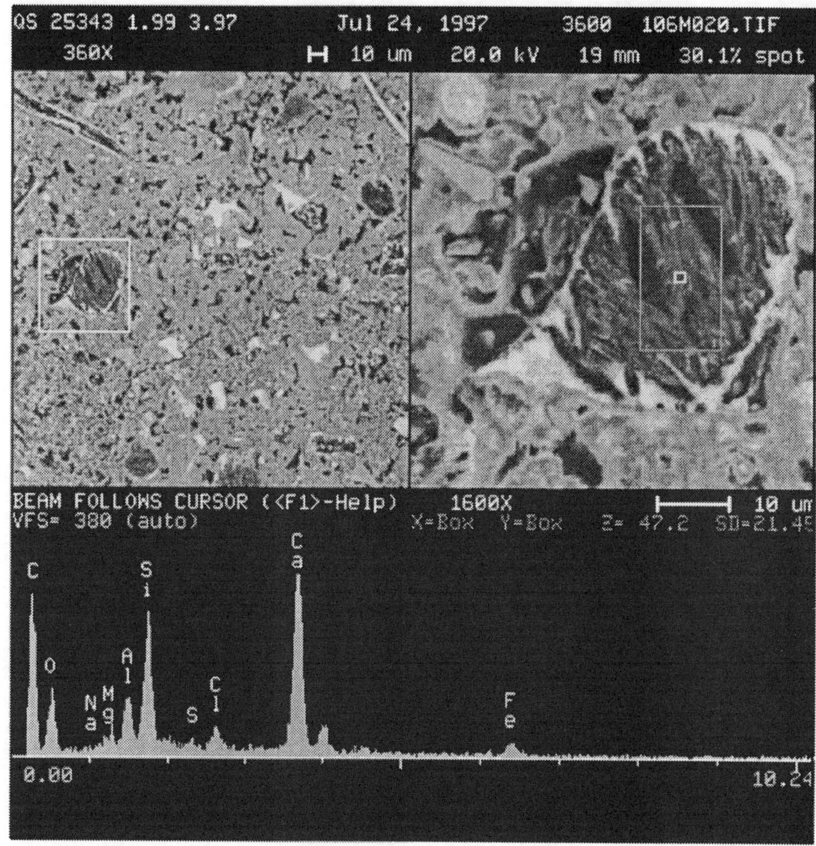

Fig. 8. Former cement grain where no unhydrated calcium silicate material is left.

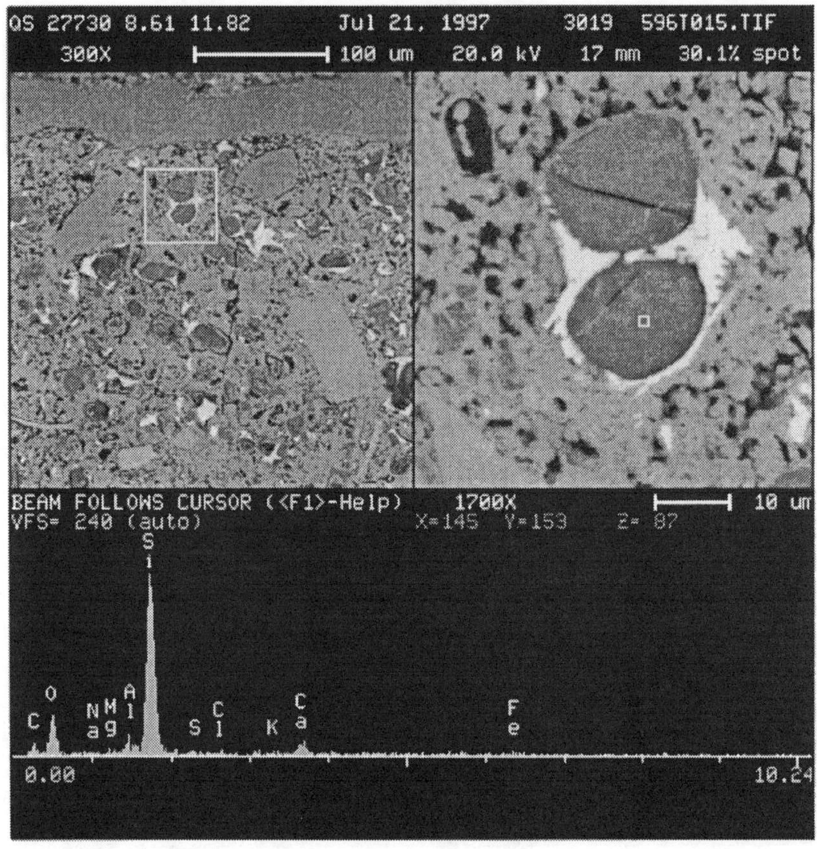

Fig. 9. Former cement grain where calcium has been selectively leached out, leaving a solid residue of silica gel.

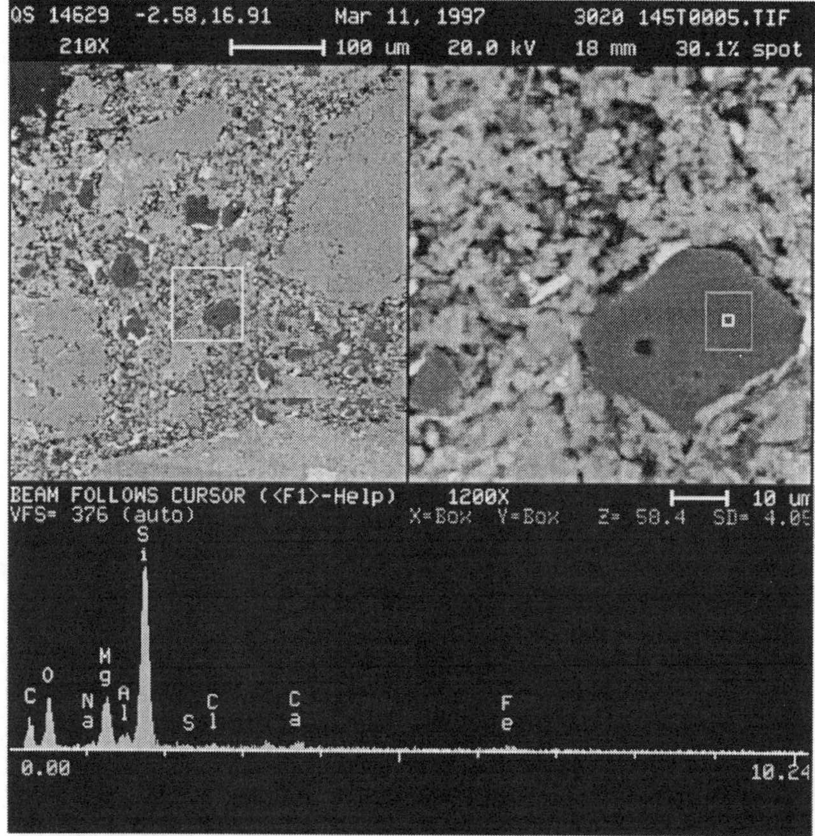

Fig. 10. Former cement grain in which calcium has been leached out, but partly replaced by magnesium.

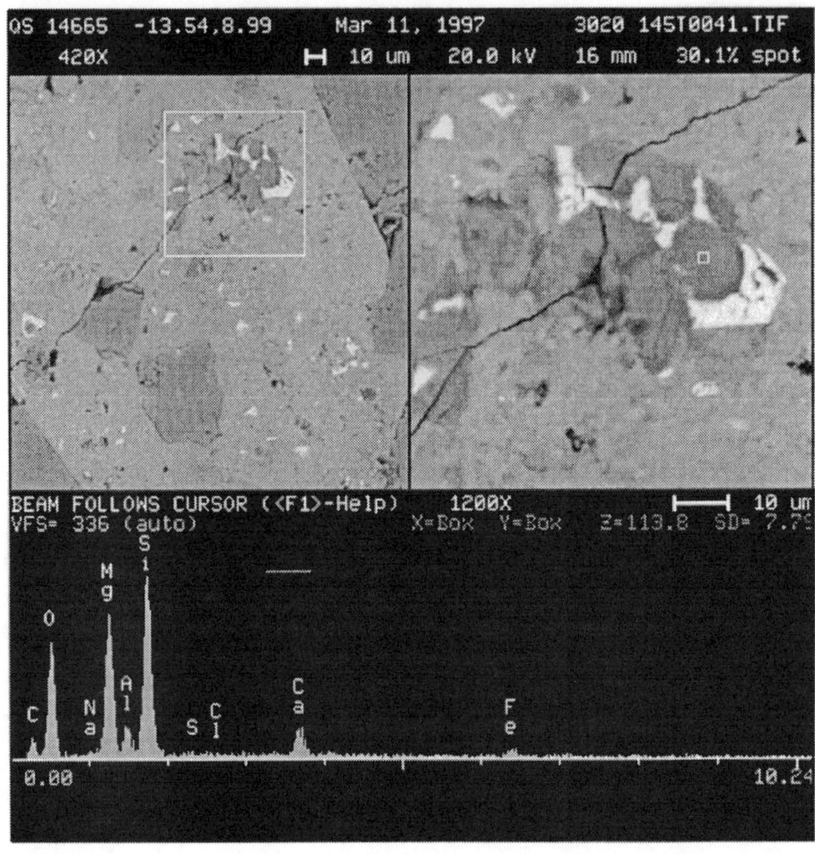

Fig. 11. Former cement grain incorporating a higher content of magnesium.

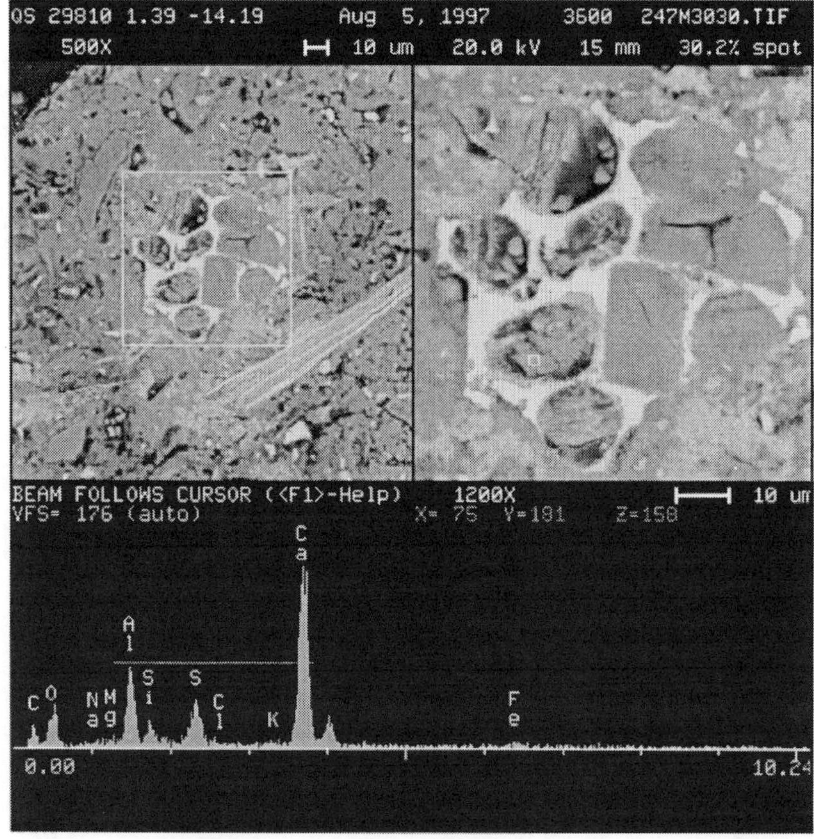

Fig. 12. Former cement grain with some of the C_2S dissolved and monosulfate deposited in the space so created.

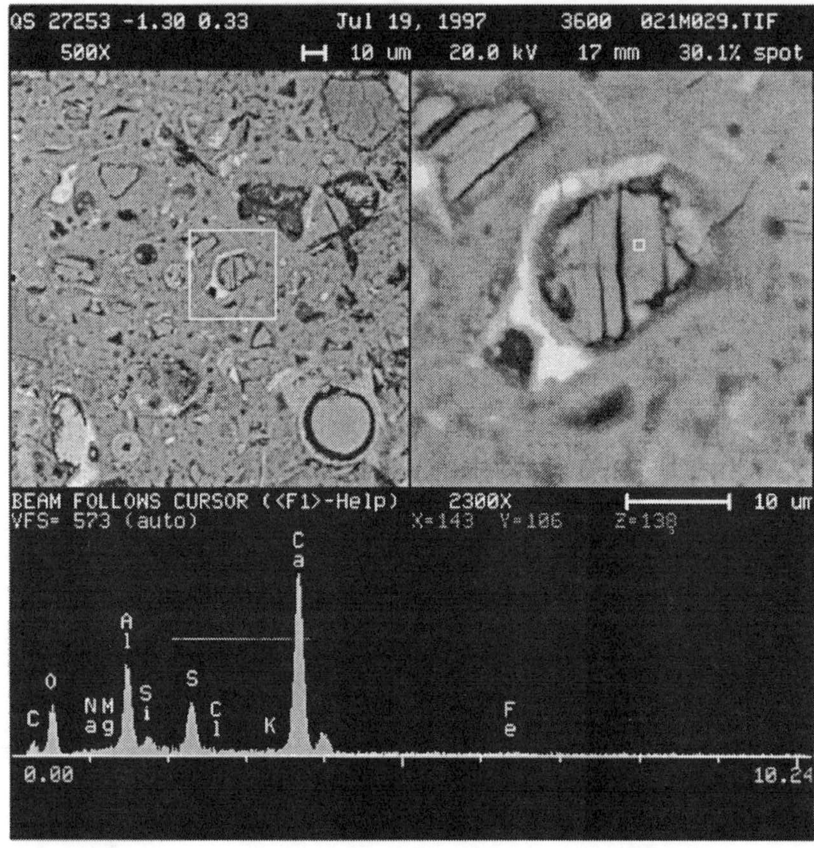

Fig. 13. Monosulfate deposit in a former cement grain in a fly ash-bearing concrete.

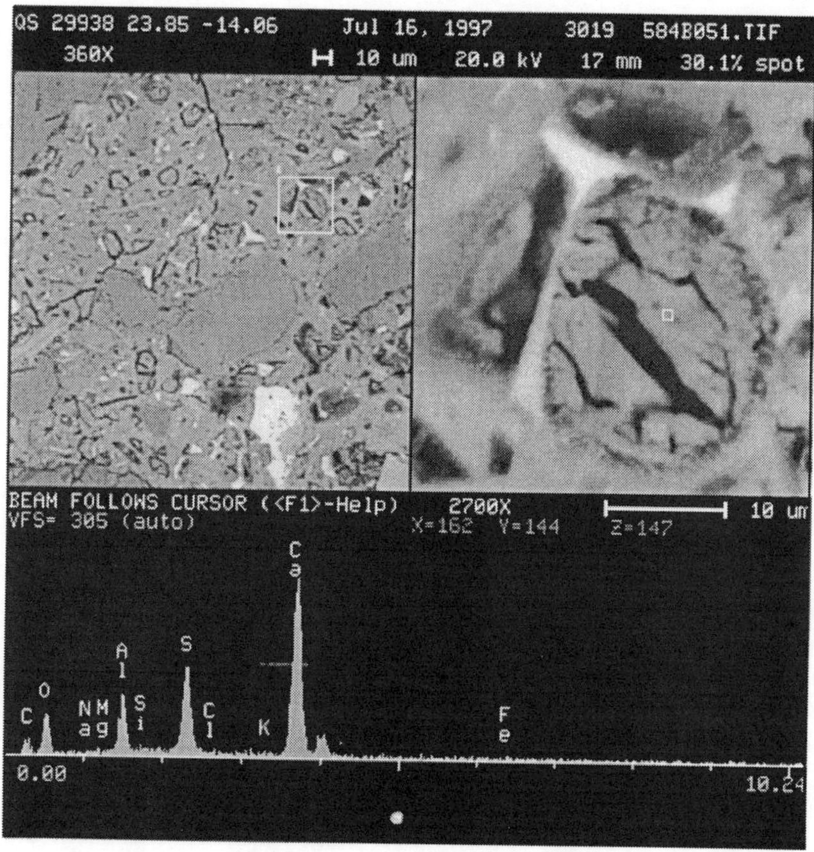

Fig. 14. Ettringite deposit in former cement grain in a plain concrete.

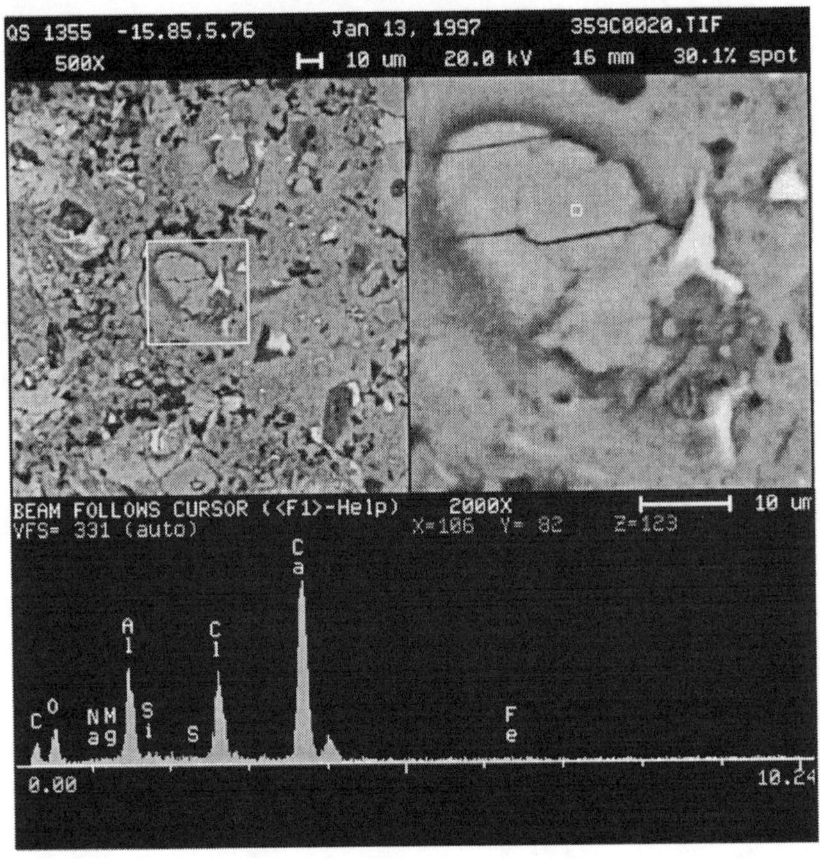

Fig. 15. Friedel's salt deposit in former cement grain in a plain concrete.

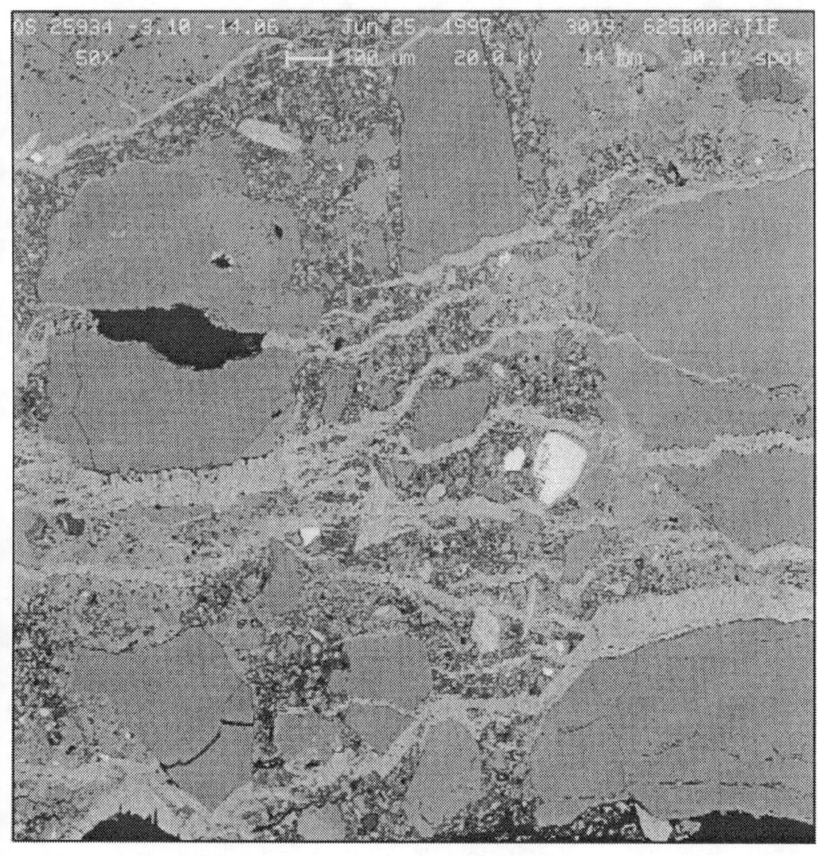

ïg. 16. Layered deposits of gypsum parallel to the surface of contact with ulfate-bearing solution. The bottom of the concrete can just be seen at the bottom f the micrograph.

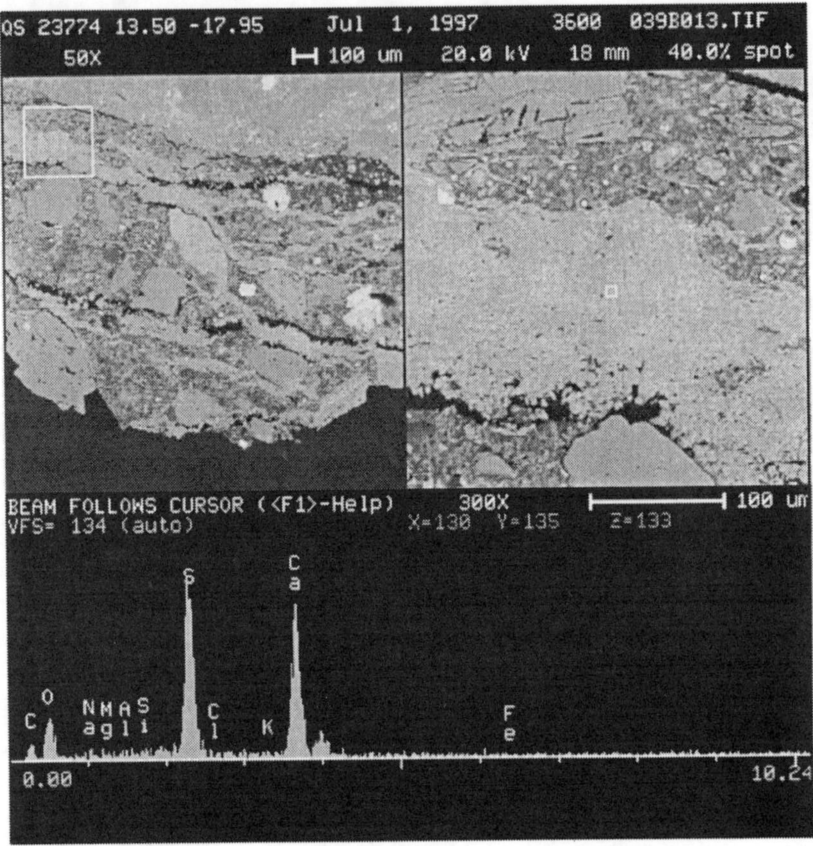

Fig. 17. Layered but nearly massive deposits of gypsum near the surface in contact with sulfate-bearing solution.

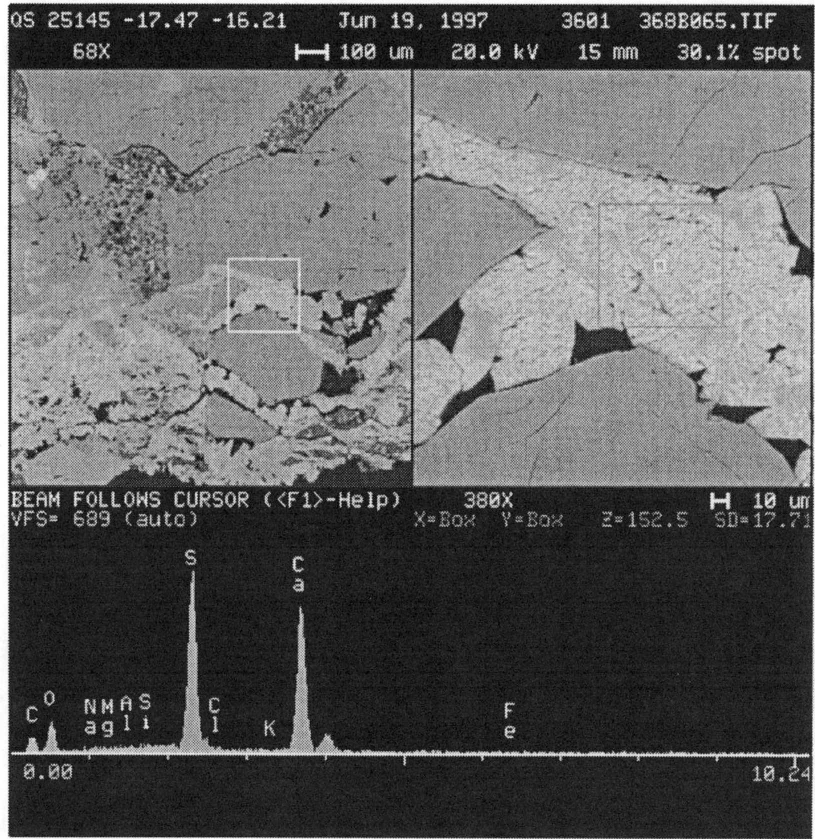

Fig. 18. Gypsum penetrating a cracked sand grain and seemingly forcing the several parts further apart.

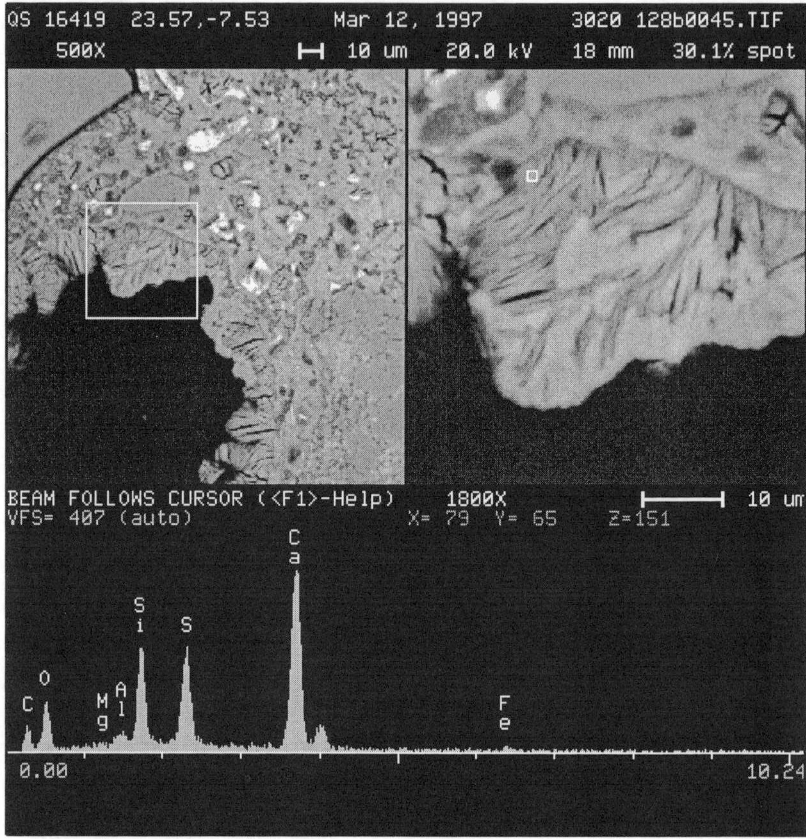

Fig. 19. Massive deposit of thaumasite at the bottom surface of concrete in contact with sulfate-bearing solution.

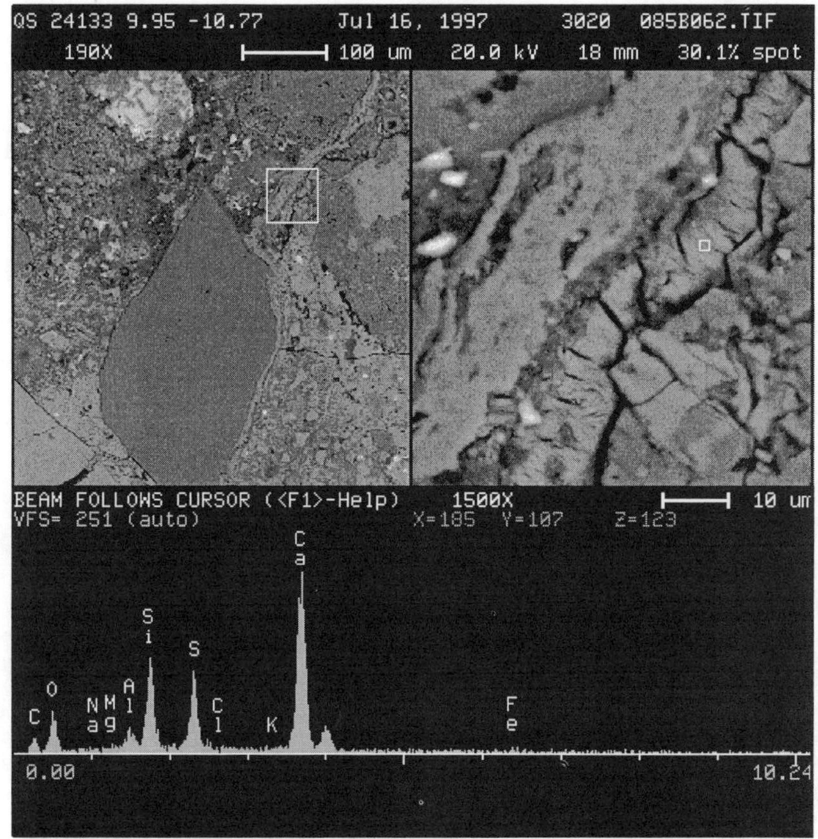

Fig. 20. Deposit of thaumasite within a concrete. Note the lower part of the deposit apparently forming an occupied rim around the sand grain to the left of it.

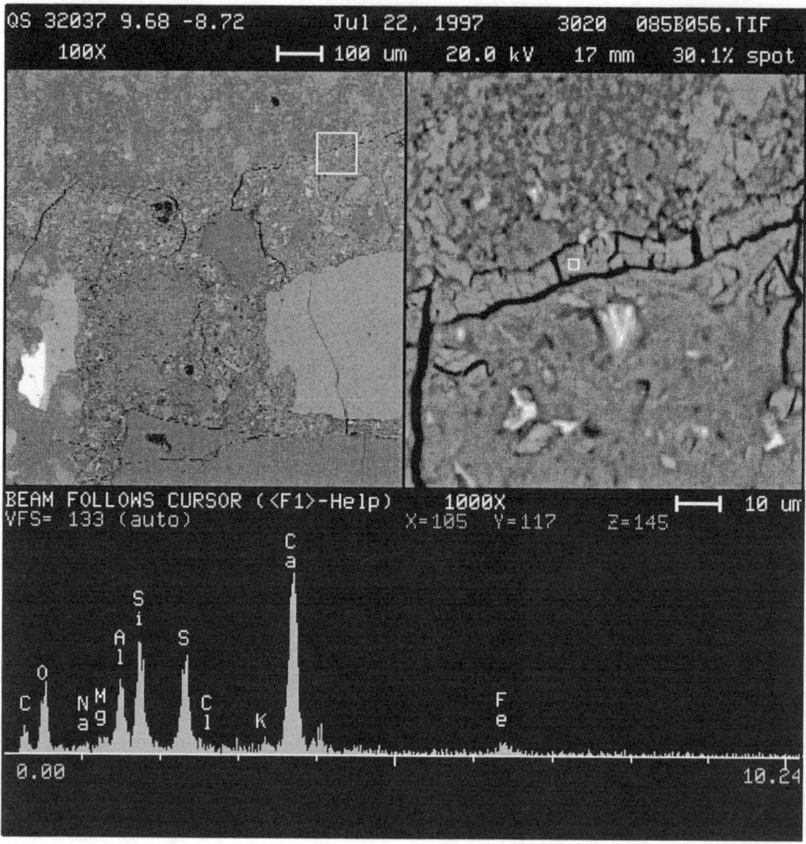

Fig. 21 A layered deposit of mixed thaumasite and ettringite (or a solid solution between them) within a sulfate-affected concrete.

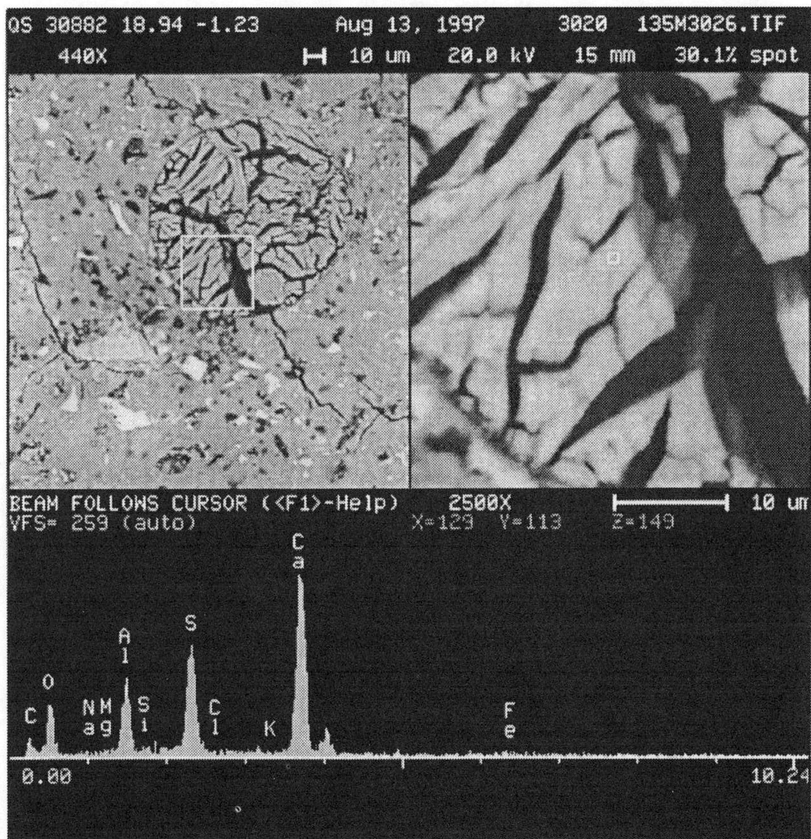

Fig. 22. A classical view of ettringite invading and filling a large air void within the concrete. The large cracks within the ettringite mass are shrinkage cracks induced by evacuation and drying of the sample.

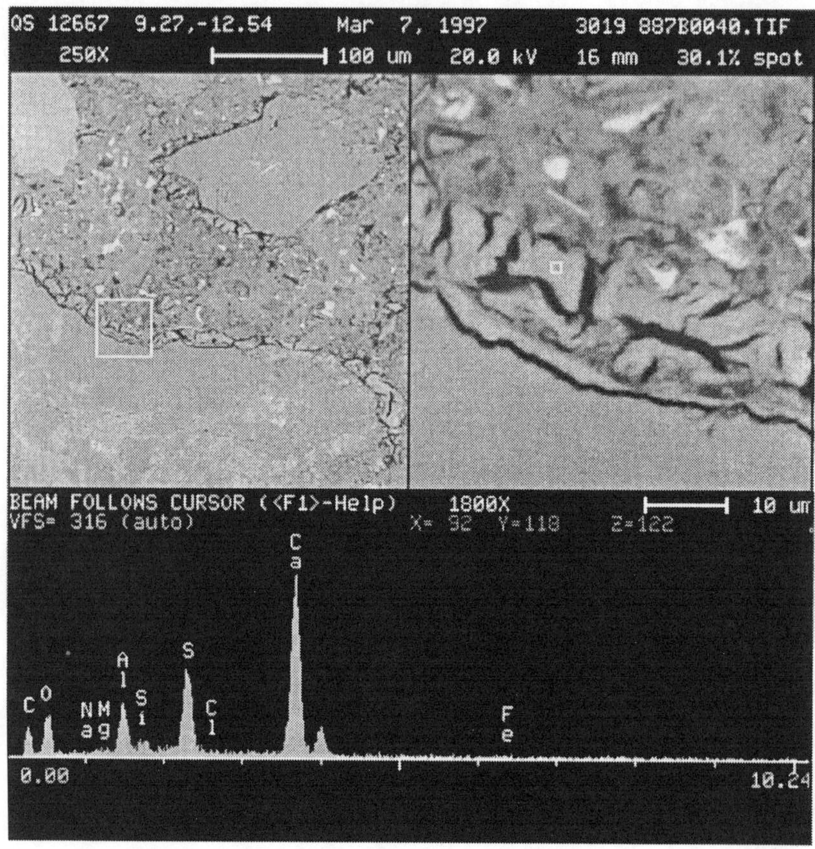

Fig. 23. Ettringite occupying a rim around a sand grain.

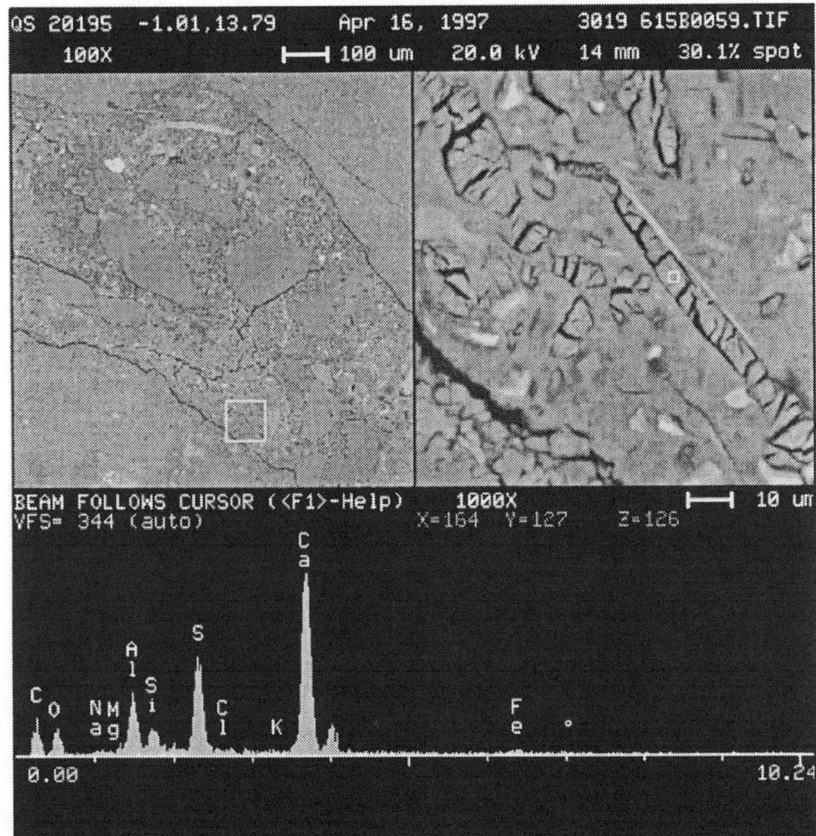

Fig. 24. Ettringite dominating the local paste morphology by insertion into the pre-existing C-S-H gel mass.

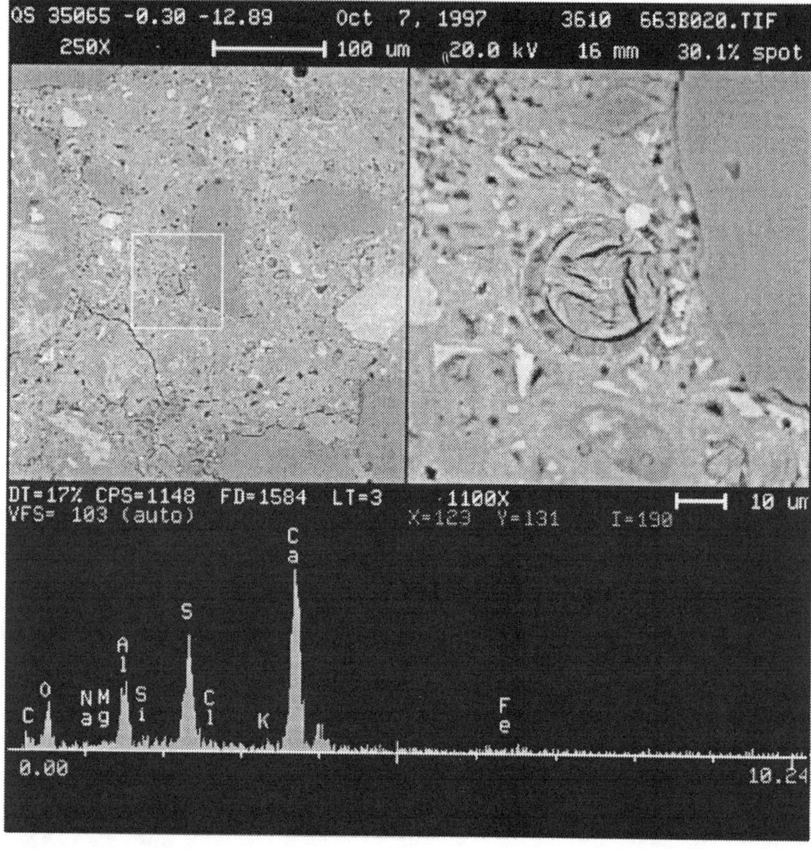

Fig. 25. Ettringite deposited in and filling the open space in a hollow fly ash cenosphere in a fly ash-bearing concrete.

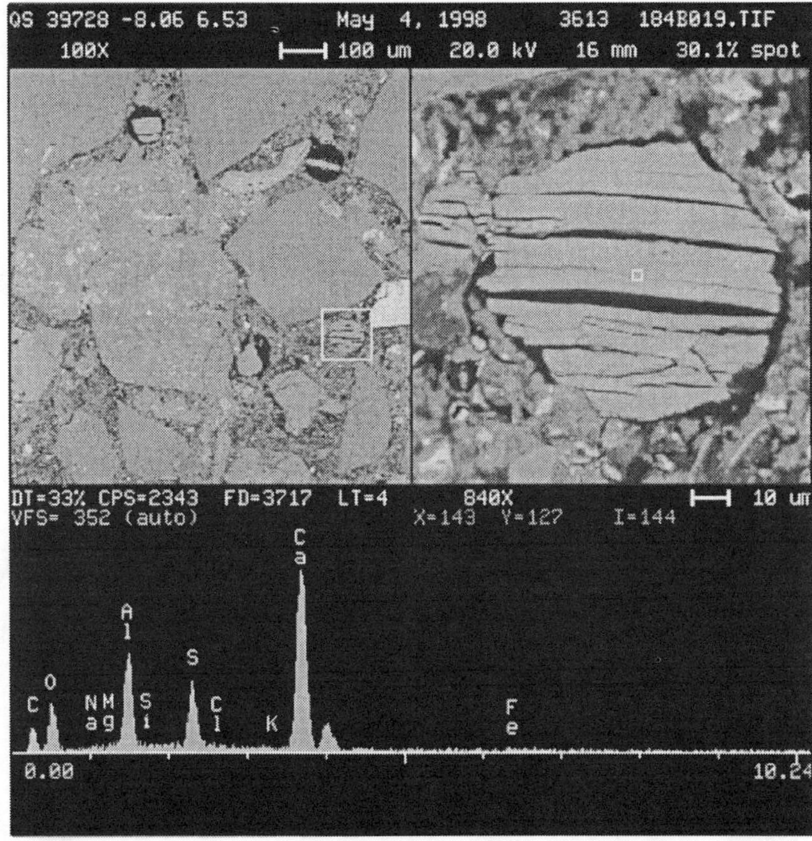

Fig. 26. A deposit of monosulfate filling an air void; such deposits are not found in concrete not exposed to sulfate attack.

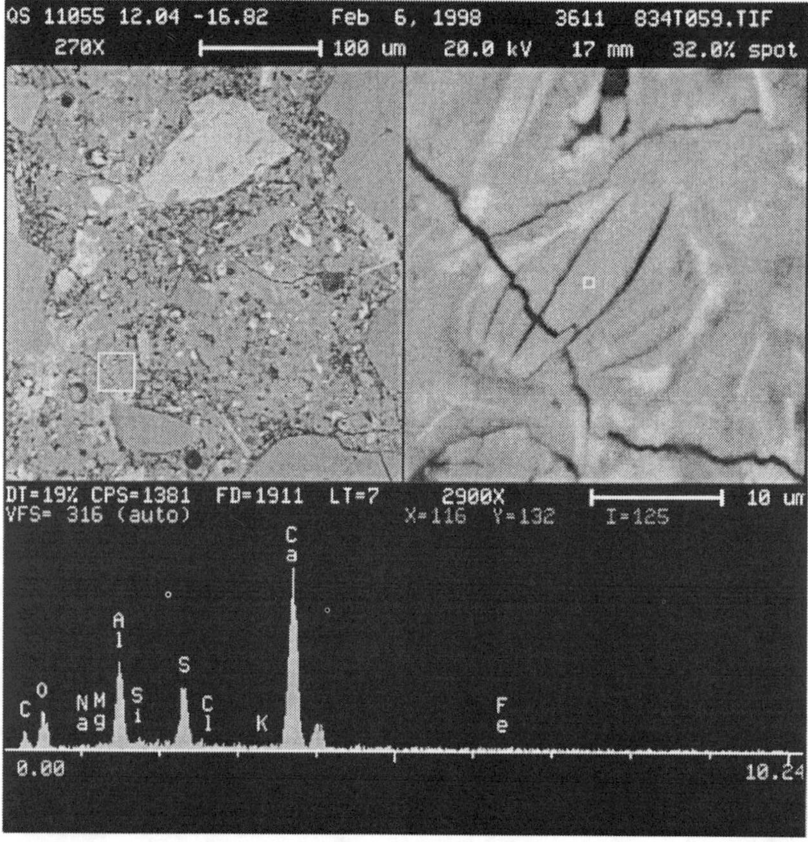

Fig. 27. Extensive monosulfate deposit mimicking the texture of "smeared out" C-S-H gel.

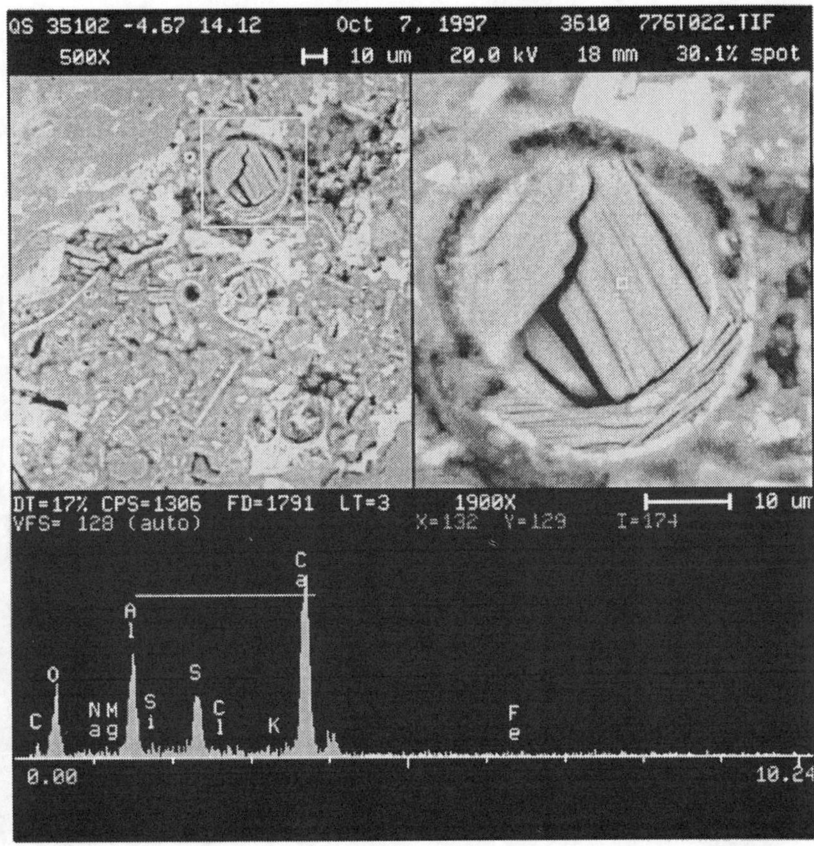

ig. 28. Area of locally ubiquitous monosulfate depsoits in a fly ash-bearing oncrete.

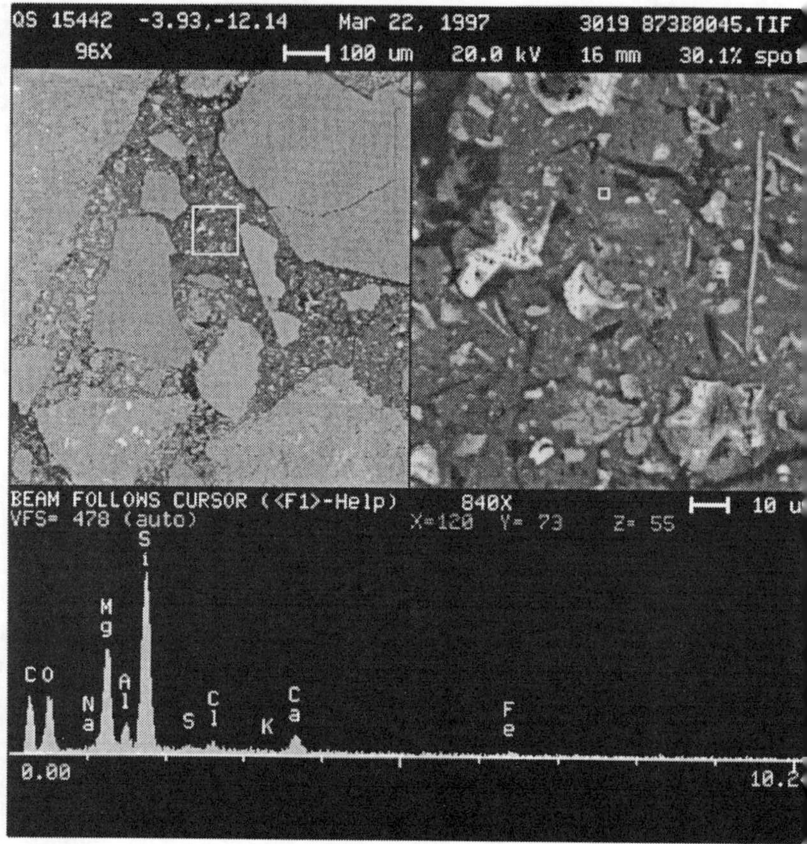

Fig. 29. Area of C-S-H gel decalcified and converted to magnesium si
hydrate.

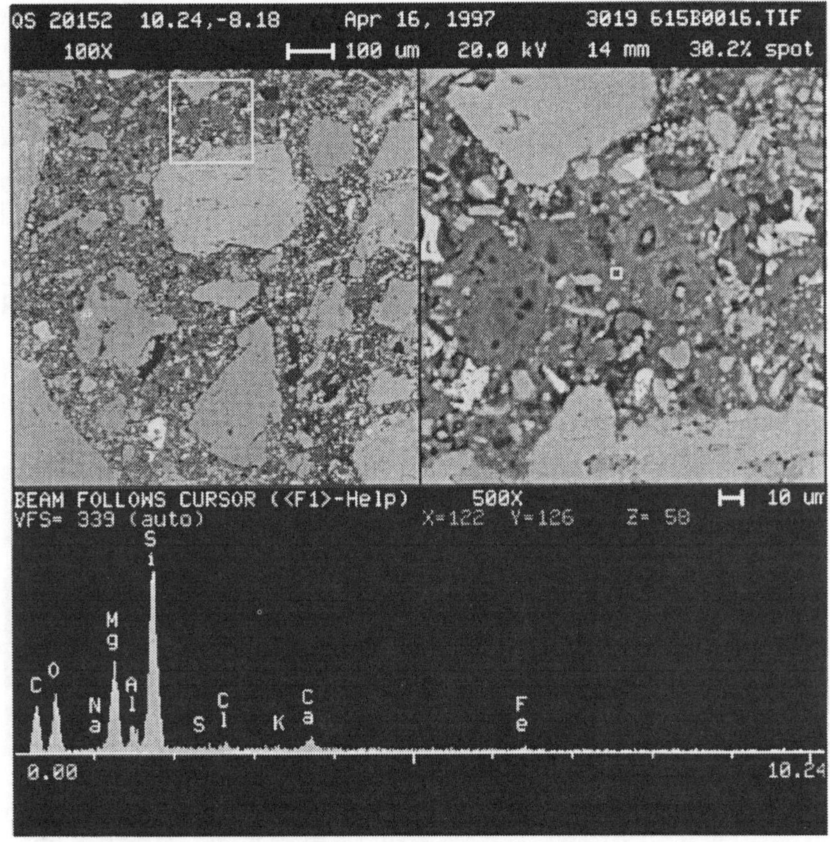

Fig. 30. Broader area of C-S-H gel converted to magnesium silicate hydrate.

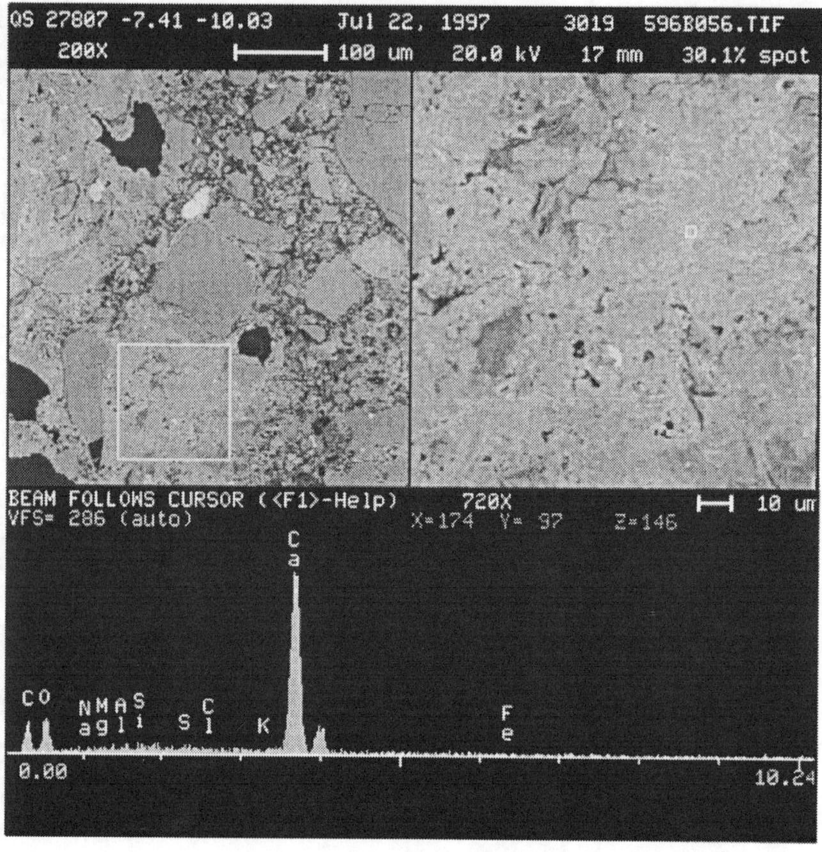

Fig. 31. Area in which only Ca is recorded by EDXA, presumably comtaining cryptocrystalline caclcium carbonate.

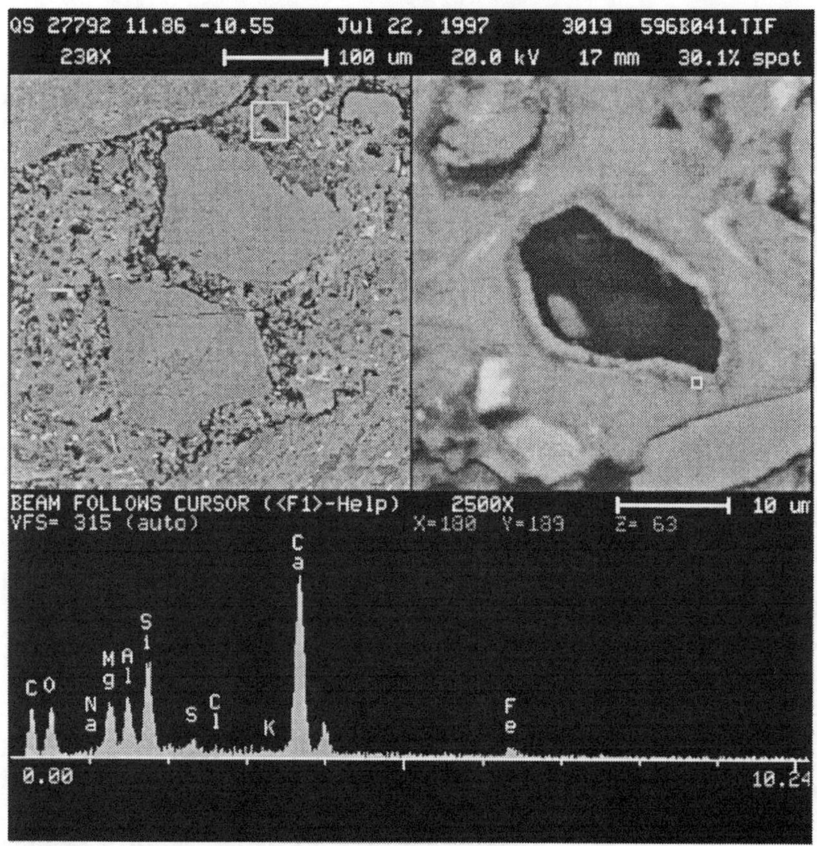

Fig. 32. Magnesium-bearing rim around an empty void.

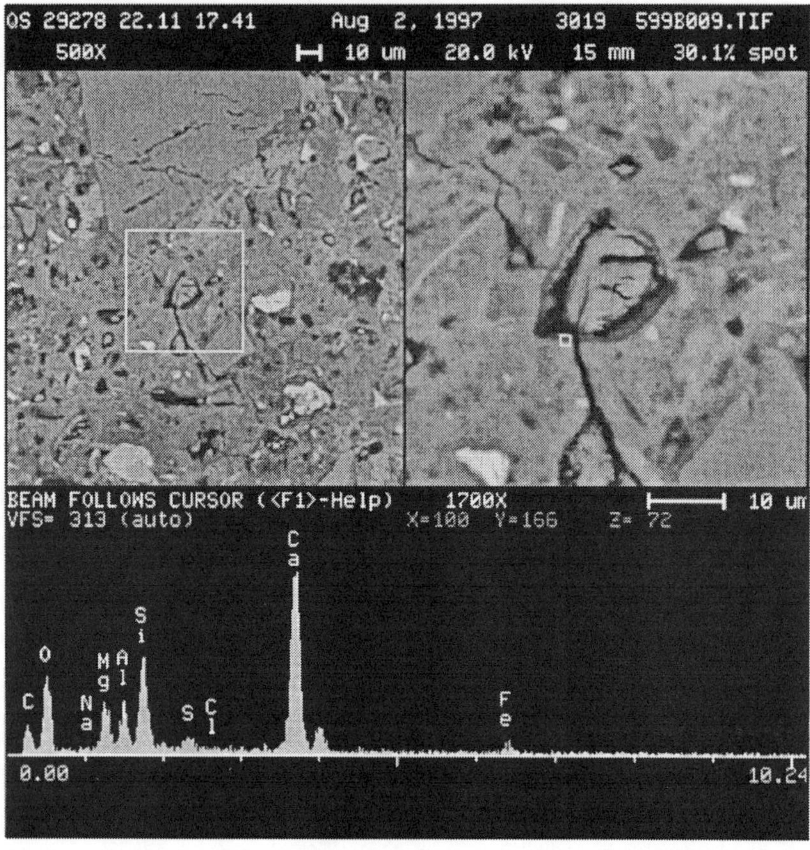

Fig. 33. Magnesium-bearing rim around an ettringite-containing void.

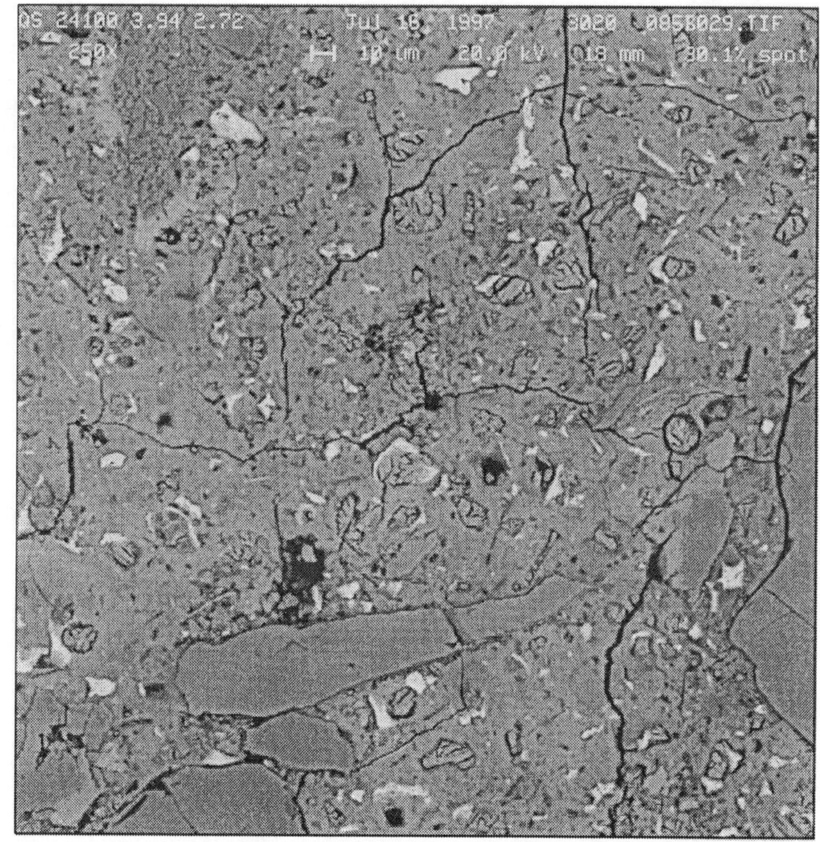

Fig. 34. Area of microcrack network formation associated with extensive local ettringite deposition.

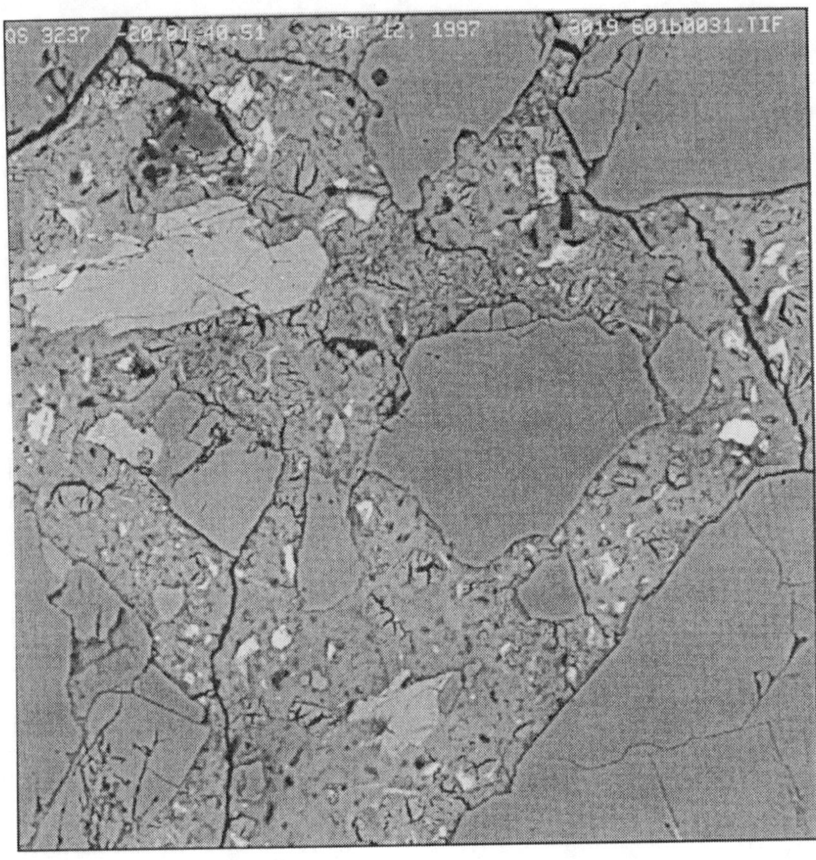

Fig. 35. Another area of extensive network cracking associated with local ettringite deposition. Note the partial debonding and open rim formation around some of the sand grains.

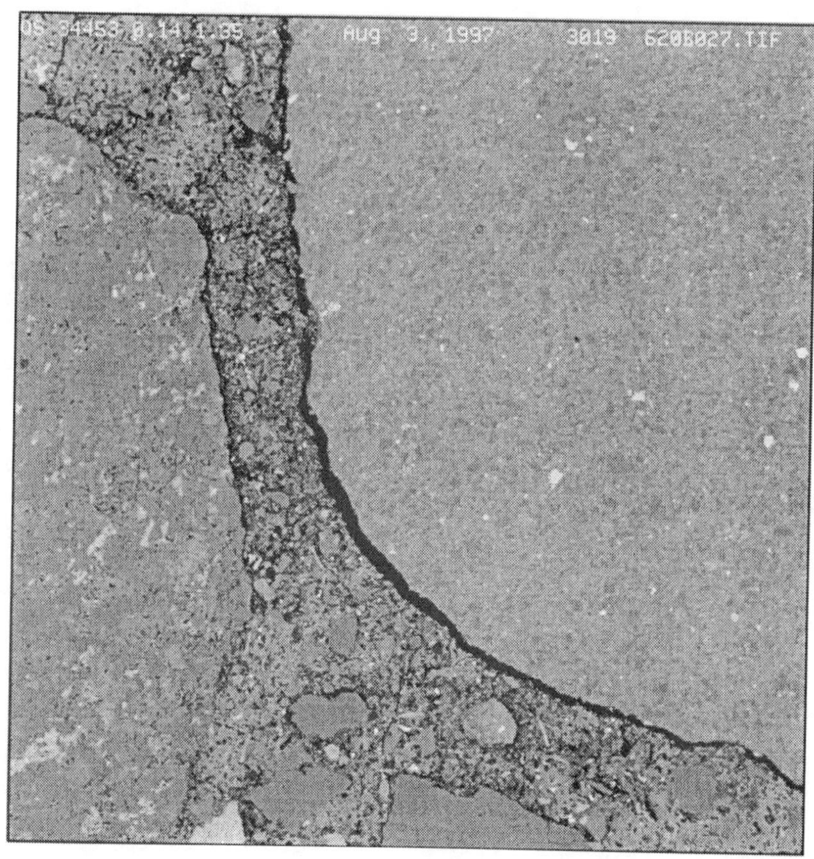

Fig. 36. An area illustrating extensive open rim formation around a coarse aggregate grain.

THE U-PHASE - FORMATION AND STABILITY

Micheline Moranville* and Guanshu Li**
*LMT, ENS de Cachan, 61 av. Président Wilson, 94230 Cachan, France
**LERM, 23 rue de la Madeleine, BP 136, 13631 Arles, France

ABSTRACT

The U-phase, a Na_2O substituted AFm first synthetized by Dosch and zur Strassen in 1967, has then been observed in cement-based materials in contact with concentrated sulfate solutions. This paper shows that its formation, possible at low NaOH concentrations, can be accelerated by temperature. Unstable at low pH, U-phase transforms into AFt, as shown by leaching tests. In presence of CH, this AFt is expansive.

INTRODUCTION

The U-phase has been detected by Dosch and zur Strassen[1] in the chemical system 4CaO. Al_2O_3. SO_3. Na_2O. H_2O in a vast range of alkaline concentrations. The U-Phase has also been characterized in cement-based systems containing high amounts of Na_2SO_4[2,3]. In Portland cements hydrated in water, 0.5 M, 1.0 M, 2.0 M, 4.5 M NaOH solutions Way and Shayan[4] found a AFm phase containing sodium, instead of ettringite. The thermodynamic investigation of the CaO-Al_2O_3-$CaSO_4$.H_2O system at 25°C by Damidot and Glasser[5] emphasized the role of Na_2O which highly modifies the stability domain of ettringite. After Brown and Bothe[6] KOH inhibits the formation of ettringite as well. The U-phase was detected by X ray diffraction in steam-cured concretes at various alkali levels in the pore solution[7]. It has recently been observed in site concretes i.e. in samples extracted from a dam[8] or in self-placed concretes made with a fly ash cement[9].

The U-phase induces deterioration of concrete through two mechanisms[2,10,11] (1) supplementary expansive formation, (2) transformation into ettringite. The kinetics of the microstructure evolution during leaching has been modelled[12], based on coupling between transport by diffusion and chemical reactions.

This paper reports the influence of both alkaline concentration in solution and temperature, on the U-phase formation. The microstructure of synthethic mixtures and cement pastes on one hand and the transformation of the U-phase under leaching on the other, are also presented.

CHARACTERISATION OF THE U-PHASE

The U-phase 4CaO. 0.9 Al_2O_3. 1.1 SO_3. O.5 Na_2O. 16 H_2O, synthetized from high alkaline solutions as 1M NaOH [13] belongs to the group of hexagonal $C_4 A \overline{S} H_x$ crystal structures[14]. After Dosch and zur Strassen[1], there is an omission of some Al^{3+} ions in the principal layer counterbalanced by the inclusion of Na^+ ions and SO_4^{2-} ions in the interlayer. So the structural formula can be written as:

$$\{Ca_4Al_{2(1-x)}(OH)_{12}[SO_4]\}^{6x-} \{yNa_2SO_4 . 6xNa. Aq\}^{6x+}$$

and the formal formula using oxides as:

$$4\ CaO . (1-x)Al_2O_3 . (1+y)SO_3 . (3x+y)Na_2O$$

The main characteristic X-ray diffraction peaks (Fig.1) are at 1.000 and 0.500 nm, corresponding to 16 molecules of H_2O. Two other hydration states at 12 and 8 moles of H_2O are identified at 0.93 and 0.81 nm respectively[14].

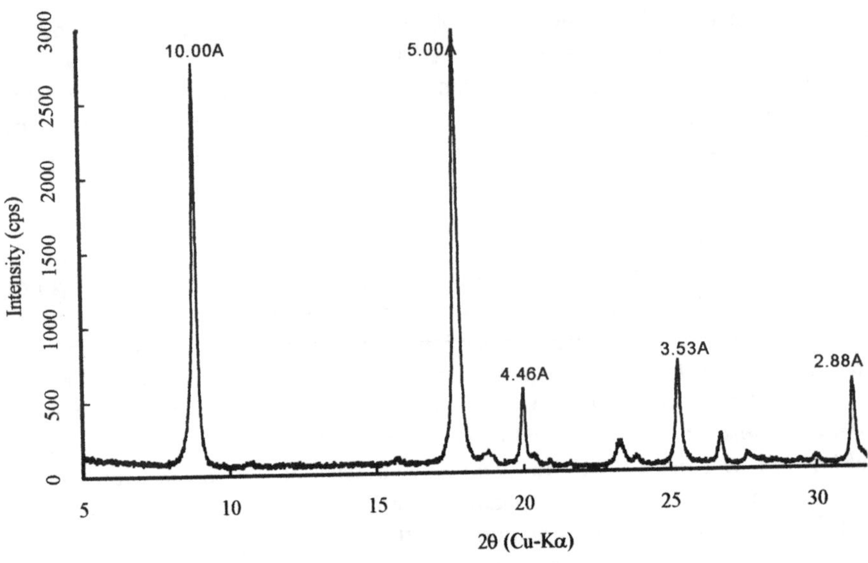

Fig.1. XRD pattern of the pure U-phase.

Under SEM, pure U-phase appears as thin hexagonal plates (Fig.2) containing Na, Al, S, Ca. U-phase is rather difficult to be distinguished from monosulfate AFm in cements and concretes, due to its low concentration of Na_2O and its intermixing with AFm. However U-phase has been clearly characterised by XRD in concretes as shown by Shayan, Quick and Lancucki[7], even if it coexists with AFm and AFt. An example of such XRD pattern is given by a mixture of C_3A and gypsum, hydrated in a 0.6 M NaOH at 80°C for 3 days (Fig.3). In the SEM photo it is impossible to distinguish U-phase among thick AFm crystals, even if they are both in hexagonal plates, besides AFt in fibers (Fig.4).

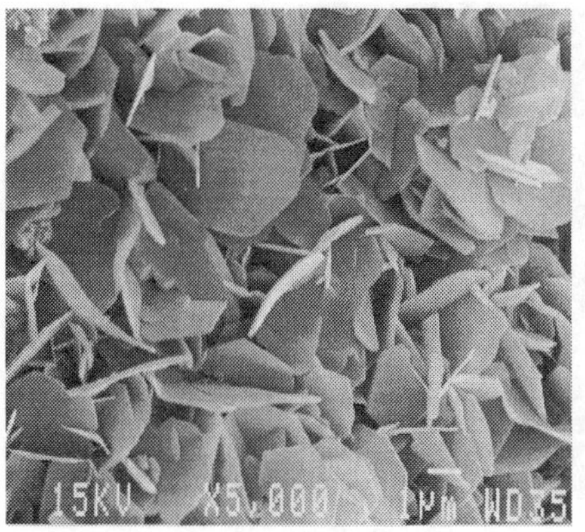

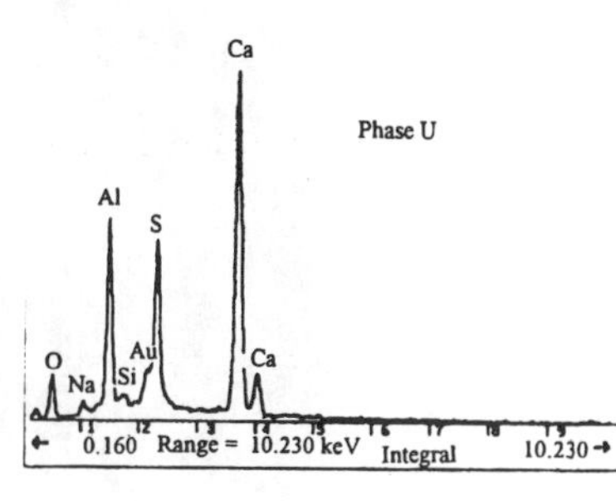

Fig.2. Pure U-phase hexagonal crystals and elementary composition by EDS.

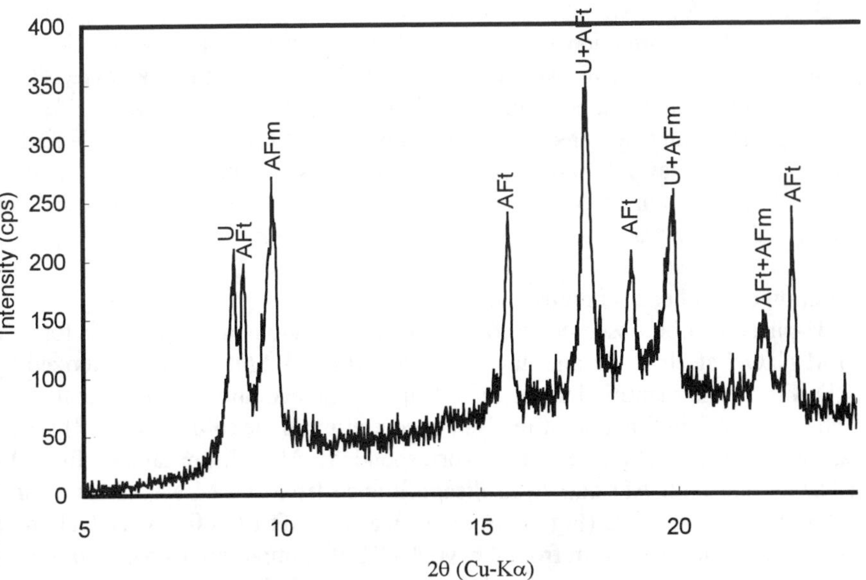

Fig. 3. XRD pattern of C_3A and gypsum in 0.6M NaOH at 80°C for 3 days: U-phase coexists with AFm and AFt.

Fig.4. SEM. Mixture of U-phase, AFt and AFm in the mixture C_3A + Gypsum in 0.6 M NaOH at 80°C for 3 days.

U-PHASE FORMATION – INFLUENCE OF ALKALINE CONCENTRATION AND TEMPERATURE

The U-phase formation has been studied in synthetic mixtures of C_3A and gypsum in function of two parameters, Na concentration and temperature. These mixtures have been prepared following the AFm chemical formula with 61% C_3A + 39% gypsum, and hydrated in alkaline solutions at 0.2, 0.6 and 1M NaOH at w/s = 1, during 1 day at two temperatures 20°C and 80°C and then 2 days at 20°C. Normal Portland cement pastes and mixtures of 90% slag + gypsum have been treated in the same way.

Evolution of synthetic mixtures

U-phase can be observed in lower alkaline solutions as expected before. In 0.6 M NaOH at 20°C for one day, U-phase, AFt and AFm are characterized by XRD (Fig.5) and identified under SEM (Fig.6). The evolution of the formation of U-phase at 20°C in function of the Na concentration in the solution (Fig. 7) shows that the increase in U-phase content corresponds to AFm disappearance from 0.2 M NaOH and both AFt and AFm disappearance from 0.6 M NaOH. The same evolution occurs at 80°C (Fig.8) with a decrease in AFt from 0.2 M NaOH and a continuous decrease in AFm from 0.6 M NaOH. It is important to point out that at 80°C U-phase is already stabilized at 0.2 M NaOH (Fig.9) and appears under SEM as very thin platelets growing from a massive mixture of AFt and AFm. At 3 days, the influence of the alkaline solution is still visible at 20°C (Fig.10) and 80°C (Fig.11), compared to results at 1 day. It is also clear that AFt had already been transformed into AFm which then decreases sharply in function of NaOH concentration.

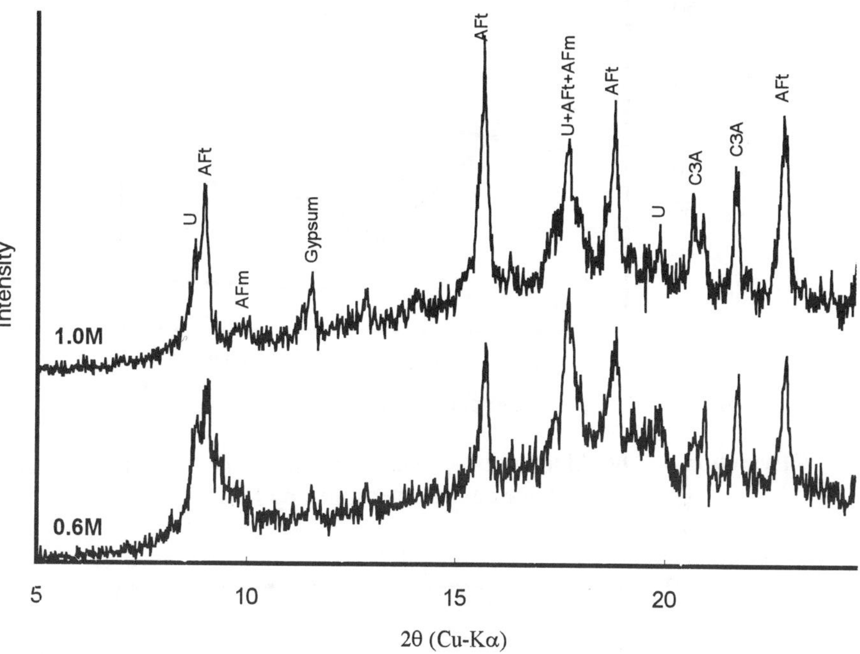

Fig.5. XRD. C_3A + gypsum in 0.6 M and 1 M NaOH at 20°C for one day.

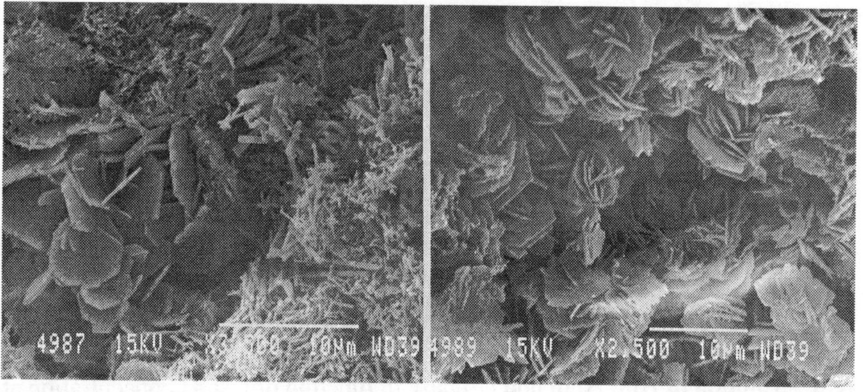

Fig.6. SEM. Fibrous AFt and hexagonal U-phase crystals in 0.6 M NaOH (left), mainly U-phase in 1 M NaOH (right).

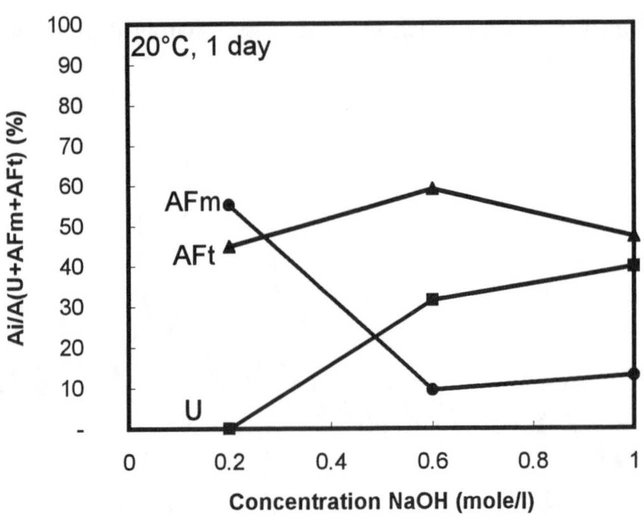

Fig.7. Evolution of AFt, AFm, and U-phase in function of NaOH concentration at 20°C for one day as determined by XRD.

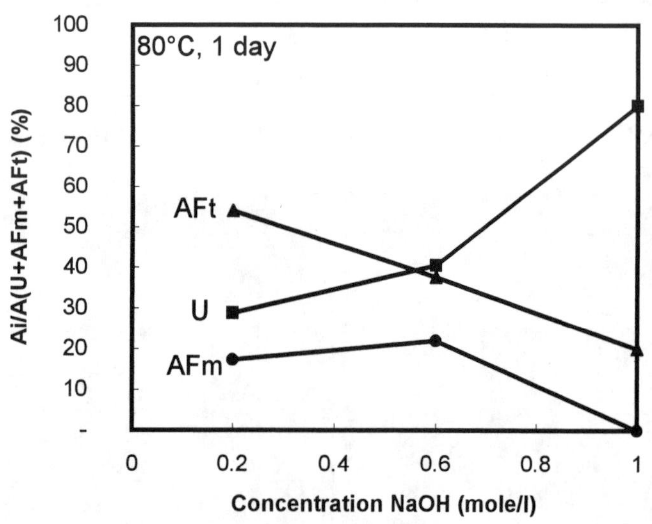

Fig.8. Evolution of AFt, AFm and U-phase in function of NaOH concentration at 80°C for one day as determined by XRD.

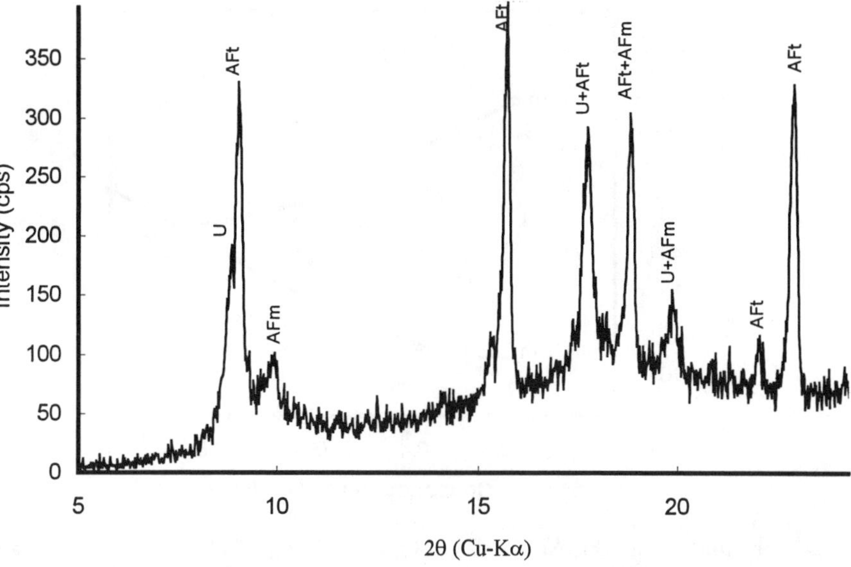

Fig.9. XRD pattern of (C_3A + G) hydrated in 0.2 M NaOH at 80°C for 1 day.

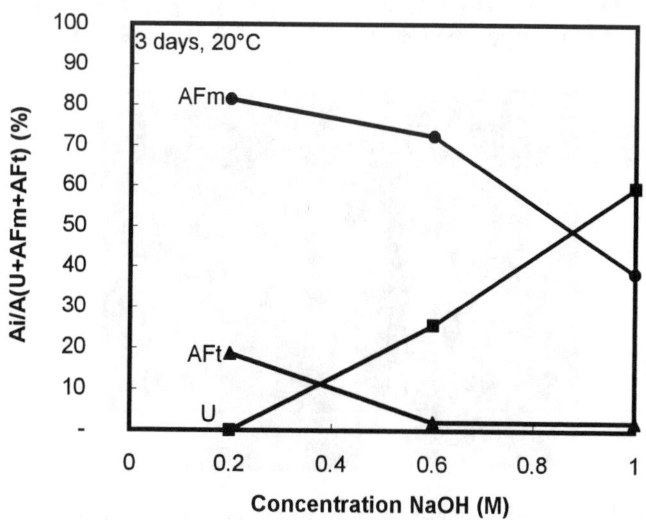

Fig.10. Evolution of AFt, AFm and U-phase in function of NaOH, at 3 days and 20°C, as determined by XRD.

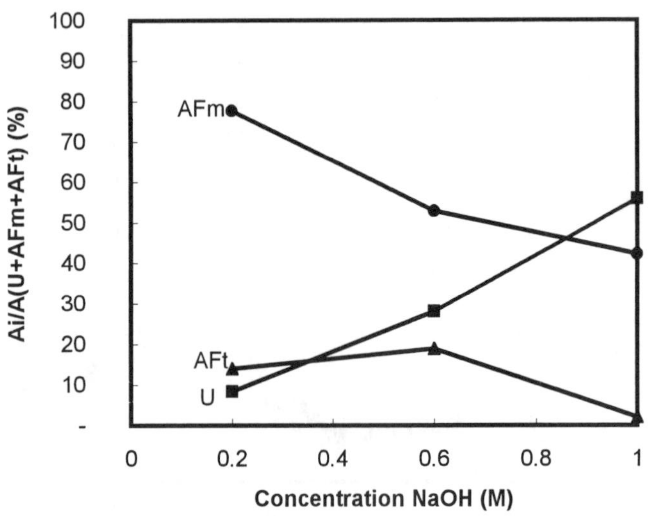

Fig.11. Evolution of AFt, AFm and U-phase, in function of NaOH, after 1 day at 80°C and 2 days at 80°C, as determined by XRD.

Evolution of Portland cement

At 20°C and 1 day, U-phase is observed in samples hydrated in the 1 M NaOH solution but in larger content at 80°C where it coexists with CH (Fig.12).

Fig.12. Thin hexagonal U-phase crystals and large thick CH plates, in an OPC hydrated in 1M NaOH solution at 80°C, for 1 day.

The U-phase appears in the XRD pattern of the OPC hydrated in 1 M NaOH at 80°C for 1 day, also coexisting with CH due to the C_3S hydration (Fig.13).

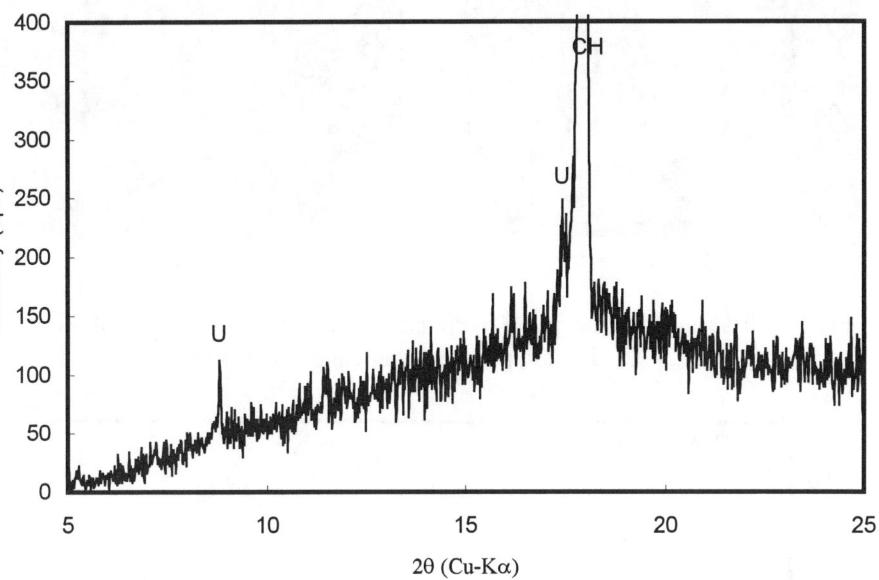

Fig.13. XRD pattern showing U-phase and CH, in the OPC after 1 day at 80°C in 1 M NaOH.

Evolution of slag + gypsum

Mixtures of blast-furnace slag and 10% gypsum have been hydrated in 1 M NaOH for 5 days. At 20°C Aft is only observed under SEM. Ettringite is well dispersed in the matrix (Fig.14). After 5 days at 80°C, U-phase coexists with a reticular C-S-H which is a typical texture of heat treated cement pastes. The XRD pattern shows after 1 day at 80°C and 4 days at 20°C, U-phase and AFt. Compared to OPC, CH is not detected (Fig.15).

Conclusion

The U-phase can exist in concretes and more particularly in steam-cured elements like concrete ties or when the temperature has reached 80°C like in massive dams. Moreover an ingress of Na_2SO_4 from outside is able to enrich the pore solution in sodium ions and especially after drying, thus favoring the U-phase stability.

Fig.14. 90% slag + 10% gypsum hydrated in 1 M NaOH for 5 days. Fibrous AFt at 20°C, U-phase plates and reticular C-S-H at 80°C.

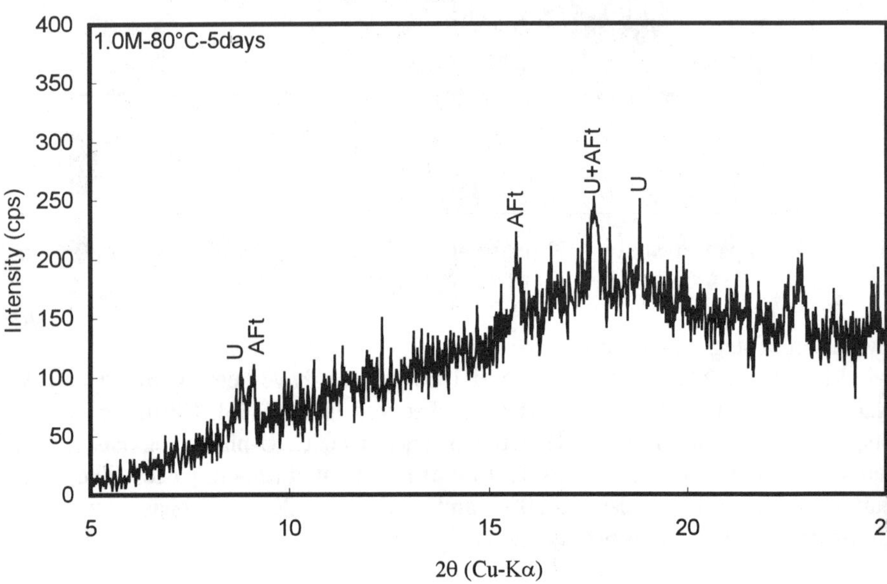

Fig.15. XRD pattern of the slag + gypsum mixture, showing the coexistence of U phase and AFt.

STABILITY OF U-PHASE UNDER LEACHING

Under leaching, the decreasing in the pore solution alkalinity results in the instability of U-phase[2]. Model cement pastes with a high content of U-phase as 50.6% C_3S + 49.4% U-phase, hydrated in 1.5 M NaOH during 1 month, have expanded and cracked under leaching using pure water at pH 7. The expansion has been related to the formation of ettringite. The evolution of the different minerals from the bulk to the surface of a cylinder has been followed by XRD on slices of 0.2 – 0.5 mm thick (Fig.16).

It has been clearly shown a zoning corresponding to four stages of phase transformations, as follows:

1. Coexistence of U-phase and CH in the sound core,
2. Disappearance of U-phase, decreasing in CH, formation of AFm and AFt (1),
3. Disappearance of AFm and CH, formation of AFt (2),
4. Disappearance of AFt (2) and softening of the superficial layer.

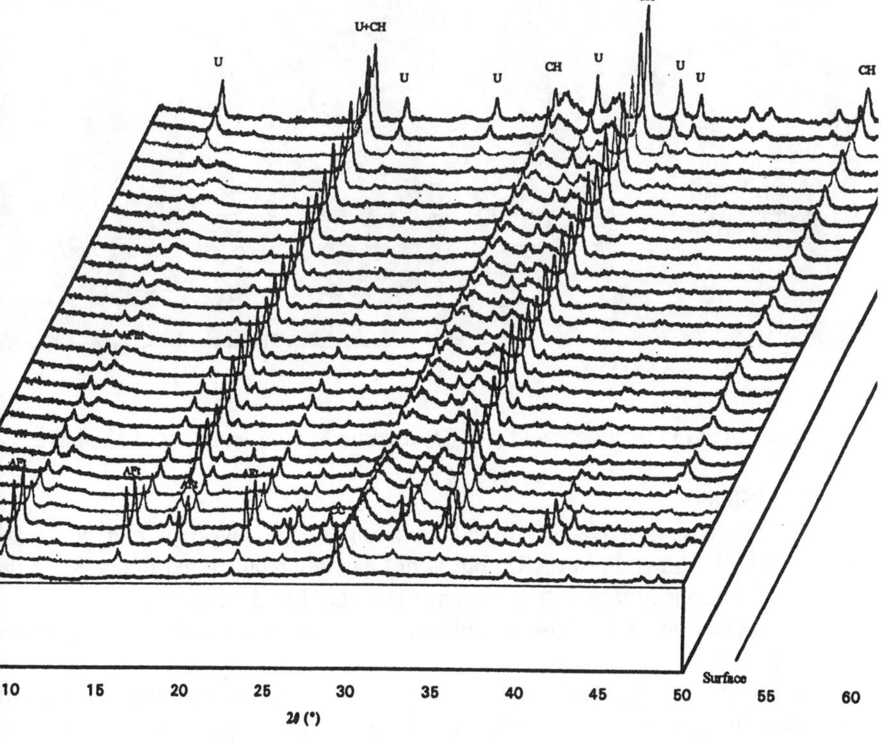

Fig.16. Layer by layer XRD analysis of hydrated C_3S + U-Phase, after leaching.

In each zone, solid phases are in equilibrium with the pore solution[1] Chemical reactions, faster than ionic diffusions, are in agreement with the produ of solubility of the four main phases i.e. p_K [$Ca(OH)_2$]=10^{-5}, p_K (U)=10^{-28}, (AFm)=10^{-29}, p_K (AFt)=10^{-40}. The p_H of the pore solution decreases from 14 in t zone 1 to 7 in the superficial layer passing through 12.5 in the zone 2 and 9 in t zone 3.

The two forms of ettringite correspond to a different morphology. Aft (which coexists with CH and AFm appears as a micro-cracked mass of den fibers (Fig.17). AFt (2) occurs in long thin fibers in the space available aft dissolution of AFt (1) in the pore solution. So the expansion and cracking a mainly due to AFt (1) formation.

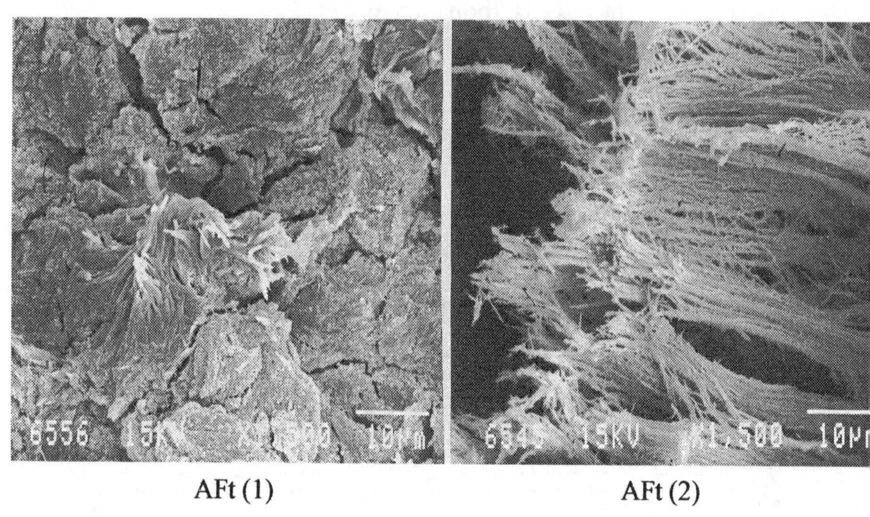

AFt (1) AFt (2)

Fig.17. AFt (1) in zone 2 and AFt (2) in zone 3, after leaching of U-phase.

CONCLUSION

The U-phase is a Na-substituted hexagonal AFm phase. It can be clearl identified by high resolution XRD but is not easily characterized under SEM du to its low Na^+ concentration. It might have been confused with AFm.

The formation of U-phase is enhanced by high alkaline concentration an temperature .

The U-phase has been observed in laboratory cement pastes and si concretes. Its presence can be related to local high alkaline concentrations. Th supersaturation of the pore solution is possible after drying or by using some additives with high alkali content.

The U-phase is unstable in low alkaline solutions as after leaching by low pH water. It transforms into ettringite which is expansive when coexisting with $Ca(OH)_2$.

REFERENCES

[1]W. Dosch and H. zur Strassen, "Ein alkalihaltiges calciumaluminat-sulfathydrat (Natrium-Monosulfat)", *Zement- Kalk-Gips,* 9, pp. 392-401 (1967).

[2]G. Li, "Etude du phénomène d'expansion sulfatique dans les bétons: comportement des enrobés de déchets radioactifs sulfatés", Thèse Ecole Nationale des Ponts et Chaussées, Spécialité Structures et Matériaux, 242 Pages, Paris, 27 Septembre 1994.

[3]G. Li, P. Le Bescop and M. Moranville, "The U-Phase Formation in Cement-Based Systems Containing High Amounts of Na_2SO_4", *Cement and Concrete Research,* **26** [1] 27-34 (1996).

[4]S.J. Way and A. Shayan, "Early Hydration of a Portland Cement in Water and Sodium Hydroxide Solutions and Nature of Solid Phases", *Cement and Concrete Research,* **19** [5] 759-769 (1989).

[5]D. Damidot and F.P. Glasser, "Thermodynamic Investigation of the CaO-Al_2O_3-$CaSO_4$-H_2O System at 25°C and the Influence of Na_2O", *Cement and Concrete Research,* **23** [1] 221-238 (1993).

[6]P.W. Brown and J.V. Bothe, "The Stability of Ettringite", *Advanced in Cement and Research,* **5** [18] 47-52 (1992).

[7]A. Shayan, G.W. Quick and C.J. Lancucki, "Morphological, Mineralogical and Chemical Features of Steamed-Cured Concretes Containing Densified Silica Fume and Various Alkali Levels", *Advances in Cement Research,* **5** [20] pp. 151-162 (1993).

[8]J.P. Bournazel, F. Boutin and N. Rafaï, "Les analogues anciens et l'analyse inverse pour la durabilité du béton", *Revue française de génie civil,* **22** [3] pp. 341-352 (1998).

[9]N. Rafaï, Private Communication.

[10]G. Li, P. Le Bescop and M. Moranville, "Expansion Mechanism Associated with the Secondary Formation of the U-Phase in Cement-Based System Containing High Amount of Na_2SO_4", *Cement and Concrete Research,* **26** [2] pp. 195-201 (1996).

[11]G. Li and Le Bescop, "Degradation Mechanisms of Cement-Stabilized Wastes by Internal Sulfate Associated with the Formation of the U Phase" *Symposium R of the Materials Research Society,* Fall Meeting, 1995.

[12]P. Lovera, P. Le Bescop, F. Adenot, G. Li, T. Tanaka and E. Owaki, "Physico-Chemical Transformations of Sulphated Compounds during Leaching of

Highly Sulphated Cemented Wastes", *Cement and Concrete Research*, **27** [10] pp. 1523-1532 (1997).

[13]G. Li, P. Le Bescop and M. Moranville, "Synthesis of the U-phase $4CaO.0.9Al_2O_3.1.1\ SO_3.0.5Na_2O.16H_2O$", *Cement and Concrete Research*, **27** [1] pp. 7-14 (1997).

[14]H.F.W. Taylor, "Hydrated Aluminate, Ferrite and Sulfate Phases" pp.161 in *Cement Chemistry*, 2nd ed. Edited by Thomas Telford, London, 1997.

[15]F. Adenot, "Durabilité du béton : caractérisation et modélisation des processus physiques et chimiques de dégradation du ciment" Thèse de Doctorat, Université d'Orléans, SpécialitéMatériaux Minéraux, Option Géochimie de l'Environnement, 239 pages, 17 décembre 1992.

W/C RATIO, POROSITY AND SULFATE ATTACK – A REVIEW

Nataliya Hearn
Department of Civil Engineering
University of Toronto
35 St. George Street
Toronto, Canada, M5S 1A4

Francis Young
Center for Advanced Cement-Based Materials
University of Illinois
Urbana, IL 61801

ABSTRACT

Analysis of the literature shows that the chemistry of cement has significant impact on the sulfate resistance of concrete down to w/c=0.45. Below this w/c ratio, the chemistry of cement becomes much less important, and at w/c=0.40 has no effect on the sulfate durability of the concrete. These findings also correlate well with theoretical analysis of the total pore volumes and the discontinuity of the pore structure. The sulfate resistance is dependent on the continuity and total volume of pores, with w/c=0.40 providing a good barrier characteristics in sulfate environment, even under limited curing conditions in the field.

INTRODUCTION

The objectives of concrete mix design are three-fold:

1. to provide sufficient strength to resist mechanical loads during the life of a structure:
2. to develop an impermeable barrier in order to reduce the rate of deterioration of the original microstructure by aggressive elements with subsequent loss of both strength and barrier performance;
3. to ensure concrete of sufficient workability for efficient placement so that the first two objectives are not compromised.

Of these three objectives the second is the most critical since performance of a structure is seldom the result of using concrete of inadequate strength for structural design. Concrete needs to resist the entry of aggressive elements into structural elements and consequent deterioration of the original microstructure, with subsequent loss of both strength and barrier performance.

The existence of a relationship between mechanical and barrier characteristics is questionable, because even though both barrier and strength depend on the mix design and subsequent curing, little correlation exists between the two. Figure 1 compiles data from various studies showing lack of strength to permeability relationship in a cross study comparison – where 30 MPa concrete has permeability values varying by five orders of magnitude. Strength is a function of total porosity and the amount of cementitious binder, while durability is related to the pore connectivity. For this reason, codes provide specifications for both strength and durability requirements.

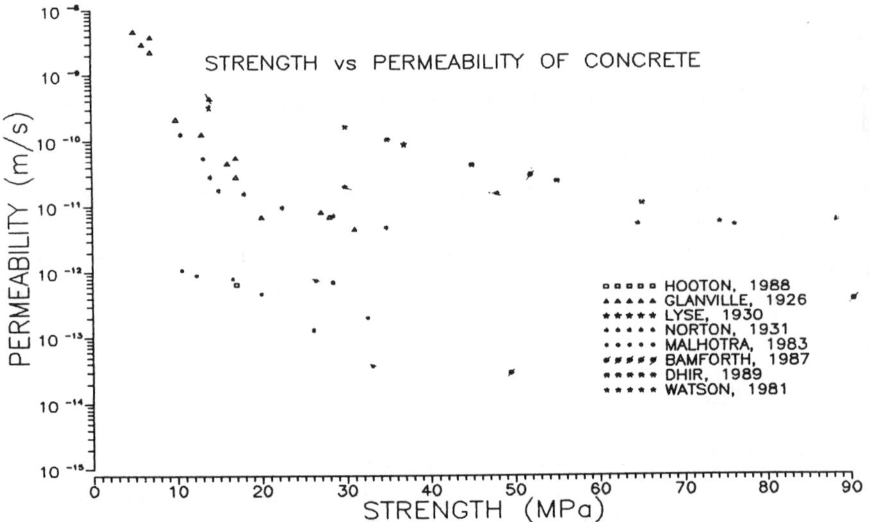

Figure 1. Permeability vs. Strength – compilation of data.

Sulfate Exposure

For concrete exposed to sulfates, current building codes rely on the recommendation of Committee ACI 201 on Durability for specification of the barrier quality. Under severe conditions of sulfate exposure, not only is a low w/c ratio prescribed, but the chemistry of cement (i.e. Type V) is also specified. Recent failures in residential construction have raised questions on the relative importance of w/c ratio and cement type. The investigations of the importance of cement chemistry and w/c in sulfate environment have been investigated for many years and have resulted in a large number of publications. A bibliography by King et al., (1925) [1] cited 700 references on sulfate attack, so that some boundary conditions had to be establish in order to obtain useful information from an extensive literature analysis. This paper reviews the parameters which determine the barrier characteristics of concrete and how these parameters relate to the sulfate resistance.

Mix Design and W/C ratio

The properties of hydrated cement paste depend on volume proportions of cement, water and cement chemistry, while the properties of concrete depend primarily on volume proportions of cement paste, air and aggregate. Therefore, although in cement pastes all the mechanical and durability characteristics are related to the w/c ratio, in concrete, the w/c ratio only defines the quality of the paste and not the overall characteristics of the composite.

The upper and lower limits of structural concrete properties are defined by the w/c ratios between 0.28 and 0.75 and water content between 120 and 240 kg/m^3. These numbers set the extreme limits of very rich to very lean mixes. In practice, concrete mix proportions and characteristics are in the range shown in Table I.

Table I. Common proportions of concrete mixes

w/c	**W (kg/m3)**	**C (kg/m3)**	**Agg. (kg/m3)**	**Initial Pores (%)**	**σ (MPa)**
0.45-0.60	160-200	140-450	660-770	16-20	30-50

RELATIONSHIP BETWEEN W/C A ND POROSITY

Categories of Pores

The amount of mixing water determines the initial porosity of a concrete mix. As th hydration process proceeds, water will be retained in the concrete in three ways.

1. Non-evaporable (or chemically bound) water as an intrinsic part of the hydratio products. When cement is fully hydrated the maximum value is about 23% by weight c original cement.
2. Gel water is held in gel pores associated the assemblage of hydration product primarily associated with C-S-H gel. This water is strongly held by surface forces and it properties differ from those of bulk water. Much of the water is held in micropore (diameters < 2.5 nm). It can be estimated by heating pastes to 105°C after equilibratio at 45% RH.
3. Capillary water is contained in residual space not occupied by the hydratio products. Capillary water has the same properties as bulk water, but with its behavic modified by its confinement in constricted space. This water can be estimated from mas loss when a saturated paste is dried at 45% RH.

The distinction between water in gel pores and capillary pores was originally propose by Powers(1960) [2] on the basis of water adsorption and desorption with pastes c varying hydration and initial w/c ratios. He set the division at 10 nm diameter. Althoug this distinction is at odds with the IUPAC (1972) [3] classification for porous material which indicates capillarity effects below 10 nm, it is a useful concept for cement paste Only capillary pores need to be considered when studying the ingress of chemical agent Gel pores are considered to be an intrinsic part of the hydration products.

Changes in Porosity with Hydration

The total porosity will decrease during hydration due to the volume of the hydratio products being approximately 2.2 times the original solid volume, and will be divide into capillary pores and gel pores. The extent of the reduction of the initial capillar porosity depends on the amount of the initial space available for hydration product (amount of mixing water) and the extent of hydration. Based on the relative volumes o hydrated and unhydrated cements and the characteristic porosity of the hydratio products of 28%, it can be determined that for w/c<0.40, there is not sufficient space t hydrate all the cement. For w/c>0.40 all of the cement in the mix has enough space t hydrate, provided that enough water is available to continue the hydration process Powers(1960) [2], estimated a relationship between evaporable (w_e) to non-evaporabl (w_n) water (chemically bound to free water) for the zone of incomplete hydratio $(0<w_{o/c}<0.40)$ to be:

$w_e/w_n = 1.170$

and for the zone of complete hydration ($0.40 < w_{o/c}$):

$w_e/w_n = 4.950w_o - 0.667$

In practice, however, even at high w/c ratios, complete hydration is rare. SEM micrographs of 0.9 and 0.7 w/c concrete cured in water for 26 years[4], show some unhydrated cement remnants.

Mills [5] derived empirical relationship for the practically achievable terminal levels of hydration from an extensive experimental study:

$\alpha^* = (1.031w_o)/(0.194 + w_o)$

Based on the terminal levels of hydration and stoiciometric relationships, Figure 2 shows volumetric distribution of solids and pores in cement paste and concrete with 75% volume of aggregate. Similar data has been calculated by Parrott (1987, 1989) [7,8], and also by Jennings and Tennis (1994) [9] who used the hydration equations for each cement compound. From Figure 2 it is obvious that pore volume in the cement paste is just a function of the initial w/c ratio, while in concrete the pore volume must be adjusted by the volume fraction of the paste, and is thus linearly related to the initial amount of the mixing water. This volumetric distribution is of significance in defining barrier characteristics of a cementitious system.

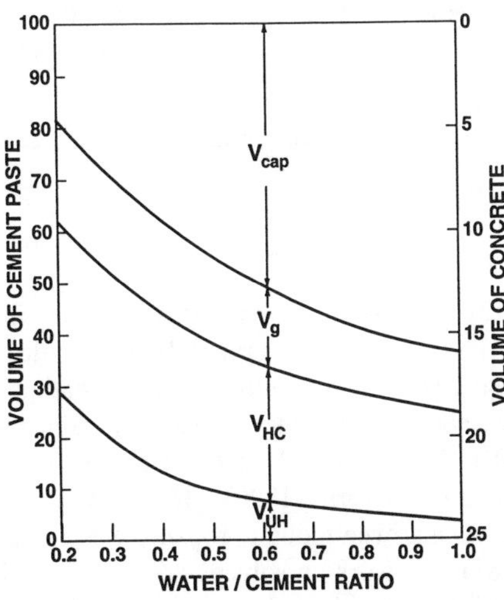

Figure 2. Volume distribution of cement pastes and concrete at terminal levels of hydration for various w/c ratios.[6]

RELATIONSHIP BETWEEN POROSITY AND PERMEABILITY

All durability aspects of concrete are related to the movement of water, which is a function of the pore structure, as defined by the relative distribution of capillary and ge pores. Because, pores in concrete span several orders of magnitude (Table II), not al porosity contributes to the movement of water through concrete and not all of the wate in the pore structure is mobile.

Table II. Sizes of the hydration products and pores in cement paste [from Mindes and Young (1981) [10] and Mehta and Montiero (1993)[11]].

Hydration products	*Typical Dimension (μm)*
C-S-H	1x 0.1
Calcium Hydroxide (CH)	1 to 7
Ettringite	10 x 0.5
Monosulfoaluminate	1x 1x 0.1

Pores	*Typical Dimension (μm)*
Interparticle spacing between C-S-H sheets	0.001 to 0.003
Capillary voids	0.01 to 5
Entrained air bubbles	1 to 50
Entrapped air voids	1000 to 3000
Micro-cracks	1 to 60

Role of Pore Sizes

Powers et al. (1958) [12] discussed the immobility of water in smaller pores during a permeability test. He proposed that if immobile water did exist, than it could be mobilized at some pressure level, thus resulting in the deviation in the linear relationship between applied pressure and the resulting permeability. Since his tests showed no such deviation up to 0.7MPa, he concluded that there was no immobile water in the cementitious pore structure. Hearn (1996) [13] showed that water and propanol permeabilities were equivalent, even though propanol was excluded from the finest pore structure. Parrott's (1981) [14] work on solvent exchange is relevant here. He showed that total porosities obtained with alcohols were always lower than those obtained by water. For instance, the apparent porosity obtained by saturation with propan-2-ol in a nine-day old moist cured hydrated alite sample with a water/solid ratio of 0.55 was 2.2%

lower than obtained with water and this difference became more pronounced with continuing hydration and the development of fine porosity. In concrete, most aggregate can be considered impervious, so that porosity fraction must be adjusted to reflect paste fraction. The mobile capillary water in concrete is defined as (Mills, 1986) [6]:

$V_{cap} = c(w/c - 0.254\alpha)$

and shows linear relationship to the amount of mixing water.

Capillary Porosity Size Distribution

The above observations lead to the conclusion that capillary porosity controls permeability and that gel pores do not contribute signficantly to the overall flow. This is to be expected since water in these smaller pores is under surface forces and will not exhibit bulk flow.

The characteristics of the capillary pore system can be probed by mercury intrusion porosimetry (MIP) in which mercury is forced into the pore system under an applied pressure. The cumulative volume of mercury intruded is plotted against an equivalent pore diameter calculated from the applied pressure assuming an ideal cylindrical pore geometry. A series of MIP curves are shown in Figure 3 for increasing degree of hydration (α) at a constant w/c. It can be noted that the curves tend to converge to a lower bound which is characteristic of the hydrated matrix. Similarly as w/c ratio is changed at constant α the intruded volume is decreased and the pores become finer.

Since mercury can only enter pores which are connected to the outside surface, the characteristics of the MIP curves reflects the connectivity of the capillary pore system and can be correlated with water permeability. Nyame and Illston (1980) [15] showed that the pore diameter at which maximum rate of intrusion of mercury could be correlated with the permeability coefficient for water, as shown in Figure 4. Manmohan and Mehta (1981) [16] have correlated the permeability coefficient of a cement paste with the onset of rapid intrusion (the threshold diameter) and the total volume of coarse capillary porosity (macropores with diameter > 0.13 μm). Mehta and Manmohan (1981) [17]and Feldman (1983) [18] also showed that the addition of mineral admixtures can reduce the size of capillary pores and hence the
permeability of a paste.

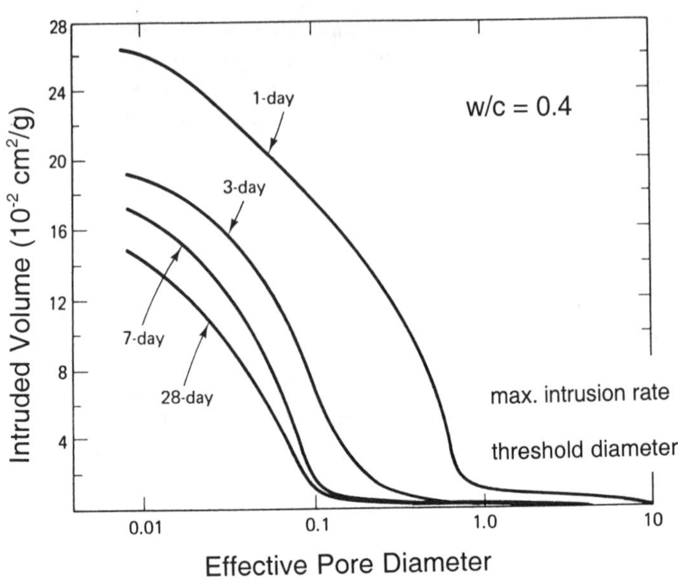

Figure 3. MIP curves of cement paste hydrated for different times (w/c = 0.4).

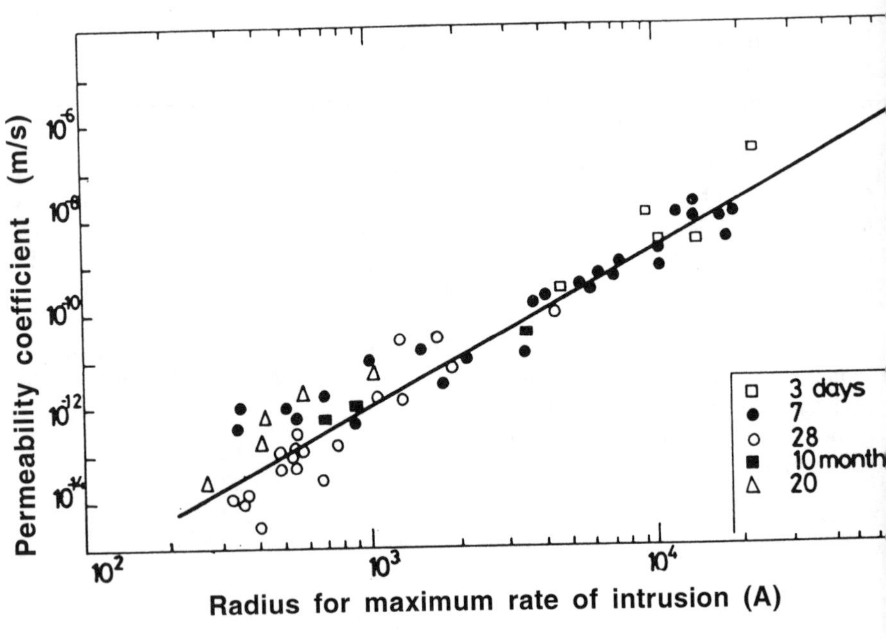

Figure 4. Correlation between the permeability coefficient of a hardened cement paste and the pore diameter at which maximum rate of mercury intrusion occurs. [After Nyame and Illston, (1980) [15]].

DISCONTINUITY OF THE CAPILLARY PORE SYSTEM

The pore structure can therefore be represented by the high porosity, low permeability model (Figure 5). The reduction of larger pores through the process of continuing hydration is also accompanied by segmentation and of the pore structure (Figure 6).

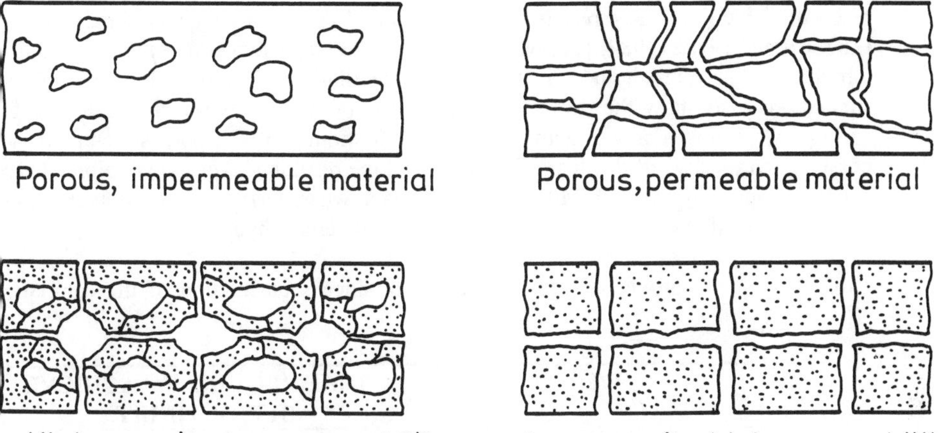

Figure 5. Illustration of various porous/permeable systems [19].

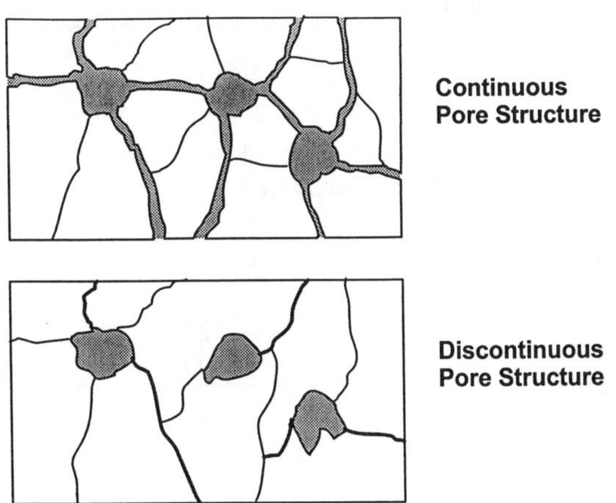

Figure 6. Discontinuity of the pore structure due to continuing hydration.

With regular cement type concretes, Powers (1959) [20] calculated the time required t achieve discontinuity in the pore structure (Table III). A w/c ratio of 0.45 will achiev discontinuity within 7 days of mixing and w/c of 0.40 in 3 days, which is of practica significance in the field concrete, where curing time is limited, and the exposure to th aggressive environment can be immediate. Work by Bonzel (1966) [21] showed tha improvement in the barrier characteristics occurred faster at the lower w/c ratios, an minimal changes in permeability are observed after a certain level of hydration i reached (Table IV). These results are also supported by recent computer simulations b Garboczi and Bentz (1989) [22] which show that the major reduction in the percolatio through capillary porosity occurrs at volume of capillaries below 40% (Figure 7). Thes simulations also show the degree of hydration required to achieve discontinuity fc various w/c ratios. For w/c above 0.6, the discontinuity of the pore structure is nc possible, irrespective of the degree of curing and the resulting level of hydration. Thes results are more in line with Mills' (1966) [5] data, where terminal level of hydration i considered, while Powers' calculations assumed 100 percent hydration.

Table III. Approximate age required to produce segmentation of the capillarie [20].

W/C	TIME REQUIRE TO ACHIEVE DISCONTINUITY
0.40	3 days
0.45	7 days
0.50	14 days
0.60	6 months
0.70	1 year
over 0.70	impossible

Table IV. Length of wet curing passed which only small changes in the barrier characteristics are possible [21].

W/C	Age (days)
0.45	7
0.60	28
0.70	90

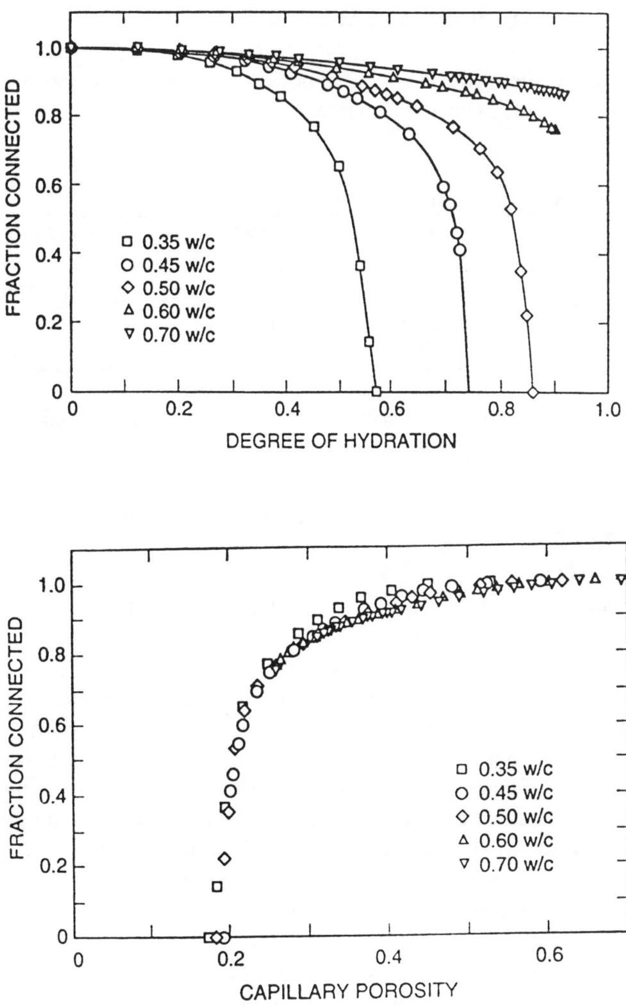

Figure 7. 3-D computer simulation of discontinuity of the pore structure at various levels of hydration for different w/c ratios; (a) versus degree of hydration; (b) versus total (capillary) porosity [22].

POROSITY AND SULFATE ATTACK

Since the water content equals the initial porosity, Table V compares the initial and final capillary porosities of paste and concrete:

Table V. Initial capillary porosities of pastes and concretes.

	W/C	WATER (kg/m3)	INITIAL CAPILLARY POROSITY
PASTE	0.45-0.60	200	60-66%
CONCRETE	0.45-0.60	160-200	16-20%

The theoretically determined curves in Figure 2 [5], are based on the following equations:

$V_{uc} = (1-\alpha)/3.22$ (S.G. of cement = 3.22)
$V_{hc} = 1.253\alpha/2.515$
$V_{gel\ water} = 0.066\alpha W$ (W= mixing water)
$V_{capillary\ water} = W - 0.254\alpha C$ (W= mixing water; C=cement content)

These calculations have been supported by the experimental work done at the same time by Wesche (1966) [23]. He found that the most important parameter in defining concrete durability in sea-water exposure is the pore volume of hardened cement paste. With pore volumes of less than 30% of the hardened cement paste the concrete will not experience measurable levels of deterioration, "no matter which of the examined binders are used". It is important to add that although Wesche's conclusions are on total pore volumes, the specified pore volume is in the range where the resulting pore structure is discontinuous. Such level of pore volume (with acceptable range of 40 to 30%) can be achieved either by maintaining w/c $\leq$ 0.40 or increasing the cement content (with recommended cement content of 400kg/m^3). If other mix requirements are maintained than higher cement content will decrease the w/c and thus improve the durability. Delineating these ranges in Figure 2 provides two important pieces of information (Figure 8):

a) for pore volumes in the paste of 30 to 40% the w/c ratio has to be between 0.30 and 0.45 (which supports Wesche results through theoretical analysis conducted by Mills); and

b) the pore volume, which determines the durability of the mix, is not represented by the total porosity, but the volume of the capillary pores only.

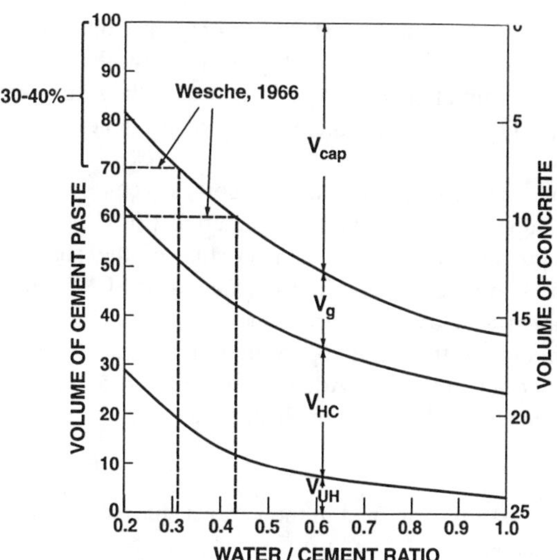

Figure 8. Mills' volume distribution curves incorporating Wesche's data.

A recent study by Stark (1991) [24] on the resistance of the various concretes to sulfate attack also supports Wesche's (1966) results. Figure 9 shows that irrespective of the cement chemistry, concretes made with w/c of 0.40 have equally high resistance to the sulfate attack.

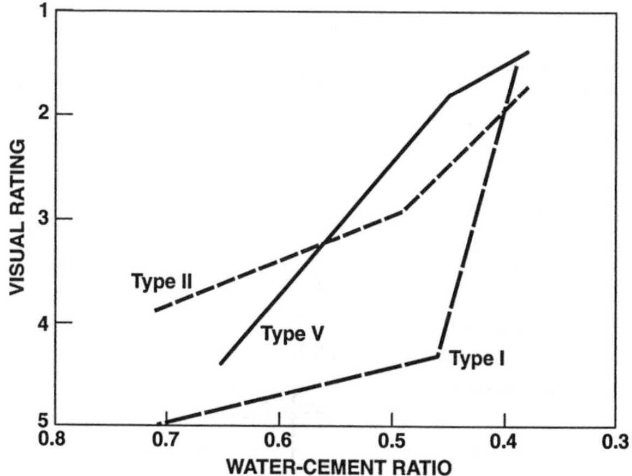

Figure 9. Sulfate resistance of concrete as a function of w/c ratio and cement chemistry [24].

In order to provide more support for the above findings, a review of the literature was conducted to determine the range of the w/c ratios tested for resistance in the sulfate environment. Table VI shows that most studies only investigated w/c ratios above 0.45. The results of Stark's study (Figure 9), however, show that at w/c=0.45 the cement chemistry has a significant effect on the performance in the sulfate environment, while at w/c=0.40 the effect of the cement chemistry is not measurable.

Such segmentation and discontinuity of the pore structure, can be also successfully achieved with the supplementary cementing materials (SCM's), where the change in the pore structure improves the barrier characteristics of the paste fraction of the mix [17,18]. The main controversy regarding usefulness of SCM's in sulfate resistance, is that their beneficial effects are more pronounced in the long term with the slow process of continuing hydration. If there is no opportunity for moist curing, the beneficial effects of SCM's may not be fully realized. Additionally some fly ashes used as SCM's may not have a suitable chemistry with respect to chemical attack.

Table VI – The W/C ratios examined in the various studies on the resistance of concrete to sulfate environments.

PAPER		W/C
Bogue [25]	1949	0.56
Day, Ward [26]	1987	0.47 0.58
Fattuhi, Hughes [27]	1988	0.4 0.5 0.6 0.7
Harrison [28]	1990	0.47 0.55 0.85
Harrison, Teychenne[29]	1981	0.50- 0.72
Hughes [30]	1985	0.47
Hughes, Grounds [31]	1985	0.47
Mathews, Baker [32]	1976	0.5
K. Mather [33]	1980	0.485
Nagano [34]	1966	0.45- 0.8
Richards [35]	1965	0.47 0.54
Steinegger [36]	1969	0.6
Wong, Poole [37]	1988	0.485

CONCLUSIONS

The most important aspect in the barrier characteristics of concrete is the quality of the paste component, which is governed by the volume and connectivity of the capillary pore system. In order to achieve satisfactory resistance to sulfate attack, the pore volumes of the paste component must be maintained below 40%, which translates into w/c< 0.45. At lower w/c ratios the connectivity of capillary pores is greatly reduced, provided the concrete is adequately moist cured. In such concrete the influence cement type becomes insignificant, because excellent barrier characteristics are achieved at the early age. However, when w/c > 0.45 the contamination of the pore structure with sulfate ions cannot be prevented, so that cement chemistry becomes important, as it slows down the rate of chemical attack; but the deleterious reactions cannot be completely suppressed.

REFERENCES

1. F.J. King, G. Ervin and O.L. Evans, "A bibliography Relating to Soil Alkalies", U.S. Department of Agricalture Bulletin 1314, pp.40 (1925).
2. T.C. Powers, "Properties of Cement Paste and Concrete", Proc. Fourth Intern. Symp. Chem. Cement, Washington, DC, II pp.577-609 (1960).
3. IUPAC "Manual of Symbols and Terminology, Appendix 2, Part 1, Colloid and Surface Chemistry", Pure Appl. Chem., 31 p.578 (1972).
4. N. Hearn, R. Detwiler and C. Sframeli, "Water Permeability and Microstructure of Three Old Concrete", Cement and Concrete Research, Vol. 24, No. 4 pp.633-640 (1994).
5. R.H. Mills, "Factors Influencing Cessation of Hydration in Water Cured Cement Pastes", Special Report 90, Highway Research Board, Washington, D.C., (1966).
6. R.H Mills, "Gas and Water Permeability of Concrete for Reactor Buildings – Small Specimens", Report INFO 0188-1, Atomoc Energy Control Board, Ottawa (1986).
7. L.J. Parrot, "Modelling of Hydration Reactions and Concrete Properties", in Materials Science of Concrete Ed. J. Skalny, I pp.181-196 (Amer. Ceram. Soc., Westerville OH) (1989).
8. L.J Parrott, "Measurement and Modelling of Porosity in Drying Cement Pastes" Mater. Res. Soc. Symp. Proc., 85 pp.91-104 (Mater. Res. Soc., Pittsburgh PA) (1987).
9. H.M.Jennings and P.D. Tennis, "Model for the Developing Microstructure in Portland Cement Pastes", J. Amer. Ceram. Soc., 77 pp.3161-3172 (1994).
10. S. Mindess and J.F. Young, "Concrete", Prince-Hall Inc., (1981).
11. P.K. Mehta, and P.J.M. Montiero, Concrete:Microstructure, Properties, and Materials p. 27 (McGraw-Hill, New York, NY) (1981).
12. T.C. Powers, L.E. Copeland, J.C. Haynes, H.M. Mann, "Permeability of Portland Cement Pastes", ACI Journal, pp.285-298 Nov. (1959).
13. N.Hearn, Comparison of Water and Propan-2-ol Permeability in Mortar Specimens", Advances in Cement Research, Vol. 8, No. 30, pp.81-86 April (1996).

14. L.J. Parrott, "An Examination of Two Methods for Studying Diffusion Kinetics in Hydrated Cements", Materials and Structures, Vol. 17, No. 98, pp.131-137 (1984).
15. B.K.Nyame, and J.M. Ilston, Capillary Pore Structure and Permeability of Hardened Cement Paste", Proc. 7th Intern. Congr. Chem. Cement, Paris, III, VI. pp.181-185 (1980).
16. P.K.Mehta, and D.Manmohan,, "Pore Size Distribution and Permeability of Hardened Cement Pastes." Proc. 7th Intern. Congr. Chem. Cement, Paris, III VIII. pp.1-5 (1980).
17. D.Manmohan, and P.K.Mehta, "Influence of Pozzolanic, Slag and Chemical Admixtures on Pore Size Distribution and Permeability of Hardened Cement Pastes." Cem. Concr. Aggr., 3 pp.63-67 (1981).
18. R.F.Feldman, "Significance of Porosity Measurements on Blended Cement Performance" in Fly Ash, Silica Fume, Slag and Other Mineral By-Products in Concrete, ed. V. M. Malhotra, SP-79 I pp.415-433 (Amer. Concr. Inst., Detroit,MI (1983).
19. R.F.M.Bakker, "Permeability of Blended Cement Concretes", ed. V.M. Malhotra ACI SP-79, Vol.2, pp.589-605 (1983).
20. T.C.Powers, L.E. Copeland and H.M. Mann, "Capillary Continuity or Discontinuity in Cement Pastes", Journal of Portland Cement Association R&D Laboratories, Vol. 1 No. 2, pp.38-48 May (1959).
21. J.Bonzel, "Der Einfluss des Zements, des W/Z Werts, des Alters und der Lagerung auf die Wasserundurchlässigkeit des Beton", Beton, No.3 pp. 379-83, No.10 pp. 417-21 (1966).
22. E.J.Garboczi, and D.P.Bentz, "Fundamental Computer Simulation Models for Cement-Based Materials" in Materials Science of Concrete, eds. J. Skalny and S. Mindess II pp.249-277 (Amer. Ceram. Soc., Westerville, OH) (1989).
23. K.Wesche, "Influence of the Pore Volume of HCP on the Resistance of Concrete Exposed to Sea-water", Bulletin RILEM No. 32, pp.291-293, Sept. (1966).
24. D.Stark, "Durability of Concrete in Sulfate-Rich Soils", PCA R&D Bulletin RD097 Skokie, Illinois, pp.1-14 (1989).
25. R.H.Bogue, "Chemistry of Portland Cement", New York, Reinhold (1955)
26. R.L.Day and M.A. Ward "Sulfate Durability of Plain and Fly Ash Mortars" Materials Research Society Symposium Proceedings Vol. 113, pp.153-161(1987).
27. N.I. Fattuhi and B.P. Hughes "SRPC and Modified Concretes Subjected to Severe Sulphuric Acid Attack" Magazine of Concrete Research, Vol. 40, No. 144, pp.159-166 (1988)
28. W.H.Harrison, "Effect of Chloride in Mix Ingredients on Sulfate Resistance of Concrete, Magazine of Concrete Research, Vol. 42, No. 152, pp.113-126(1990).
29. W.H.Harrison and D.C. Teychenne, "Sulfate Resistance of Buried Concrete; Second Interim Report on Long-Term Investigation at Northwick Park", Department of the Environment, Building Research Establishmant, Garston, Watford, England (1981).
30. D.C.Hughes, "Sulfate Resistance of OPC, OPC/Fly Ash and SRPC Pastes: Pore Structure and Permeability", Cement and Concrete Research, Vol. 15, pp.1003-1012 (1985).

31. D.C.Hughes and T. Grounds, "The Use of Beams with a Single Edge Notch to Stuy the sulphate Resistance of OPC and OPC/pfa Pastes", Magazine of Concrete Research, Vol. 37, No. 131, pp.67-74 June (1985).
32. J.D.Mathews and R.S. Baker, "An Investigation of the Comparative Sulfate Resistance of Ordinary and Sulfate-Resisting Portland Cement and Their Blends with Fly Ash and Blast Furnace Slag", Department of the Environment, BRE Note N116/76, (1976).
33. K.Mather, "Factors Affecting Sulfate Resistance of Mortars", Proc. Of the 7th Interantional Congress on the Chemistry of Cement, Paris, Vol.IV, pp. 580-588, (1980).
34. R.Nagano, "A Ten Year Investigation of the Resistance of Concrete Sulfate Solution by Means of the Ultrasonic Method", Zement-Kalk-Gips, Vol. 9, No. 10, pp.478-486 (1966)
35. J.D.Richards, "The Effect of Various Sulfate Solution on the Strength and Other Properties of Cement Mortars at Temperatures up to 80°C", Magazine of Concrete Research, Vol. 1, No.51, pp.69-76 (1965).
36. H.Steinegger, "Testing and Assesment of the Sulfate Resistance of Cements", RILEM International Symposium on Durability of Concrete", Vol.2, Prague, pp.C379-C395 (1969).
37. G.S.Wong and T. Poole, "The Effect of Pozzolans and Slag on the Sulfate Resistance of Hydraulic-Cement Mortars", Concrete Durability, ed. J.M. Scanlon, K. and B. Mather International Conference, Vol. 2, ACI SP-100, pp.2121-2134 (1987).

A NOTE ON THE HISTORY OF TYPE V CEMENT DEVELOPMENT

by Eugene D. Hill, Jr.
Openaka Corporation inc.
3910 South Hillcrest Drive
Denver, Colorado 80237, USA

The evolution of sulfate-resisting cements and concretes has occurred over a period of years, and many outstanding researchers have contributed to it. Before the turn of the nineteenth century, le Chatelier suggested substituting ferric oxide for alumina in cement manufacture. Erz cement was made in Germany not long after le Chatelier's suggestion. In about 1920, Ferrari cement was made in Italy. Both of these cements had very low ratios of alumina to iron oxide.

In 1933, Miller and Manson[1] reported the results of tests of over 50,000 specimens exposed to sulfates in the laboratory and in Medicine Lake, South Dakota. A major observation from this study was that concrete and mortar made with cements resistant to sulfate solutions may last 10 times as long as that made with cements of low resistance. In later work[2] reported in 1951, they showed that the maximum C_3A for the most sulfate resistant portland cements was 5.6 percent. They also pointed out the importance of low permeability concrete, well consolidated.

Work was done in Canada by Thorvaldson and his associates as early as the 1920s. In 1930, Canada Cement made the first portland cement in North America deliberately designed to be sulfate-resisting. It was the cement around which the original ASTM specification for Type V cement was designed, and it had low C_3A.

In 1940, ASTM adopted a tentative specification[3] that provided for five types of portland cement, including Type V for use where high sulfate resistance is required. In Type V cement, C_3A was limited to 5 percent. There were also limits on alumina, ferric oxide and the ratio of Al_2O_3 to Fe_2O_3. Over the years, there have been various ways of limiting the C_4AF. In 1978, a mortar bar expansion test was offered as an option to the limit on the sum of twice the C_3A plus the C_4AF.

The Portland Cement Association initiated the Long-Time Study of Cement Performance in Concrete[4] in the early 1940s. In exploring sulfate resistance, they very wisely included all five ASTM types of cement, and tested them in concretes at three different cement contents and water-cement ratios. Figure 1 summarizes their results.

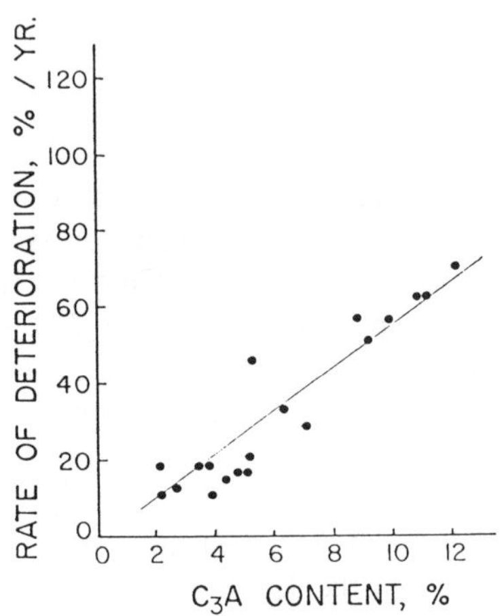

Figure 1 - Effect of cement content of concrete and average C_3A contents of different types of cement on rate of deterioration in suphate soil exposure (basin 1) [Courtesy of Portland Cement association][5]

In lean mixes, there are great differences in the sulfate resistance of the different types of cements (and C_3A levels). In rich mixes, the differences between types diminishes, and the sulfate resistance of the concrete increases. The importance of low water-cement ratio is apparent. The average water-cement ratios for these mixtures were as follows:

Sacks per Cubic Yard	W/C
4	0.74
5.5	0.51
7	0.42

The work of Kalousek, Porter and Harboe[6], reported in 1976, is a very interesting addition to our understanding of sulfate resistance. The Bureau of Reclamation had observed considerable differences in the sulfate resistance of concrete pipe made with Type V cements, and suspected that porosity might be a principal cause of these differences. Many concrete pipe are molded with earth-moist concrete mixtures that result in concrete structures much more porous that cast-in-place concrete. Water-cement ratios of 0.30 to 0.45 are found, but absorptions as high as 10 percent by mass are acceptable. Ten percent absorption by mass is roughly equivalent to 20 percent by volume. Kalousek and his colleagues made test specimens with a water-cement ratio of 0.35 and a slump of zero, using various types of cement. Absorption was varied by using different degrees of vibration and compaction of the specimens. With 6.2 percent absorption, the number of days to 0.5 percent expansion was 1240. With 8.5 percent absorption, the number of days to the same expansion was only 140, a drop of almost 90 percent. As absorptions increased further, performance declined even more. They stated:

"In the range of porosities between about 8 and 13% Type I, Type II, and Type V cement concretes showed the same poor resistance to sulfate attack."

Pastes with high surface to volume ratios failed in exposure to sulfate solutions in spite of a water-cement ratio 0.35.

Kalousek's work is a dramatic demonstration of the need for a tight concrete structure that resists the penetration of aggressive solutions. His data as well as other historical data show that low permeability is a primary precondition for sulfate-resisting concrete. No combinations of adequate cement content, low water-cement ratio, or sulfate-resisting cement can make concrete capable of resisting sulfate attack unless the concrete also has low permeability and high resistance to the penetration of attacking solutions.

References

[1] Miller, D. G. and Manson, P. W., "Laboratory and Field Tests of Concrete Exposed to the Action of Sulphate Waters," United States Department of Agriculture Technical Bulletin No. 358, Washington, D.C., June, 1933.

[2] Miller, D. G. and Manson, P. W., "Long-Time Tests of Concretes and Mortars Exposed to Sulfate Waters," Univ. of Minnesota Agr. Exp. St. Tech. Bull. No. 194, University of Minnesota, Minneapolis, MN, May, 1951.

[3] "Tentative Specification for Portland Cement," ASTM C 150-40 T, ASTM, West Conshohocken, PA, 1940.

[4] McMillan, F. R., Stanton, T. E., Tyler, I. L., and Hansen, W. C., "Long-Time Study of Cement Performance in Concrete, Chapter 5, Concrete Exposed to Sulfate Soils," Research Bulletin 30, Portland Cement Association, Skokie, IL, December, 1949.

[5] Verbeck, G. J., "Field and Laboratory Studies of the Sulphate Resistance of Concrete," Research Bulletin 227, Portland Cement Association, Skokie, IL, 1967.

[6] Kalousek, G. L., Porter, L. C., and Harboe, E. M., "Past, Present, and Potential Developments of Sulfate-Resisting Concretes," Journal of Testing and Evaluation, ASTM, West Conshohocken, PA, Vol. 4, No. 5, Sept. 1976, pp. 347-354.

MODELING MICROSTRUCTURAL ALTERATIONS OF CONCRETE SUBJECTED TO EXTERNAL SULFATE ATTACK

J. Marchand[1-2], É. Samson[1-2] and Y. Maltais[1-2]

(1) Centre de Recherche Interuniversitaire sur le Béton
Université Laval, Sainte-Foy, Canada, G1K 7P4

(2) SIMCO Technologies inc.
1400, boul. du Parc Technologique, Québec, Canada, G1P 4R7

ABSTRACT

The main features of a numerical model developed to predict the microstructural alterations of concrete subjected to external sulfate attack are presented. The model accounts for the transport by diffusion and advection of five different ionic species. The main originality of the numerical model presented in the following sections lies in the fact that it takes into account the electrical coupling between the various ionic species in solution. Such an approach yields a reliable description of the ionic transport process in the material. The model also considers the chemical interaction of ions with the cement paste hydrated phases and the effects of the chemically-induced microstructural alterations on the transport properties of the material. Examples of the application of the model to typical external sulfate degradation cases are given.

INTRODUCTION

Concrete degradation mechanisms generally involve the penetration of external ions (such as sulfate, sodium and magnesium) into the material porosity and/or the dissolution of chemical species from its hydrated and unhydrated phases. The penetration of ions in concrete can be associated with the flow of liquid under a capillary potential gradient. The phenomenon is usually called advection. The penetration of ions can also be associated with an ionic drift originating from a chemical potential gradient. This phenomenon is generally referred to as diffusion. In most practical cases, ions are transported through the concrete pore structure by a combined advection/diffusion process.

The following paragraphs are devoted to the presentation of a global mathematical model that combined the ionic transport originating from both capillary and chemical potential gradients. The main originality of the model presented in the following sections lies in the fact that it takes into account the electrical coupling between the various ionic species in solution. Such an approach yields a reliable description of the ionic transport process in the material. In addition to the transport of ions by diffusion and advection, the mathematical model accounts for the various ionic interactions with the solid phases. It can predict the formation of new chemical products as well as for the dissolution of existing solid phases.

Given its complexity, the system of equations at the basis of the mathematical model has to be solved numerically. A section of this paper is thus devoted to the brief description of the finite-element model that has been used to solve the system of equations. Finally, in last part of this paper, two typical examples of the application of the model to external sulfate degradation cases are given.

THEORETICAL CONSIDERATIONS PERTAINING TO ION TRANSPORT IN CEMENT-BASED MATERIALS

As can be seen in Figure 1, in any cement-based materials, ions can be found in three different states:

- they can be free in the pore solution;
- they can be physically bound to the pore wall (or possibly trapped in the interlayer spaces);
- or chemically bound to the hydration products.

The presence of bound ions arises from the fact that many ionic species tend to strongly interact with the hydrated cement paste. The interaction can be both chemical and physical in nature. For instance, many studies indicate that sulfate ions can physically interact with the calcium silicate hydrates (Kalousek and Adams, 1951; Fu, 1996; Li, 1995). Sulfates can also chemically react with the unhydrated and hydrated phases to form new compounds (Damidot and Glasser,

1993; Fu, 1996; Li, 1995). The mechanisms of ion binding will be further discussed in a following section.

In this global mathematical transport model, the mass conservation equation is used to describe the movement of ions in (saturated or unsaturated) porous media. It is applied on a representative elementary volume of concrete, i.e. on a piece of concrete large enough so that the material can be considered isotropic throughout the entire volume.

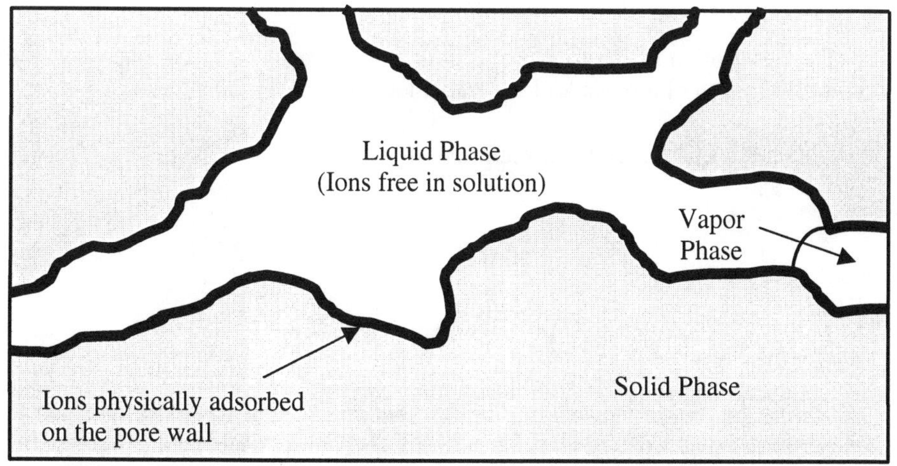

Figure 1 – Various phases that can be found in a porous medium

For the representative elementary volume shown in Figure 1, the total concentration of an ionic species i is given by the total number of mole N divided by the total volume V_t:

$$c_{i\,tot} = \frac{N}{V_t} \tag{1}$$

However, in order to properly use the mass conservation equation, one has to distinguish the concentration of each phase separately. Knowing that the total concentration of ions in the solid is the sum of the concentration of free and bound ions (physically adsorbed or chemically bound), one can write:

$$c_{i\,tot} = \frac{N_f + N_a + N_s}{V_t} \qquad (2)$$

Where the subscripts f, a and s stand for the free, adsorbed and solid phases respectively. Likewise, by separating V in its different parts, the latter equation can be developed as follows:

$$c_{i\,tot} = \frac{N_f}{V_f}\frac{V_f}{V_t} + \frac{N_a}{V_a}\frac{V_a}{V_t} + \frac{N_s}{V_s}\frac{V_s}{V_t} \qquad (3)$$

The description of the transport of ions through cement-based materials requires the definition of a few additional variables, such as:

- The free ion concentration (mol/m^3):

$$c_i = \frac{N_f}{V_f} \qquad (4)$$

- The adsorbed ion concentration (mol/m^3):

$$c_{ia} = \frac{N_a}{V_a} \qquad (5)$$

- The ion concentration in the solid phase (mol/m^3):

$$c_{is} = \frac{N_s}{V_s} \qquad (6)$$

- The free water content (m^3/m^3):

$$\theta_w = \frac{V_f}{V_t} \qquad (7)$$

- The adsorbed water content (m^3/m^3):

$$\theta_a = \frac{V_a}{V_t} \qquad (8)$$

- The solid phase content (m^3/m^3):

$$\theta_s = \frac{V_s}{V_t} \tag{9}$$

The porosity ϕ of a porous medium being defined as the fraction of voids in a given volume, one can write:

$$\theta_s = 1 - \phi \tag{10}$$

The adsorbed water content can be related to a well-known quantity: the specific surface area of the concrete α (expressed in m^2/kg of solid phase). By multiplying α by the solid phase density ρ_s (expressed in kg/m^3 of solid phase), one can get the surface area per m^3 of solid. If the assumption is made that the adsorbed water layer on the pore wall is of constant thickness κ (m), the adsorbed water content can be written as:

$$\theta_a = \rho_s \, \alpha \, \kappa \tag{11}$$

By combining all the previous definitions, the total ionic concentration is given by the following equation:

$$c_{i\,tot} = (1-\phi)\, c_{is} + \rho_s \alpha \, \kappa \, c_{ia} + \theta_w c_i \tag{12}$$

In order to apply the mass conservation equation over an elementary representative volume of concrete, the total ionic flow crossing this material has to be defined ($\mathbf{F}_{i\,tot}$ – mol/m^2s). As for the definition of the ionic concentration, the total flow is given by the sum of the flows of each individual phase, weighted by the volumetric fraction of that phase (Bear and Gilman, 1995; Borkovec et al., 1996; Porcelli et al., 1994; Silberbush et al., 1993):

$$\mathbf{F}_{i\,tot} = (1-\phi)\, \mathbf{F}_{is} + \rho_s \alpha \, \kappa \, \mathbf{F}_{ia} + \theta \, \mathbf{F}_i \tag{13}$$

Where $\mathbf{F}_{is}$, $\mathbf{F}_{ia}$ and $\mathbf{F}_i$ are the flow of ion i in the solid, adsorbed and free water phases respectively.

Using the definition of the total ionic concentration (equation 12) and the tota flow (equation 13), the mass conservation equation for the whole concret volume is given by:

$$\frac{\partial c_{i\,tot}}{\partial t} = -\operatorname{div}\left(\mathbf{F}_{i\,tot}\right) \tag{14}$$

In this model, it is assumed that the movement of the ions can only occur in th liquid phase. Thus, the ionic flow in the solid and adsorbed phases can b neglected. Hence, the equation describing the total ionic flow is reduced to:

$$\mathbf{F}_{i\,tot} = \theta_w \mathbf{F}_i \tag{15}$$

Substituting equation (12) and (15) in equation (14), the mass balance equation i now given by:

$$(1-\phi)\frac{\partial c_{is}}{\partial t} + \rho_s \alpha \kappa \frac{\partial c_{ia}}{\partial t} + \frac{\partial(\theta_w c_i)}{\partial t} + \operatorname{div}(\theta_w \mathbf{F}_i) = 0 \tag{16}$$

As previously emphasized, in most practical cases, concrete elements are kept i an unsaturated state, and the external ions can be transported in the material by combined advection-diffusion process. In order to account for both phenomena the ionic flow $\mathbf{F}_i$ is mainly driven by two forces: the capillary and the chemica potential. Accordingly, the total free ionic flow can be written as:

$$\mathbf{F}_i = \mathbf{F}_{iu} + \mathbf{F}_{iv} \tag{17}$$

Where the subscripts *u* and *v* respectively stand for the chemical (diffusion) an capillary (advection) potential applied to the i species.

TRANSPORT BY IONIC DIFFUSION

The Gibbs free energy (G) of an ionic species (i) in solution is equivalent to it electrochemical potential (μ_i) that can be calculated on the basis of the followin equation:

$$\mu_i = \mu_{io} + RT \ln(\gamma_i c_i) + z_i FV \qquad (18)$$

In the previous equation, μ_{io} is the standard chemical potential (i.e. for very diluted solutions), R the ideal gas constant (8.3143 J/mol/K), T the temperature (°K), γ_i the activity coefficient of the ionic species, c_i its concentration in solution (mol / m^3), F the Faraday constant (9.64846 x 10^4 C/mol) and V the local electrical potential (Volt).

When a gradient of electrochemical potential is present , the ions in solution will tend to move in the direction opposite to the gradient. This can be written as:

$$\mathbf{F}_{iu} = -\frac{D_{0i}}{RT}\,\mathbf{grad}\,(\mu_i) \qquad (19)$$

By replacing equation (18) into equation (19), one finds:

$$\mathbf{F}_{iu} = -D_{0i}\left(\mathbf{grad}\,c_i + c_i\,\mathbf{grad}\ln\gamma_i + \frac{Fz_i}{RT}c_i\,\mathbf{grad}\,V\right) \qquad (20)$$

where (D_{i0}) is the intrinsic diffusion coefficient (m^2/s) of the species. For a given ionic species, the coefficient (D_{i0}) is a constant and corresponds to the diffusion coefficient of the ion in a very dilute solution. In free solutions, (D_{i0}) is therefore an intrinsic property of the ionic species. In a porous media, (D_{i0}) has to be

corrected by some geometrical factors that accounts for the tortuosity and the porosity of the pore system (Bear and Bachmat, 1991; Samson et al., 1998a). In that case, the transport of ions in a porous medium can be described by:

$$\mathbf{F}_{iu} = -D_i\left(\mathbf{grad}\,c_i + c_i\,\mathbf{grad}\ln\gamma_i + \frac{Fz_i}{RT}c_i\,\mathbf{grad}\,V\right) \qquad (21)$$

where D_i stands for the diffusion coefficient of the ionic species in the porous solid. It is a characteristic of both the solid and the species.

In the literature, equation (21) is often called the extended Nernst-Planck equation. It accounts for the particular feature of the ionic diffusion mechanisms. As previously discussed, ions are, contrary to molecules, charged particles. The most important feature that distinguishes ion diffusion from molecular diffusio is the electrical coupling of the various ionic flows. In an ionic solution, the loca electroneutrality shall be preserved at any point. The conservation o electroneutrality requires that the flows of all diffusing species should be couplec During the diffusion process, all ions are not drifting at the same speed. Som ions tend to diffuse at higher rates than others. However, any excess charg transferred by the faster ions builds up a local electric field (called the diffusio potential). This potential slows down the faster ions, and reciprocally accelerate the slower ions. This local potential can be calculated on the basis of the Poisso equation (Marchand et al., 1998; Samson et al., 1998b, Helfferich, 1962):

$$\nabla^2 V + \frac{\rho}{\varepsilon_0 \varepsilon_r} = 0 \qquad (22)$$

In this equation, ∇^2 represents the Laplacian operator, ρ is the charge distributio (Coulomb/m^3), ε_0 is the permittivity of vacuum and ε_r is the dielectric constant th medium in which the ions are drifting (in our case the medium is water). Sinc the ions bear electrical charges, one can relate ρ to the ionic concentration usin the following relation (Helfferich, 1962):

$$\rho = F \sum_{i=1}^{n} z_i \, c_i \qquad (23)$$

where n is the number of ionic species present in the system.

The complete description of the diffusion process in saturated porous material can be done through the Nernst-Planck/Poisson set of equations, i.e. on the basi of equations (21) (that should be written for each ionic species present i solution) and equation (22). The resolution of this system of non-linea differential equations can be quite difficult particularly for systems made o numerous polyvalent species. Over the years, numerous analytical solutions hav been developed for simple cases (electrolytic solutions made of monovalent ions ...). However, the treatment of most practical cases requires the development o numerical solutions. The inherent difficulties of the numerical integration of thes equations are discussed in references Samson et al., 1998b; Samson an Marchand, 1998.

TRANSPORT BY ADVECTION

General considerations

The mass conservation equation (in a representative unit or elementary volume) can be used to describe the evolution of the moisture content (θ_m) (Jacobsen et al. 1996; Friaa et al. 1987; Bazant and Najjar, 1972). In its most general form, the mass balance equation for the moisture transport can be simply written as (Nilsson et al., 1996):

$$\frac{\partial \theta_m}{\partial t} + \mathrm{div}\left(\mathbf{F}_m\right) = 0 \tag{24}$$

In this equation, $\mathbf{F}_m$ represents the total moisture flow in the representative elementary volume of concrete. However, the transport of moisture can take place as liquid and vapor flows. Thus, to account for both types of flow, the previous equation should be modified as follows:

$$\frac{\partial \theta_w}{\partial t} + \frac{\partial \theta_v}{\partial t} + \mathrm{div}\left(\mathbf{F}_m\right) = 0 \tag{25}$$

where θ_W is the liquid water content (m^3 of liquid water / m^3 of concrete) and θ_V is the vapor water content (m^3 of vapor water / m^3 of concrete).

As Luikov (1966) emphasized in his comprehensive review of heat and mass transport in porous media, the transport of moisture in a capillary-porous material can take place simultaneously in the liquid and vapor phases. Hence, the moisture flow vector ($\mathbf{F}_m$) can be written as:

$$\mathbf{F}_m = \mathbf{F}_w + \mathbf{F}_v \tag{26}$$

In the next paragraphs, the liquid and the vapor flow vectors will be treated separately. This approach is necessary since, in cement-based materials, the ionic species are solely transported in the liquid phase (Nilsson et al. 1996). Accordingly, the transport by liquid flow will be distinguished from the transport by vapor flow.

Liquid Transport

The mass transport equation of liquid water can be written as (Pel, 1995; Luikov 1966):

$$\frac{\partial \theta_w}{\partial t} = -\operatorname{div}\left(\mathbf{F}_w\right) + I_w \tag{27}$$

In this equation, the term I_W represents the volumetric capacity of the source (or sink) of liquid, dependent on the phase change.

The mass flow of liquid water in a porous medium ($\mathbf{F}_w$) can be described by the so-called extended Darcy law. If one takes into account gravity effects and the capillary potential, the total mass flow can be written as follows (Perrin e Bonnet, 1995; Hall, 1994, Friaa et al., 1987):

$$\mathbf{F}_w = -K\left(\theta_w\right) \operatorname{grad} \Phi = -K\left(\theta_w\right) \operatorname{grad}\left(\Psi + Z\right) \tag{28}$$

where Φ is the hydraulic potential, Ψ is the capillary potential, Z is the gravitational potential, θ_W is the water content (m^3 of water / m^3 of concrete) and $K(\theta_W)$ is the water conductivity. For a capillary-porous body (such as concrete) the capillary potential is significantly greater than the gravity potential ($\Psi >> Z$) (Luikov, 1966). In this case, the action of gravity on the capillary liquid can be neglected and the water flow equation can be reduced to:

$$F_w = -K\left(\theta_w\right) \mathbf{grad}\, \Psi \tag{29}$$

By stating that, for a constant temperature, the capillary potential is only a fonction of the water content, on can write:

$$\mathbf{grad}\, \Psi = \left(\frac{\partial \Psi}{\partial \theta_w}\right) \mathbf{grad}\, \theta_w \tag{30}$$

If the following quantity is defined as the water diffusivity:

$$D_w = K(\theta_w)\left(\frac{\partial \Psi}{\partial \theta_w}\right) \tag{31}$$

Equation (29) becomes:

$$\mathbf{F}_w = -D_w \ \mathbf{grad}\,\theta_w \tag{32}$$

This equation indicates that the flow of liquid water is directly linked to the gradient of water content $\theta_w(x,t)$ or to the gradient of relative humidity H(x,t) in the pore system. The development of a model to predict the flow of moisture in any porous medium can be based on any of these two variables (water content or relative humidity).

In many instances, researchers have chosen to express the moisture flow in terms of the gradient of the pore relative humidity (Xi et al., 1994a, Xi et al., 1994b). At first sight, the choice of the relative humidity as the main variable may appear more practical. However, as will be seen later, the pore relative humidity is extremely sensitive to any variations of the pore solution chemistry. In that respect, the choice of expressing the moisture flow in terms of the gradient of water content simplifies the mathematics and the experimental validation of the model.

The amount of ions transported by an advective flow is proportional to the water flow and the concentration of ions in the pore solution (Nilsson et al., 1996). Since the ions are solely transported in the liquid phase, the mass transport process under an advective flow can be described by:

$$\mathbf{F}_{iv} = c_i \ \mathbf{F}_w \tag{33}$$

Diffusion coefficient of water

In order to solve equation (33), the diffusion coefficient (D_w) of water in concrete must be determined. Recently, numerous authors have shown that the water diffusion coefficient (which varies strongly with the water content of the material) can be determined on the basis of water content distributions measured at some precise elapsed times (Carpenter et al., 1993; Hall, 1994, Pel, 1995, Hazrati, 1998).

In a recent investigation, Hazrati (1998) has measured water profiles in a wide range of mortars using a nuclear magnetic resonance imaging (NMRI) technique. With an apparatus especially designed for porous building materials (such as fired-clay brick, sand-lime brick and mortar) the evolution of the water conten can be determined very precisely. This experimental method is non-destructive Furthermore it is characterized by a relatively good spatial resolution (Pel, 1995) The complete description of the nuclear magnetic resonance imaging technique and the experimental set-up can be found in Hazrati (1998).

The results obtained by Hazrati (1998) have clearly confirmed that the water diffusivity of concrete varies strongly with the water content. According to the author, the variation of the transport coefficient over the entire range of the material water contents (i.e. from a fully dry state to a saturated state) can be modeled by a double-exponential function. However, when the water content of the material ranges from 50% to 100% (i.e. full saturation), the variation can be approximated by a simple exponential function such as:

$$D_w = A\exp(B\theta_w) \quad (m^2/sec) \tag{34}$$

Where A and B are material parameters that can be easily defined on the basis of NMRI water profiles.

Boundary conditions

In the previous equations, the transport of water is expressed in terms of a gradient of water content. However, in most practical cases, it is much easier to express the boundary conditions in terms of relative humidity. Thus, in order to evaluate the water distribution of a given concrete subjected to a drying process, one has to relate the water content of the material to the relative humidity. In this context, a desorption isotherm should be used.

Figure 2 gives a typical the desorption isotherms obtained by Hazrati (1998) for a 0.60 water/cement ratio mortar mixture moist cured for a minimal period of one year.

For a given concrete mixture, the water content of the material exposed to a given relative humidity can be calculated on the basis of the following equation:

$$\theta_w = (\Gamma_{concrete}\,\phi)\,\Omega(H) \tag{35}$$

In this equation, $\Gamma_{concrete}$ represents the paste content of the concrete, ϕ is the porosity of the cement paste, $\Omega(H)$ is the relative water content at a given relative humidity (H).

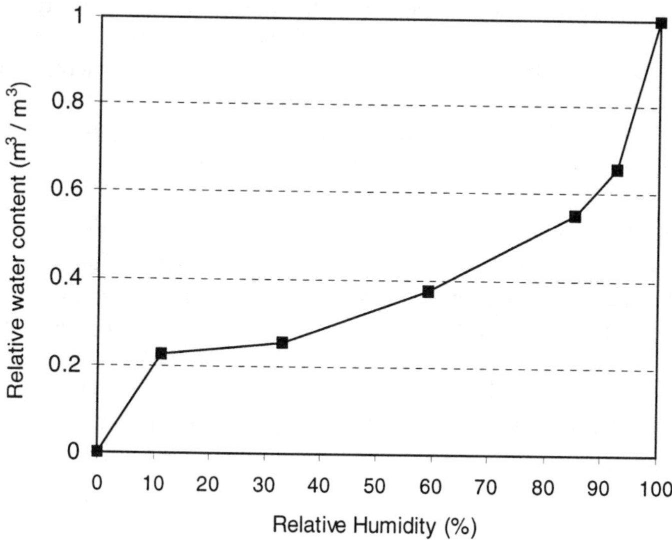

Figure 2 – Desorption isotherm for a 0.60 water/cement ratio mortar mixture

IONIC TRANSPORT BY DIFFUSION AND ADVECTION

Coupling the diffusion and advection ionic flows, one finds:

$$\mathbf{F}_i = \mathbf{F}_{iu} + \mathbf{F}_{iv} \tag{36}$$

Incorporating equations (21) and (33) in equation (36), the coupled ionic flow becomes in one-dimension:

$$\mathbf{F}_i = -D_i\left(\frac{\partial c_i}{\partial x} + \frac{z_i F}{RT} c_i \frac{\partial V}{\partial x}\right) - c_i\left(D_w \frac{\partial \theta_w}{\partial x}\right) \tag{37}$$

Now, if the expression of $\mathbf{F}_i$ is included in the mass balance equation (16), one gets (for a given ionic species):

$$(1-\phi)\frac{\partial c_{is}}{\partial t}+\rho_s \alpha \kappa \frac{\partial c_{ia}}{\partial t}+\frac{\partial(\theta_w c_i)}{\partial t}-\frac{\partial}{\partial x}\left(\theta_w D_i\left(\frac{\partial c_i}{\partial x}+\frac{z_i F}{RT}c_i\frac{\partial V}{\partial x}\right)+\theta_w c_i\left(D_w\frac{\partial \theta_w}{\partial x}\right)\right)=0 \tag{38}$$

For a given ionic transport problem, this advection-diffusion transport equation has to be solved for all the ionic species considered in a particular system. Considering the nature of the external sulfate attack of concrete, the number of ionic species to be considered can be limited to four, namely : sulfate (SO_4^{2-}), sodium (Na^+), calcium (Ca^{2+}) and hydroxyl ions (OH^-). In many instances, a fifth species (magnesium – Mg^{2+}) has to be accounted for.

In order to solve equation (38), the terms $\partial c_{is}/\partial t$ and $\partial c_{ia}/\partial t$ must be defined for each ionic species considered. In the model, the term $\partial c_{is}/\partial t$ is used to take into account the formation or the dissolution of numerous solid phases upon an external sulfate attack. For instance, these terms can be used to model the formation of ettringite or gypsum or they can also be used to model the dissolution of portlandite or the decalcification of the C-S-H. On the other hand, the term $\partial c_{ia}/\partial t$ is used to model the physical interaction of ions on (or near) the pore wall. The complete mathematical treatment of the terms $\partial c_{is}/\partial t$ and $\partial c_{ia}/\partial t$ is described in the next section.

MATHEMATICAL MODELING OF ION INTERACTION IN UNSATURATED CONCRETE

A simple approach to model the ion interaction mechanisms in cement-based materials is presented in the following paragraphs. According to this approach, it is proposed to model the modifications of the pore solution chemistry by a series of source and sink terms. For instance, the generation of hydroxyl and calcium ions upon the dissolution of portlandite or the decalcification of the C-S-H is taken into account by a source term, while the consumption of sulfates (upon the formation of ettringite or gypsum) is modeled by a sink term.

In the general set of transport/degradation equations, the modifications of the pore solution chemistry are considered through the chemical interaction terms ($\partial c_{is}/\partial t$). Accordingly, a chemical interaction term must be defined for each ionic species that is assumed to interact with the solid matrix (i.e. SO_4^{2-}, OH^-, $Ca^{2+}$and eventually Mg^{2+}). If the chemical reactions are assumed to be fast enough as compared to the transport process, the ionic binding mechanism can be treated as an equilibrium problem and the chemical interaction terms can be written:

(1) for sulfate ions:

$$\frac{\partial c_{sulfate-s}}{\partial t} = \frac{\partial c_{sulfate-s}}{\partial c_{sulfate}} \frac{\partial c_{sulfate}}{\partial t} \tag{39}$$

(2) for hydroxyl ions:

$$\frac{\partial c_{hydroxyl-s}}{\partial t} = \frac{\partial c_{hydroxyl-s}}{\partial c_{hydroxyl}} \frac{\partial c_{hydroxyl}}{\partial t} \tag{40}$$

(3) for calcium ions:

$$\frac{\partial c_{calcium-s}}{\partial t} = \frac{\partial c_{calcium-s}}{\partial c_{hydroxyl}} \frac{\partial c_{hydroxyl}}{\partial t} \tag{41}$$

Equations (39) to (41) require the determination of a new term: $\partial c_{i\text{-}s}/\partial c_i$. This term has to be evaluated for each ionic species using an *ionic interaction isotherm* (i.e. a relationship that relates the free ion concentration in solution to the bound species concentration in the solid). In the following paragraphs, the interaction isotherm for the four ionic species considered in the binding model will be established on the basis of theoretical considerations.

Elaboration of the Sulfate Interaction Isotherm

The sulfate interaction isotherm can be designed on the assumption that the mass of sulfate bound to the solid phase is a function of (1) the free sulfate concentration in the pore solution and (2) the chemical composition of the cement. On the basis of the phase diagrams developed by Damidot and Glasser (1993) and assuming that the only hydrated phases present are hydrogarnet, portlandite, ettringite and gypsum, the free sulfate concentration can be divided into three intervals:

		Existing solid phases
(interval I)	$0 \leq \{SO_4^{2-}\} \leq \{SO_4^{2-}\}^{ett.\ free}$	CH, C_3AH_6
(interval II)	$\{SO_4^{2-}\}^{ett.\ free} \leq \{SO_4^{2-}\} \leq \{SO_4^{2-}\}^{gyp.\ free}$	Ett.,CH,C_3AH_6
(interval III)	$\{SO_4^{2-}\}^{gyp.\ free} \leq \{SO_4^{2-}\}$	Gypsum, CH

These intervals are shown on Figure 3. On this figure, a minimum sulfate concentration is required to form both ettringite and gypsum. In fact, when the concentration of the sulfate ions in the pore solution is low (interval I), no sulfate-bearing products can be formed. However, if the concentration in sulfate increases, ettringite and gypsum will be produced in intervals II and III respectively.

The critical concentrations in sulfate ions required to form ettringite and gypsum can be estimated on the basis of the phase diagrams developed by Damidot and Glasser (1993). They tend to vary with the type of cement and the concrete mixture characteristics.

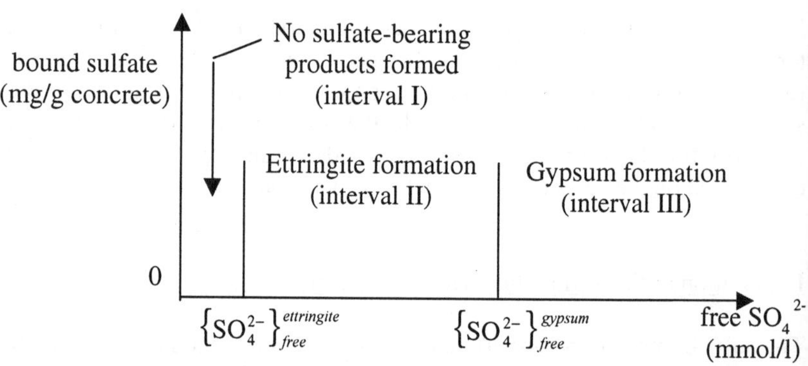

Figure 3 – Elaboration of the theoretical sulfate interaction isotherm (treatment of the free sulfate ions in the concrete pore solution)

The last step to establish the interaction isotherm is to evaluate the mass of ettringite and gypsum formed at different concentrations of sulfate. In first approximation, for a concentration of sulfate comprised in the interval of ettringite formation, it is assumed that all the hydrogarnet phase will react to form ettringite. It is also assumed, for a concentration of sulfate comprised in the interval of gypsum formation, that all the remaining portlandite (after the formation of ettringite) will be dissolved to produce gypsum. On the basis of these assumptions, it can be shown that the maximum mass of sulfate bound as ettringite and gypsum can be calculated on the basis of equations (42) and (43) respectively:

$$\left\{SO_4^{2-}\right\}_{bound}^{ettringite} = \left[\frac{\{C\}}{\{C\}+\{W\}+\{A\}}\right]\cdot\left[1.067\cdot\left[\left(\{C_3A\}-1.12\{SO_3\}\right)+0.78\{C_4AF\}\right]\right] \tag{42}$$

$$\left\{SO_4^{2-}\right\}_{bound}^{gypsum} = \left[\frac{\{C\}}{\{C\}+\{W\}+\{A\}}\right]\cdot\left[1.297\left[0.487\cdot\{C_3S\} + 0.215\cdot\{C_2S\} - 0.822\{C_3A\}\right]\right] \tag{43}$$

where :

$\{C_2S\}$ = initial mass of C_2S in the cement (expressed in g per g of cement)
$\{C_3S\}$ = initial mass of C_3S in the cement (expressed in g per g of cement)
$\{C_3A\}$ = initial mass of C_3A in the cement (expressed in g per g of cement)
$\{C_4AF\}$ = initial mass of C_4AF in the cement (expressed g per g of cement)
$\{SO_3\}$ = initial mass of SO_3 in the cement (expressed g per g of cement)

$\{C\}$ = mass of cement (expressed in kg per m^3 of concrete)
$\{W\}$ = mass of water (expressed in kg per m^3 of concrete)
$\{A\}$ = mass of aggregate (expressed in kg per m^3 of concrete)

Figure 4 shows the relationship between the maximum mass of sulfate bound to the solid phase and the concentration of free sulfate.

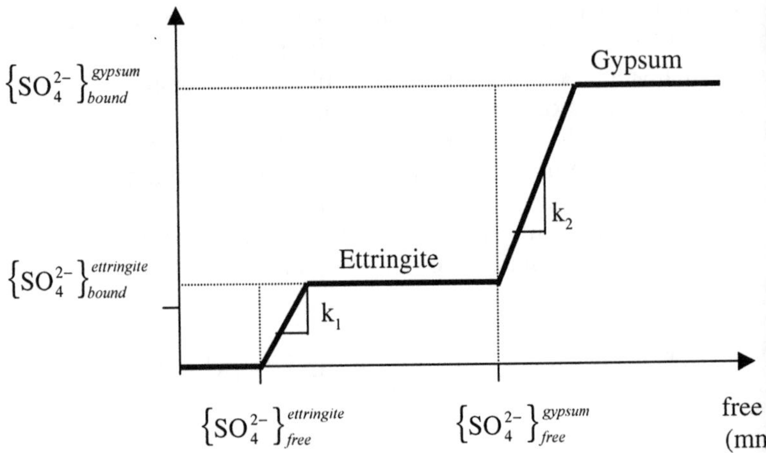

Figure 4 – Theoretical sulfate interaction isotherm

Modeling the hydroxyl and calcium interaction isotherms for the dissolution of portlandite

The general approach used to model the hydroxyl and calcium interaction isotherms is similar to that presented in the previous section. In fact, assuming that a mature concrete is subjected to a sulfate attack, the mass of hydroxyl and calcium ions bound as portlandite will be calculated before and after the ettringite formation. Furthermore, the level of hydroxyl ions required to dissolve the portlandite phase will also be established.

- Stage I – Evaluation of the Mass of Hydroxyl and Calcium Bound as Portlandite

The amount of portlandite (expressed in g / g of cement) is given by the following relation:

$$\{Ca(OH)_2\}_{initial} = 0.487 \cdot \{C_3S\} + 0.215 \cdot \{C_2S\} \qquad (44)$$

In this equation, the terms 0.487 and 0.215 account for the mass of calcium hydroxide produced by the hydration of 1g of tricalcium and dicalcium repectively.

It can be demonstrated that, during an external sulfate attack, a part of the portlandite is used to form ettringite. In order to take into account this phenomenon, the residual mass of portlandite (upon ettringite formation) can be estimated using the following equation:

$$\{Ca(OH)_2\}_{residual} = 0.487 \cdot \{C_3S\} + 0.215 \cdot \{C_2S\} - 0.822 \cdot \{C_3A\} \quad (45)$$

From this equation, the mass of hydroxyl ions bound as portlandite (expressed in g/g of cement) can be written:

$$\{OH^-\}_{bound}^{portlandite} = 0.459 \cdot [0.487 \cdot \{C_3S\} + 0.215 \cdot \{C_2S\} - 0.822 \cdot \{C_3A\}] \quad (46)$$

In this equation, 0.459 represents the weight ratio between the hydroxyl ions and the calcium hydroxide phase. The mass of calcium ions bound as portlandite (expressed in g/g of cement) can be calculated as follows:

$$\{Ca^{2+}\}_{bound}^{portlandite} = (1 - 0.459) \cdot [0.487 \cdot \{C_3S\} + 0.215 \cdot \{C_2S\} - 0.822 \cdot \{C_3A\}] \quad (47)$$

As for the sulfate interaction isotherm, the mass of the bound species has to be expressed in <u>gram of bound portlandite per gram of concrete</u>. Thus, equation (46) and (47) respectively become:

$$\{OH^-\}_{bound}^{portlandite} = \left[\frac{\{C\}}{\{C\}+\{W\}+\{A\}}\right] \cdot .459 \cdot [.487 \cdot \{C_3S\} + .215 \cdot \{C_2S\} - .822 \cdot \{C_3A\}] \quad (48)$$

$$\{Ca^{2+}\}_{bound}^{portlandite} = \left[\frac{\{C\}}{\{C\}+\{W\}+\{A\}}\right] \cdot (1 - .459) \cdot [.487 \cdot \{C_3S\} + .215 \cdot \{C_2S\} - .822 \cdot \{C_3A\}] \quad (49)$$

- Stage II – Evaluation of the Critical Hydroxyl Level to Dissolve the Portlandite Phase

Calcium hydroxide is the first hydrated phase to be influenced by the pH level of the concrete pore solution. Accordingly, the amount of portlandite present in the solid will be assumed to be a function of the free hydroxyl concentration.

- Stage III – Elaboration of the Hydroxyl and Calcium Interaction Isotherms for Portlandite

The hydroxyl and calcium ion interaction isotherms presented in this section have been essentially designed on the basis of the following assumptions:

- First, the mass of calcium hydroxide presents in the solid phase (when the dissolution process begins) is assumed to be equal to $\{Ca(OH)_2\}_{residual}$;
- Second, if the pH of the pore solution is higher than 13.0, then no calcium or hydroxyl ions are released into the pore solution;
- Third, if the pH of the pore solution is between 12.4 and 13.0, then the hydroxyl and calcium bound to the portlandite crystal are assumed to be released in the pore solution at a molar ratio of 2.0;
- Finally, if the pH of the pore solution is lower than 12.4, all the portlandite phase is assumed to be dissolved.

The hydroxyl and calcium interaction isotherms established with these assumptions are shown on Figures 5 and 6 respectively.

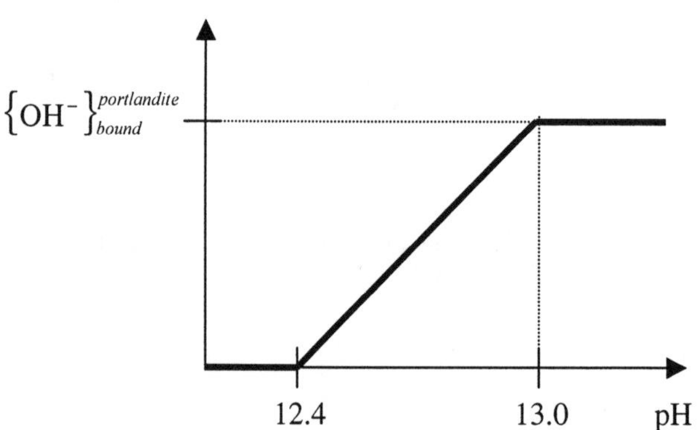

Figure 5 – Hydroxyl interaction isotherm for the dissolution of the portlandite crystal

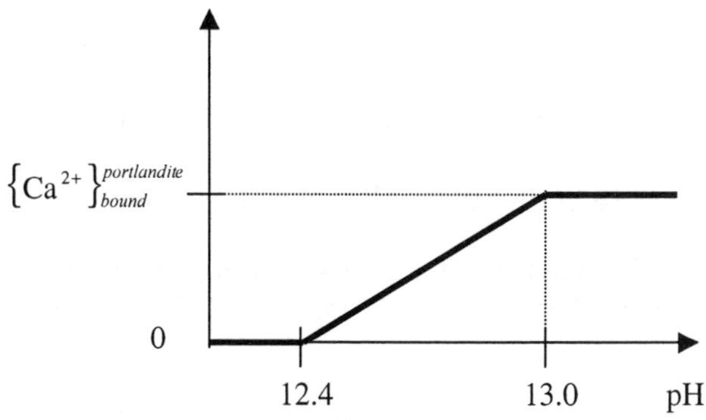

Figure 6 – Calcium interaction isotherm for the dissolution of the portlandite crystal

Modeling the hydroxyl and calcium ion interaction isotherms for the decalcification of C-S-H

The hydroxyl and calcium ion interaction isotherms for the C-S-H are based on similar assumptions than the ones used for the dissolution of the portlandite phase. As for the elaboration of the previous interaction isotherms, the modeling work is divided in three stages.

- Stage Ia – Evaluation of the Mass of Calcium Bound in C-S-H

In order to evaluate the mass of calcium bound in C-S-H, it is essential to determine the amount of C-S-H produce by the hydration of 1g of Portland cement. This mass can be calculated using the following equation:

$$\{C\text{-}S\text{-}H\} = 0.75 \cdot \{C_3S\} + 0.994 \cdot \{C_2S\} \tag{50}$$

In this equation, the terms 0.75 and 0.994 represent the mass of C-S-H produced by the hydration of 1g of tricalcium and dicalcium silicate respectively.

Assuming that C_3-S_2-H_3 is the final product of hydration of both C_3S and C_2S, the mass of calcium bound to the solid phase can be calculated on the basis of the following equation:

$$\{Ca^{2+}\}_{bound}^{C\text{-}S\text{-}H} = 0.35 \cdot [0.75 \cdot \{C_3S\} + 0.994 \cdot \{C_2S\}] \qquad (51)$$

The 0.35 coefficient represents the fraction of solid calcium in the C-S-H.

- Stage Ib – Evaluation of the Mass of Hydroxyl Bound in C-S-H

The mass of hydroxyl bound to the C-S-H can be established if one assumes that the dissolution of calcium (Ca^{2+}) from the C-S-H also liberate an equivalent amount of hydroxyl ions.

From the previous assumption, the maximum mass of hydroxyl ions liberated in the pore solution (from the C-S-H) can be calculated using the following relation:

$$\{OH^{2-}\}_{bound}^{C\text{-}S\text{-}H} = 0.15 \cdot [0.75 \cdot \{C_3S\} + 0.994 \cdot \{C_2S\}] \qquad (52)$$

The 0.15 coefficient represents the fraction of solid hydroxyl in the C-S-H.

- Stage II – Evaluation of the Critical Hydroxyl Level to Dissolve the C-S-H Phase

The decalcification of the C-S-H only occurs when the amount of available calcium hydroxide is not sufficient to restore the pore solution equilibrium. When all the available portlandite has been dissolved (pH < 12.4), the equilibrium of the system is then controlled by the C-S-H themselves that undergo partial decalcification.

- Stage III – Elaboration of the Hydroxyl and Calcium Interaction Isotherms for C-S-H

The hydroxyl and calcium interaction isotherms presented in this section have been essentially designed on the basis of the following assumptions:

  - First, the mass of the C-S-H present in the solid phase (prior to the dissolution process) is assumed to be equal to {C-S-H};
  - Second, if the pH of the pore solution is higher than 12.4, no calcium or hydroxyl ions (from the C-S-H) are released in the pore solution.

- Third, if the pH of the pore solution is lower than 12.4, then C-S-H undergo partial decalcification. The calcium and hydroxyl ions are assumed to be released in the pore solution at a molar ratio of 1.0.
- Finally, if the pH of the pore solution is lower than 11.5, all the C-S-H phase is assumed to be totally decalcified.

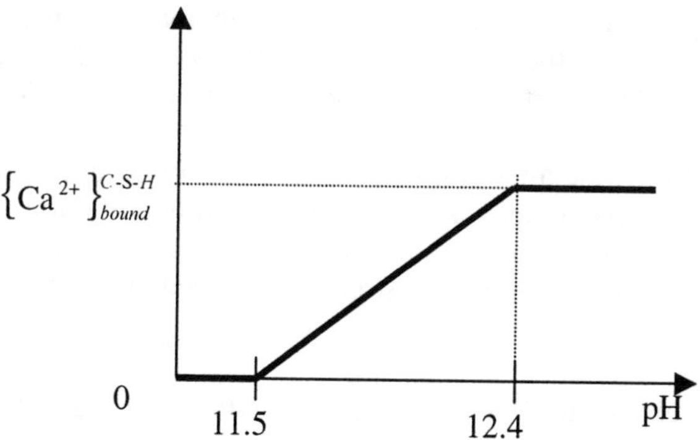

Figure 7 – Calcium interaction isotherm for the decalcification of the C-S-H

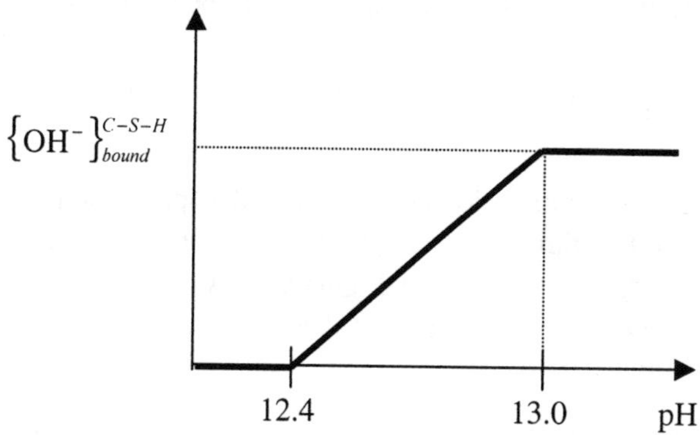

Figure 8 – Hydroxyl interaction isotherm for the decalcification of the C-S-H

The interaction isotherms established with these assumptions are shown on Figures 7 and 8.

Modeling the Influence of Microstructural Damage on the Material Transport Properties

As emphasized in the previous chapters, the rate at which these species can diffuse through the concrete pore structure largely determines the evolution of the degradation process. However, as the external sulfate attack progresses, external ions react with the hydrated cement paste matrix. These chemical reactions significantly alter the microstructure of the solid by favoring, for instance, the dissolution of the portlandite and the decalcification of C-S-H.

The effect of these alterations on the ionic diffusion coefficients are discussed in the following paragraphs. It will be explained how the ionic and water diffusion coefficients determined on undamaged samples are modified to account for the alterations of the material pore structure induced by the external sulfate attack.

Influence of the material pore structure on its transport properties

According to Garboczi and Bentz (1992a, 1992b), the diffusion coefficient (D_i) of a given ionic species in an undamaged hydrated cement paste can be directly linked to the material capillary porosity (ϕ) by the following relationship:

$$\frac{D_i}{D_{io}} = 0.001 + 0.07\phi^2 + H(\phi - 0.18) \cdot 1.8 \cdot (\phi - 0.18)^2 \qquad (53)$$

where D_{i0} is the diffusion coefficient of the species in pure water and H(x) is the Heaviside function. According to this function, $H(x = \phi - 0.18) = 1$ for $x > 0$ and $H(x = \phi - 0.18) = 0$ for $x \leq 0$. Figure 9 shows the variation of the ionic diffusivity of a hydrated cement paste (calculated on the basis of equation 53) as a function of its capillary porosity.

Determination of the material porosity after portlandite dissolution

As previously emphasized, calcium hydroxide is the first hydrated phase to be influenced by a reduction of the pH level of the concrete pore solution. The dissolution process starts when the pH level goes under 13.0 and is followed by a massive dissolution of the portlandite once the pH level of the pore solution drops below 12.4.

Knowing the density of the various phases initially present in the solid, one can calculate the variation of the capillary porosity due to the dissolution of portlandite. On the basis of equation (53), the variation of the capillary porosity can then be used to calculate the modification of the diffusion coefficient of the each ionic species associated to the dissolution of portlandite.

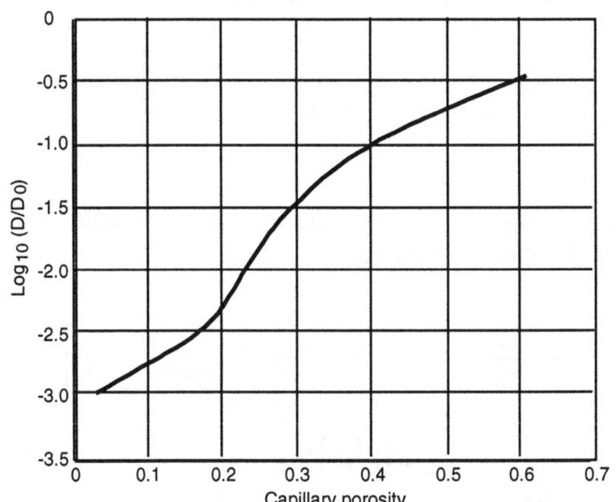

Figure 9 – Logarithm of the relative diffusivity D/D_0 of an hydrated cement paste as a function of its capillary porosity (adapted from Garboczi and Bentz, 1992a).

In order to determine the effect of the dissolution process on the material porosity (noted $\phi_{\text{cement paste}}$), one has to estimate the volume occupied by the amount of portlandite initially present in the solid. To do so, the following equation can be used:

$$\{Ca(OH)_2\}_{\text{initial}} = \mu_1 \cdot \{C_3S\} + \mu_2 \cdot \{C_2S\} \tag{54}$$

where

$$\mu_1 = 0.487 \text{ and } \mu_2 = 0.215. \tag{55}$$

This equation gives the amount of portlandite (expressed in g/g of cement) produced by the hydration of a unit mass of portland cement. However, this equation must be slightly modified in order to express the volume of portlandite on a unit volume of cement paste basis. The volume of portlandite (expressed in m^3 of portlandite per m^3 of cement paste) can be obtained with the following equation:

$$V_{portlandite} = \frac{\{Ca(OH)_2\}_{initial}}{\rho_{portlandite}} \left(\frac{1000\{C\}}{\{C\}/3.15 + \{W\}} \right) \tag{56}$$

In this equation, the terms {C} and {W} stand for the mass of cement and the mass of water respectively (expressed in kg per m^3 of cement paste). On the other hand, the term $\rho_{portlandite}$ refers to the density of portlandite ($\rho_{portlandite}$ = 2240E03 g/m^3).

The total porosity can now be calculated as follows:

$$\phi_{\substack{cement\ paste\ with \\ no\ portlandite}} = \phi_{cement\ paste} + V_{portlandite} \tag{57}$$

Using the previous equation, one can finally calculate the ionic diffusion coefficient for a portlandite-depleted cement paste:

$$\frac{D_{\substack{i-portlandite \\ dissolved}}}{D_{io}} = 0.001 + 0.07\left(\phi_{\substack{cement\ paste\ with \\ no\ portlandite}}\right)^2 + H\left(\phi_{\substack{cement\ paste\ with \\ no\ portlandite}} - 0.18\right)\cdot 1.8 \cdot \left(\phi_{\substack{cement\ paste\ with \\ no\ portlandite}} - 0.18\right)^2 \tag{58}$$

Determination of the material porosity after ettringite formation

During an external sulfate attack, the chemical reactions between the external sulfate ions and the concrete hydrated and unhydrated phases can produce ettringite. The formation of this phase can have an influence on the material microstructure and on the ionic transport properties.

One can calculate the variation of the capillary porosity due to the formation of ettringite. On the basis of equation (53), the variation of the capillary porosity can then be used to calculate the diffusion coefficient of each ionic species in the altered material.

The maximum volume of ettringite formed upon the reaction between the aluminate phase and the external sulfates ions can be calculated using the following equation:

$$V_{ettringite} = \frac{\left\{SO_4^{2-}\right\}_{bound}^{ettringite}}{\rho_{ettringite}}\left(\frac{1000\{C\}}{\{C\}/3.15 + \{W\}}\right) \tag{59}$$

In this equation, the term $\{SO_4^{2-}\}$ stands for the mass of sulfate bound as ettringite calculated using equation (42). This mass is expressed in g/g of cement. On the other hand, the term $\rho_{ettringite}$ refers to the density of ettringite (1717E03 g/m^3). Finally, the terms {C} and {W} stand for the mass of cement and the mass of water (expressed in kg per m^3 of cement paste).

The capillary porosity can now be calculated as follows:

$$\phi_{\substack{cement\ paste\ upon\\ ettringite\ formation}} = \phi_{cement\ paste} - V_{ettringite} \tag{60}$$

Using the previous equation, one can finally calculate the ionic diffusion coefficient for a given cement paste upon ettringite formation:

$$\frac{D_{\substack{i-upon\ ettringite\\ formation}}}{D_{io}} = 0.001 + 0.07\left(\phi_{\substack{cement\ paste\ upon\\ ettringite\ formation}}\right)^2 + H\left(\phi_{\substack{cement\ paste\ upon\\ ettringite\ formation}} - 0.18\right)\cdot 1.8\cdot\left(\phi_{\substack{cement\ paste\ upon\\ ettringite\ formation}} - 0.18\right)^2 \tag{61}$$

Determination of the material porosity after C-S-H decalcification

As pointed out in the first part of this report, the solubility of the calcium silicate hydrates appears to be significantly increased below a critical pH level of approximately 11.5. In other words, the calcium silicate hydrate decalcification becomes significant when the amount of available calcium hydroxide is not sufficient to restore the pore solution equilibrium.

The decalcification of C-S-H contributes to markedly increase the material porosity. When the pH level of the concrete pore solution is lower than 11.5, the porosity of the material is assumed to be increased in such a way than the ionic diffusion coefficient is approximately ten times higher than that of the undamaged material.

Evolution of the transport properties of concrete subjected to external sodium sulfate attack

On the basis of these assumptions, the effect of the various chemical alterations on the ionic diffusion coefficients can now be estimated. The evolution of the ionic diffusion coefficient, for a concrete mixture subjected to a sodium sulfate attack, is presented in Figure 10. This figure illustrates the variation of the ionic diffusion coefficient with the pH level of the pore solution (interval A: pH ≥ 13.0; interval B: 12.4 ≤ pH ≤ 13.0; interval C: 11.5 ≤ pH ≤ 12.4).

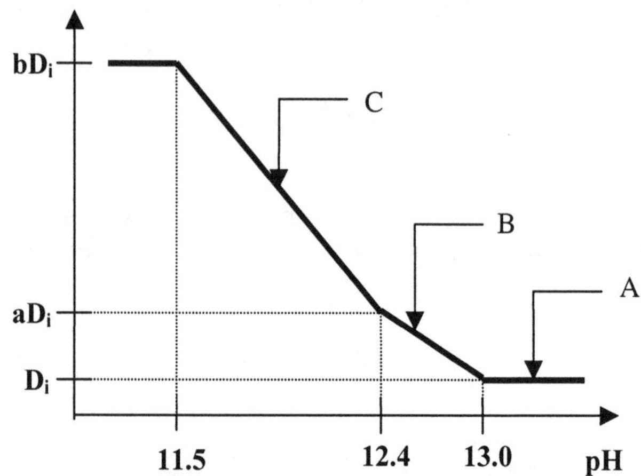

Figure 10 – Evolution of the ionic diffusion coefficient: (A) no porosity changes; (B) dissolution of portlandite; (C) decalcification of C-S-H.

- In the first interval (i.e. for pH values higher than 13.0), the ionic diffusion coefficient is set equal to D_i (see equation 53). In this interval, the reduction of the porosity due to the formation of ettringite is compensated by the consumption of approximately one third of the portlandite crystal.

- In the second interval (i.e. for pH values in between 12.4 and 13.0), the diffusion coefficient of the material slightly increases. This modification of the diffusion coefficient is related to the progressive dissolution of portlandite. At pH = 12.4, the ionic diffusion coefficient is set equal to aD_i.

- Finally, for interval C (i.e. for a pH values lower than 12.4), all the portlandite is dissolved and the subsequent increase of the ionic diffusion coefficient is solely due to the decalcification of the calcium silicate hydrate. At pH = 11.5, the value of the diffusion coefficient is assumed to be bD_i.

The value of the ionic diffusion coefficients D_i, aD_i and bD_i will be determined for some specific concrete mixtures in the last section of this paper.

NUMERICAL IMPLEMENTATION OF THE MODEL

To solve such a complex system of non-linear equations, a numerical algorithm must be used. Based on the work of Samson et al. (1998b), all the equations are solved simultaneously. The spatial discretization of this coupled system is performed through the finite element method. The spatial discretization is performed according to the standard Galerkin procedure (Zienkiewicz and Taylor, 1991). An Euler implicit scheme is used to discretize the transient part of the model (Zienkiewicz and Taylor, 1991).

The nonlinear set of equations is solved with the Newton-Raphson algorithm (Zienkiewicz and Taylor, 1991). This second order algorithm gives a good convergence rate and is robust enough to handle the electrical coupling between the ionic species as well as the nonlinearities coming from the coupling between the ionic flux and the water movement. All the other nonlinearities, i.e. the chemical reactions and the variation of the diffusion coefficients, induced steep variation in some variables. Numerical tests have shown that these terms are better handled with the substitution method (Zienkiewicz and Taylor, 1991) than with the Newton-Raphson method.

NUMERICAL SIMULATIONS – EXTERNAL SODIUM SULFATE ATTACK

This section presents the results of the numerical simulations carried out in order to predict the penetration of sulfate and magnesium ions in a concrete slab over a 50-year period. The simulations give the evolution in the formation of ettringite, the dissolution portlandite and the decalcification of C-S-H in the concrete slab after being exposed to different sodium and magnesium sulfate solutions. In order to predict the formation or the dissolution of these phases, four ionic species were considered in the simulations: SO_4^{2-}, Na^+, Ca^{2+}, OH^-.

Two series of simulations are presented. In the first series, the ionic species are assumed to be transported through a reference concrete produced with a water/cement ratio equal to 0.45. In the second series, the ionic transport phenomenon is assumed to take place through a 0.65 water/cement ratio concrete mixture. Each simulation is performed over a 50 years span.

The complete composition of the concrete mixtures considered in the simulations is presented in the following paragraphs. The transport coefficients (D_w and D_i) associated to each concrete mixtures are also given. Finally, the initial and boundary conditions required to solve the set of transport equations are defined.

Mixture characteristics

The characteristics of the mixtures considered for the simulations are given in Table 1. The design of these mixtures is similar to what can be found in practice. In the present work, it has been assumed that both mixtures had been produced with the same ASTM Type V cement. The chemical and mineralogical compositions of the cement are given in Tables 2 and 3.

Table 1 – Characteristics of the concrete mixtures considered in the numerical simulations

	W/C = 0.45	W/C = 0.65
Cement (kg/m^3)	450	270
Water (kg/m^3)	180	175
Aggregate (kg/m^3)	1825	1920

Table 2 – Chemical composition of the cement

Constituents	ASTM Type V (%)
SiO_2	22.78
Al_2O_3	3.24
Fe_2O_3	3.50
CaO	63.35
MgO	1.9
Mn_2O_3	0.55
Na_2O	1.33
K_2O	0.15
SO_3	2.12
Loss on ignition	2.00

Table 3 – Mineralogical composition of the cement

Mineralogical composition (Bogue)	ASTM Type V (%)
C_3S	50.0
C_2S	27.6
C_3A	3.6
C_4AF	10.7

Transport coefficients

Two transport coefficients are required to run the numerical model. First of all, one has to know the coefficient of water diffusion (D_w) of the considered concrete mixture. As previously emphasized, this coefficient mainly depends on the microstructure and the water content of the concrete. The ionic diffusion coefficient (D_i) of each ionic species considered in the simulation is also required.

- Water diffusion coefficients of the undamaged concrete

The coefficients of water diffusion for the two mixtures can be estimated on the basis of equations (62) and (63). These equations were established on the basis of the NMRI data obtained by Hazrati (1998) for mortar mixtures. Corrections were made to the equations obtained by Hazrati in order to account for the paste content of the concrete mixtures.

For the 0.45 water/cement ratio concrete:

$$D_w = 4.10E-11 \exp(38.7\,\theta_w) \tag{62}$$

For the 0.65 water/cement ratio concrete:

$$D_w = 2.08E-12 \exp(87.3\,\theta_w) \tag{63}$$

Where D_w is expressed in m^2/sec and θ_w in m^3 of water / m^3 of concrete.

- Ionic diffusion coefficients of the undamaged concrete

The ionic diffusion coefficients of both concrete mixtures were determined on the basis of migration experiments according to a procedure described in Samson (1998). The ionic diffusion coefficients of both concrete mixtures are summarized in Table 4.

Table 4 – Ionic diffusion coefficients

Ions	W/C = 0.45 (x 10^{-11} m^2/s)	W/C = 0.65 (x 10^{-11} m^2/s)
Hydroxyl	5.06	15.00
Sodium	1.28	3.80
Calcium	7.69	2.95
Sulfate	1.05	3.54

Evolution of the transport coefficients upon external sulfate attack

As previously emphasized, the effect of the external sulfate attack on the concrete transport properties can be evaluated on the basis of the evolution of the material porosity. The porosity of the paste fraction of a concrete mixture subjected to external sulfate attack can be calculated on the basis of the following equations:

$$\phi_{\substack{\text{cement paste upon ettringite}\\ \text{formation and porlandite}\\ \text{dissolution}}} = \phi_{\text{cement paste}} - V_{\text{ettringite}} + V_{\text{portlandite}} \tag{64}$$

It can be easily determined that the initial (undamaged) porosity of the paste fraction of a 0.45 water/cement ratio mixture is approximately equal to 37%. For that mixture, equation (64) thus yields:

$$\phi_{\substack{\text{cement paste upon ettringite}\\ \text{formation and porlandite}\\ \text{dissolution}}} = 0.37 - 0.079 + 0.175 = 0.47 \tag{65}$$

Using equation (53), the increase of the ionic diffusion coefficient (due to porosity changes) can be calculated as follows:

$$\frac{\left.\dfrac{D_i}{D_{io}}\right|_{\phi=47\%}}{\left.\dfrac{D_i}{D_{io}}\right|_{\phi=37\%}} = 2.2 \tag{66}$$

On the basis of the previous result, the coefficient (a) (see Figure 10) can be set to 2.2 and the value of the coefficient (b) is set to 10.

The initial (undamaged) cement paste fraction of a concrete produced at a water/cement ratio of 0.65 is of approximately 52%. On the basis of equation (64), one finds:

$$\phi_{\substack{\text{cement paste upon ettringite}\\ \text{formation and porlandite}\\ \text{dissolution}}} = 0.52 - 0.062 + 0.140 = 0.60 \tag{67}$$

The increase of the ionic diffusion coefficient (due to porosity changes) can be calculated using equation (53):

$$\frac{\left.\dfrac{D_i}{D_{io}}\right|_{\phi=60\%}}{\left.\dfrac{D_i}{D_{io}}\right|_{\phi=52\%}} = 2.8 \tag{68}$$

On the basis of the previous result, the coefficient (a) can be set to 2.8 and the value of the coefficient (b) is set to 10.

<u>Boundary and initial conditions</u>

In the next paragraphs, the boundary and initial conditions considered in the simulations are briefly discussed.

- *Description of the Concrete Slab*

All simulations were carried out assuming that the ionic species are transported through a concrete slab of 11.5 cm (see Figure 11). The bottom part of the slab was assumed to be directly in contact with the soil (containing SO_4^{2-}, Na^+ and Mg^{2+} ions) and the top part exposed to the external environment. The concrete slab was assumed to be initially undamaged.

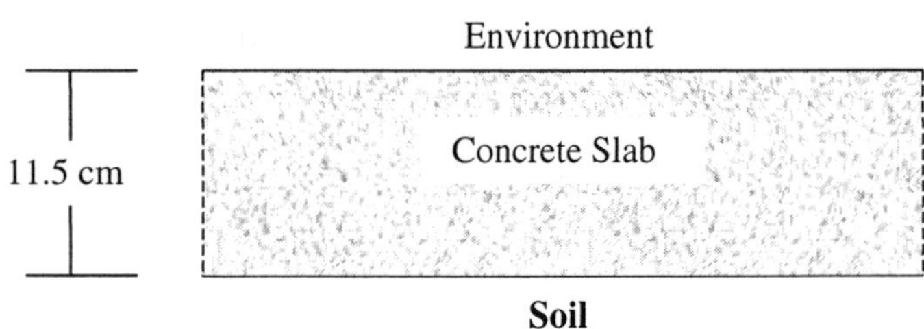

Figure 11 – Schematic representation of the concrete slab considered in the numerical simulations

- *Ionic concentrations of the soil solution*

Most of the simulations presented in the following paragraphs were performed assuming that the sulfate concentration of the ground water surrounding the bottom part of the concrete slab was equal to 4500 ppm. Some other simulations were also performed using a sulfate concentration equal to 0 and 7600 ppm.

For each simulation, the concentration in sodium of the soil solution was adjusted to balance the sulfate and the magnesium ionic content of the ground water. The concentrations in calcium and hydroxyl ions of the soil water were assumed to be equal to zero.

- *Relative Humidity*

In most simulations, the relative humidity of environment was fixed to 72%, and that of the soil was set equal to 85%. The conversion of the relative humidities of the air (on top of the slab) and the soil to the water content in the concrete was carried out according to equation (35):

For the 0.45 water/cement ratio concrete, equation (35) yields:

- Bottom part of the concrete slab:

$$\theta_w = (\Gamma_{concrete}\,\phi)\,\Omega(H = 85\%) = (0.315 \cdot 37\%) \cdot 0.60 = 0.0699 \ m^3/m^3 \qquad (69)$$

- Top part of the concrete slab:

$$\theta_w = (\Gamma_{concrete}\,\phi)\,\Omega(H = 72\%) = (0.315 \cdot 37\%) \cdot 0.50 = 0.0583 \ m^3/m^3 \qquad (70)$$

For the 0.65 water/cement ratio concrete, equation (35) yields:

- Bottom part of the concrete slab:

$$\theta_w = (\Gamma_{concrete}\,\phi)\,\Omega(H = 85\%) = (0.26 \cdot 50\%) \cdot 0.55 = 0.0715 \ m^3/m^3 \qquad (71)$$

- Top part of the concrete slab:

$$\theta_w = (\Gamma_{concrete}\,\phi)\,\Omega(H = 72\%) = (0.26 \cdot 50\%) \cdot 0.46 = 0.0598 \ m^3/m^3 \qquad (72)$$

In some cases, the concrete was assumed to be totally saturated throughout its service life.

- Initial conditions

The simulations were performed assuming that the total porosity of the 0.45 and 0.65 water/cement ratio concretes were 11.7% and 13.0% respectively.

In all simulations, the concrete was assumed to be initially fully water saturated. Further more, all the simulations have been carried out under isothermal conditions (T = 25° C).

The initial composition of the pore solution for both concrete mixtures were evaluated using the computer code developed by Reardon (1992). The initial pore solution compositions calculated on the basis of that code are given in Table 5.

Table 5 – Initial composition of the concrete pore solution

Species	W/C = 0.45 (mmol/l)	W/C = 0.65 (mmol/l)
Calcium	0.84	1.33
Hydroxyl	760	402
Sulfate	40.9	5.96
Sodium	246	122
Potassium	598	290

In order to predict the evolution of the amount of portlandite and calcium silicate hydrate in concrete upon the degradation process, the initial amount of each phase (for a fully hydrated concrete mixture) must be determined. This can be done on the basis of the equations previously presented. The amounts of portlandite and calcium silicate in a fully hydrated concrete are presented in Table 6.

Table 6 – Initial amount of C-S-H and portlandite in both concrete mixtures

Solid phase	W/C = 0.45 (mg/g concrete)	W/C = 0.65 (mg/g of concrete)
C-S-H	108.9	74.1
Portlandite	50.5	34.4

The maximum amount of sulfate that can be bound to the solid as ettringite can be calculated using equation (42). Considering the cement composition given in Table 2, one can calculate that the maximum mass of sulfate that can be bound is 103.4 mg/g of cement. This mass can be easily transformed on an unit mass of concrete basis:

- for the 0.45 water/cement ratio: 17.4 mg of bound sulfate/g of concrete
- for the 0.65 water/cement ratio: 11.8 mg of bound sulfate/g of concrete

Presentation of the numerical simulations

Twelve simulations were performed according to the conditions previously described:

1. The 0.65 water/cement ratio concrete with no sulfates in unsaturated conditions.
2. The 0.45 water/cement ratio with no sulfates in unsaturated conditions.
3. The 0.65 water/cement ratio concrete with 4500 ppm of sulfates in unsaturated conditions.
4. The 0.45 water/cement ratio with 4500 ppm of sulfates in unsaturated conditions.
5. The 0.65 water/cement ratio concrete with 7600 ppm of sulfates in unsaturated conditions.
6. The 0.45 water/cement ratio with 7600 ppm of sulfates in unsaturated conditions.
7. The 0.65 water/cement ratio concrete with no sulfates in saturated conditions.
8. The 0.45 water/cement ratio with no sulfates in saturated conditions.
9. The 0.65 water/cement ratio concrete with 7600 ppm of sulfates in saturated conditions.
10. The 0.45 water/cement ratio with 7600 ppm of sulfates in saturated conditions.

Numerical results and discussion

- General comments

Typical degradation curves for the two concrete mixtures are given in Figure 12. As can be seen, the various simulations clearly confirm that the mechanisms of sulfate attack yield to a major reorganization of the hydrated cement paste microstructure. In addition to the formation of the new sulfate-bearing products, the penetration of external ions markedly accentuates the dissolution of portlandite and the decalcification of C-S-H. As can be seen in the figure, the model predicts the progression of « fronts » (more or less precisely defined) from the bottom portion of the concrete (in contact with the sulfate-contaminated soil) to the core of the element.

The reorganization of the paste and more particularly the dissolution of Ca(OH) and the decalcification of C-S-H contribute to significantly increase the porosity of the material. Such an increase has numerous severe consequences on the concrete properties. First of all, it reduces the ability of the material to act as an effective barrier and favors the penetration of moisture and aggressive ions. The significant increase in porosity also affects the mechanical properties of the concrete. Considering that the degradation process progresses from the external surfaces to the core of the slab, the pore structure alterations should be particularly detrimental for the flexural resistance of the material. For a concrete slab subjected to a bending moment, the maximum stresses are developed at the extremities of the section. i.e. where the degradation process is more important.

It should be emphasized that the decalcification of C-S-H is, most probably, the most severe degradation that can happen to a concrete mixture. The gradual leaching of calcium from C-S-H leaves a residual silica gel. This gel is extremely porous and permeable and, it has no binding capacity.

The numerical simulations clearly underline that the kinetics of the degradation process is predominantly affected by the rate of transport of the various ions within the material. This can be easily visualized in Figure 13. As can be seen, the rate of penetration of the various fronts within the concrete is initially quite rapid and tends to be more or less attenuated with time according to the type of concrete considered. The rapid progression of the penetration fronts during the first years of exposure is related to the fact that the ionic concentration gradients are initially more important. It should be emphasized that the gradual reduction of the penetration rates of the various fronts are greatly dependent of the type of concrete considered. This point will be further discussed in the following paragraphs.

It is worth noting that none of the simulations presented in this paper predict the formation of gypsum in the concrete. This can be explained by the fact that the sulfate concentrations considered for the simulations are below the critical concentration required for gypsum formation.

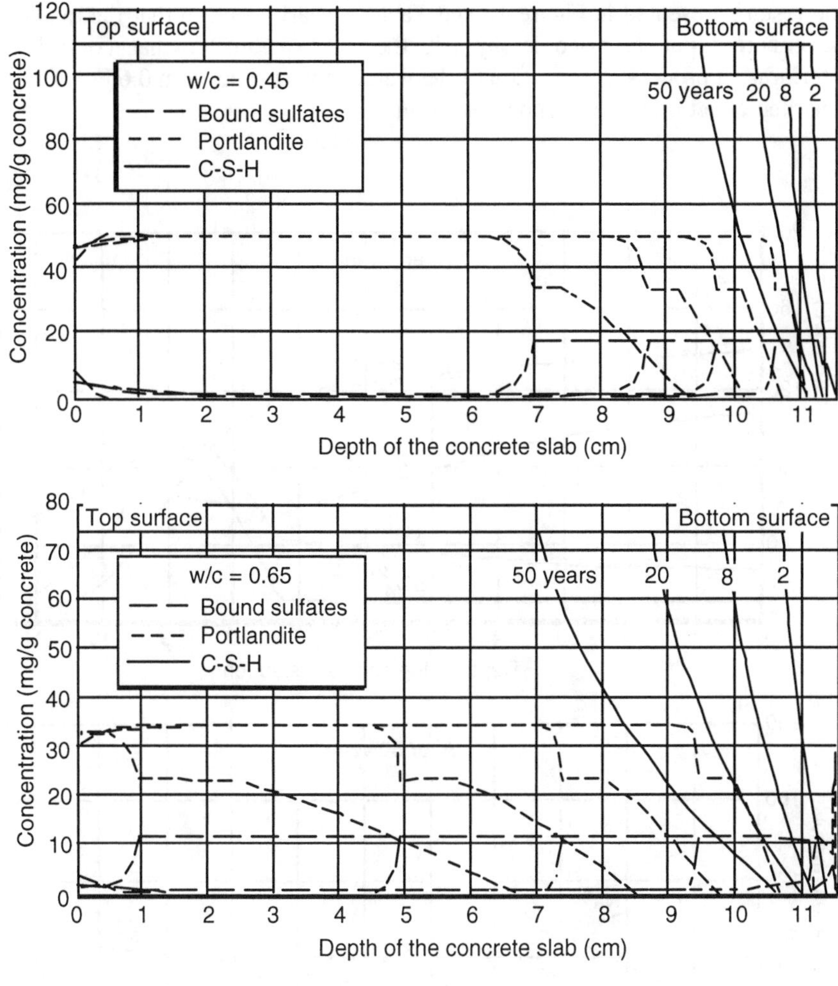

Figure 12 – Penetration of the degradation « fronts » in concrete

- Influence of the type of concrete

The numerical simulations can be used to investigate the influence of various parameters on the degradation process. The effect of the concrete quality and the concentrations in sulfate and magnesium ions in the soil water are briefly discussed in the following paragraphs.

The results presented in Figure 12 and Table 7 clearly confirm that the quality of the concrete put in place is, by far, the main parameter that controls the deterioration process. A reduction of the water/cement ratio from 0.65 to 0.45 has a marked effect on the rate of deterioration.

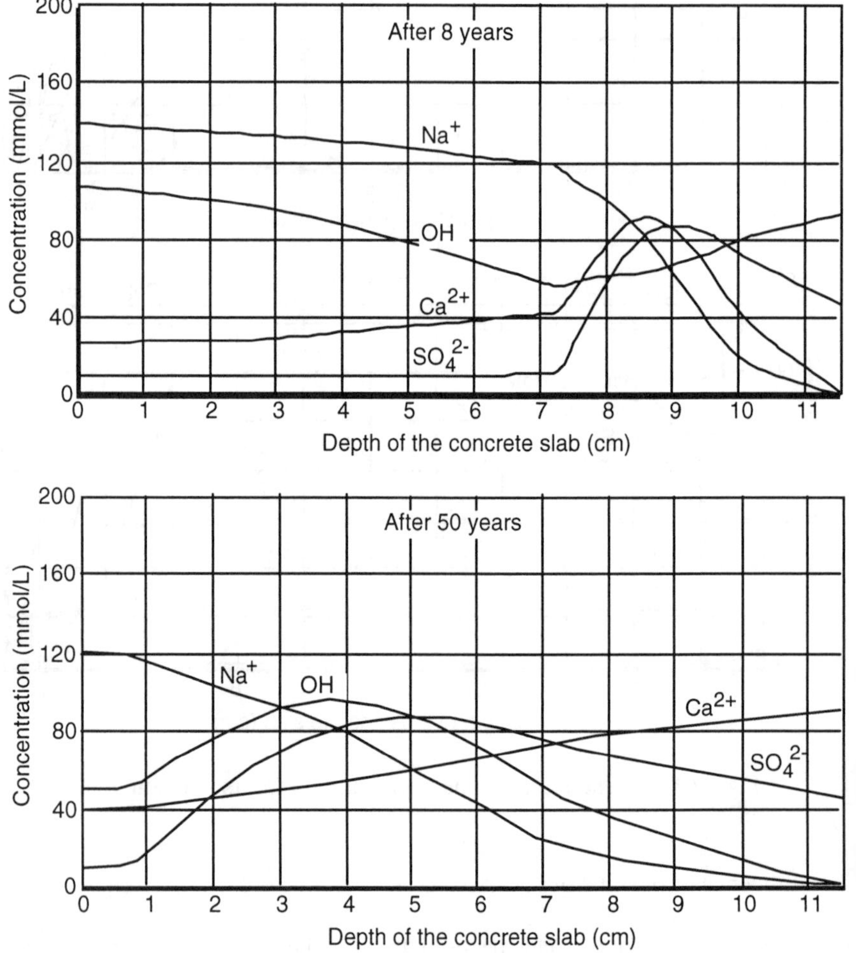

Figure 13 – Evolution of the pore solution chemistry of the 0.65 water/cement ratio concrete after 8 and 50 years

The numerical simulations clearly indicate that the degradation "fronts" tend to penetrate much more quickly in the 0.65 water/cement ratio concrete than in the 0.45 water/cement ratio reference mixture. As can be seen, the quality of the concrete put in place not only affects the intensity of the degradation but it also has a strong influence on the kinetics of degradation. While the degradation predicted for the reference (0.45 water/cement ratio) mixture tends to level off after a certain exposure period, that of the 0.65 concrete progresses more or less linearly with time.

Table 7 – Depth of penetration* (in mm) predicted by the model after 8 and 50 years of exposure

n°	W/C	External Sulfates Concentration	Water State	Time of exposure	Ettringite Formation	Portlandite Dissolution	C-S-H Dissolution
1	0.65	0	Unsaturated	8 years	n/a	30	12
				50 years	n/a	82	30
2	0.45	0	Unsaturated	8 years	n/a	10	6
				50 years	n/a	35	13
3	0.65	4500	Unsaturated	8 years	35	35	14
				50 years	85	85	40
4	0.45	4500	Unsaturated	8 years	16	16	8
				50 years	45	45	20
5	0.65	7600	Unsaturated	8 years	42	42	16
				50 years	105	105	40
6	0.45	7600	Unsaturated	8 years	20	20	8
				50 years	50	50	20
7	0.65	0	Saturated	8 years	n/a	45	17
				50 years	n/a	115	40
8	0.45	0	Saturated	8 years	n/a	14	7
				50 years	n/a	52	17
9	0.65	7600	Saturated	8 years	65	65	24
				50 years	115	115	60
10	0.45	7600	Saturated	8 years	29	29	10
				50 years	75	75	27

* Beginning of the « front » (the depths of penetration are measured from the bottom part of the concrete slab)

The particular behavior of the 0.65 water/cement ratio concrete is directly related to its high water/cement ratio that favors the transport of moisture and ions within the material. As previously discussed, the kinetics of the degradation process is predominantly influenced by the ability of the material to act as an effective barrier. In that respect, the simulations obtained for the reference concrete clearly support the requirements included in the various building codes (see for instance the Unified Building Code) that limits at 0.45 the maximum water/cement ratio for any concrete exposed to a sulfate environment.

It should finally be underlined that the reduction of the water/cement ratio from 0.65 to 0.45 should contribute to markedly improve the service life of the concrete slab. According to the curves showed at the end of the chapter, for the field concretes, the beginning of the $Ca(OH)_2$ dissolution "front" should the center of the slab after a few years of exposure (i.e approximately 15 years according to the exposure conditions). At this point, the damage done to the material is quite important. The dissolution of $Ca(OH)_2$ contributes to increase the porosity of the material and thus reduces its ability to act as an effective barrier. This increase in porosity should also detrimentally affect the mechanical properties of the material. As can be seen in the figures, in the case of the reference concrete mixtures, the progression of the $Ca(OH)_2$ dissolution front is markedly reduced.

- Influence of the sulfate concentration

In order to determine the influence of the sulfate content of the soil water on the degradation process, numerical simulations were carried out for three different sulfate concentrations: 0 ppm, 4500 ppm and 7600 ppm.

As can be seen in Figure 14, the results clearly underlines the very detrimental influence of sulfate ions on the degradation process. An increase of the sulfate concentration of the soil solution from 0 to 4500 ppm has a significant influence on the position of the portlandite dissolution « front ». Similar effects are also noted for the bound sulfate and the C-S-H decalcification « fronts ». This phenomenon is mainly related to the fact that the presence of sulfates ions contributes to the portlandite dissolution by forming new sulfate bearing phases such as ettringite and (eventually) gypsum.

The presence of sulfate ions in solution also seems to favor the leaching of hydroxyl ions. It should be emphasized that the model accounts for the electroneutrality effect by coupling the transport of all ionic species. The penetration of an external species in the concrete pore solution tends to significantly affect the local equilibrium and may well accelerate the leaching of another species. In that respect, it should also be underlined that the leaching of hydroxyl ions further contributes to accelerate the dissolution of Ca(OH)2 (and eventually the decalcification of C-S-H).

The increase in concentration from 4500 ppm to 7600 ppm appears to have a slight (but much less marked) influence on the degradation kinetics. The main reason for this phenomenon is that the concentration of 4500 ppm is already quite sufficient to create severe damage to the solid. As previously mentioned, the value of 7600 ppm is below the critical concentration above which gypsum can be formed in the solid.

- Influence of the soil water content

The influence of the of the water content of the soil is also illustrated in Figure 14. As can be seen, it appears that the worst case of degradation is found for the concrete kept in saturated conditions (i.e. when there is no water transport by advection). These results tend to indicate that the flow of water by capillary suction contributes to favor the penetration of sulfate ions but also reduces the kinetics of hydroxyl ion leaching. The overall effect is a reduction of the degradation kinetics. These results also tend to indicate that any protection or repair strategy that would contribute to keep the concrete fully saturated without totally impeding the transport of ions in the material will only increase the kinetics of degradation.

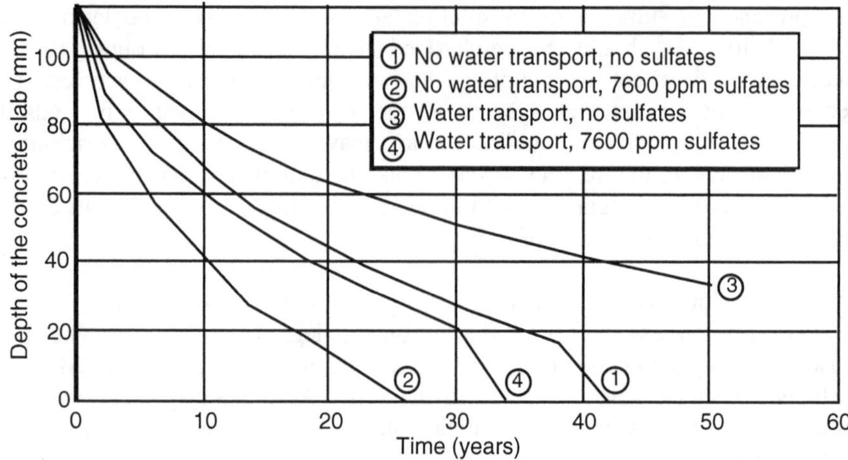

Figure 14 – Penetration of the $Ca(OH)_2$ dissolution « front » for various exposur conditions

CONCLUDING REMARKS

- The main features of a numerical model developed to predict th microstructural alterations of concrete subjected to external sulfate attac were presented. The model accounts for the coupled transport by diffusio and advection of four different ionic species.

- Numerical simulations yielded by the model confirm that the sulfate attac yields to a major reorganization of the hydrated cement paste microstructur In addition to the formation of the new sulfate-bearing products, th penetration of external ions markedly accentuates the dissolution c portlandite and the decalcification of C-S-H.

- The microstructural alterations induced by the degradation process contribut to significantly increase the porosity of the material.

- Numerical simulations clearly confirm that the quality of the concrete put i place is, by far, the main parameter that controls the deterioration process. A reduction of the water/cement ratio from 0.65 to 0.45 has a very significa effect on the rate of deterioration. The degradation fronts tend to penetrat much more quickly in the poor quality concrete than in the reference 0.45 water/cement ratio mixture.

ACKNOWLEDGMENTS

The authors are grateful to the Natural Sciences and Engineering Research Council of Canada and to Simco Technologies inc., for their financial support for this project. The authors would also like to thank V. Johansen for very fruitful discussions on the mechanisms of sulfate attack.

REFERENCES

Bazant, Z.P., Najjar, L.J., (1972). Nonlinear water diffusion in non-saturated concrete, Materials and Structures, Vol. 5, N° 25, pp. 3-20.

Bear, J., Bachmat, Y., (1991). Introduction to modeling of transport phenomena in porous media, Kluwer Academic Publishers, Netherlands, 553p.

Bear, J., Gilman A., (1995). Migration of salts in the unsaturated zone caused by heating, Letters in Mathematical Physics, Vol. 19, pp. 139-156.

Bentz, D.P., Garboczi, E.J., (1992a). Computer simulation of the diffusivity of cement-based materials, Journal of Material Science, Vol. 27, pp. 2083-2092.

Bentz, D.P., Garboczi, E.J., (1992b). Modelling the leaching of calcium hydroxide from cement paste: Effects on pore space percolation and diffusivity, Materials and Structures/Matériaux et Constructions, Vol. 25, pp. 523-533.

Borkovec, M., Burgisser, C.S., Cernik, M., Glatti, U., Sticher, H., (1996). Quantitative description of multicomponent reactive transport in porous media: an empirical approach, Transport in Porous Media, Vol. 25, pp.193-204.

Carpenter, T.A., Davies, E.S., Hall, C., Hall, L.D., Hoff, W.D., Wilson, N.A., (1993). Capillary water in roch: process and material properties examined by NMR imaging, Materials and Structure, Vol. 26, pp. 286-292.

Damidot, D., Glasser, F.P., (1993). Thermodynamic investigation of the CaO-Al_2O_3-$CaSO_4$-H_2O system at 25° C and the influence of Na_2O, Cement and Concrete Research, Vol. 23, N° 1, pp. 221-238.

Friaa, A. Mensi, R., Acker, P., (1987). Study of the simultaneous transfert of water and heat in concrete, Durability of Construction Materials, proceedings of the first RILEM congress, Chapman and Hall, pp. 1266-1273.

Fu, Y., (1996). Delayed ettringite formation in Portland cement products, Ph.D. Thesis, Department of Civil Engineering, University of Ottawa, 297 p.

Hazrati, K., (1998). Études des mécanismes de transfert d'humidité dans les matériaux cimentaires. Ph. D. Thesis, Université Laval, Canada. (in preparation - in French)

Helfferich, F., (1962). Ion exchange, McGraw-Hill, New York, USA, 624 p.

Jacobsen, S., Gérard, B. Marchand, J., (1996). Prediction of short time drying for OPC and silica fume concrete frost/salt scaling testing, Durability of Building Materials and Components VII (volume one), Published by E&FN Spon, London, pp 401-410.

Kalousek, G.L., Adams, M. (1951), Journal of the American Institute Proceedings, Vol. 48, pp. 77-89.

Li, G. (1994). Study of expansion mechanisms of concrete under sulfate attack - Behavior of radioactive wastes containing sulfates. Ph.D. Thesis, École Nationale des Ponts et Chaussées, France, 242 p., (in French).

Luikov, A.W., (1966). Heat and mass transfert in capillary-porous bodies, Pergamon press, Headington Hill Hall, Oxford.

Marchand, J., Gérard, B., Delagrave, A., (1998). Ion transport mechanisms in cement-based materials, Materials Science of Concrete, Vol. V, Edited by J. Skalny and S. Mindess, American Ceramic Society, pp. 307-400.

Nilsson, L.O., Poulsen, E., Sandberg, P., Sosenren. H.E., Klinghoffer, O., (1996). HETEK, Chloride penetration into concrete, State of the art: Transport processes, Corrosion initiation, Test methods and Predicting models, Report N° 53, Road Directorate, Denmark, 151 p.

Pel, L., (1995). Moisture transport in porous building materials, Ph. D Thesis, Technishche Universiteit Eindhoven, Netherlands, 127 p.

Perrin, B., Bonnet, S., (1995). Experimental results concerning combined transport of humidity and chloride in non steady state, International Workshop on Chloride Penetration into Concrete, October 15-18, Saint-Rémy-les-Chevreuse, France, 13 p.

Porcelli, P.C., Bidner, M.S., (1994). Simulation and transport phenomena of a ternary two-phase flow, Transport in Porous Media, Vol. 14, pp. 101-122.

Samson, E. (1998). Modeling the transport of ions in porous materials. Ph.D. Thesis, Department of Civil Engineering, Université Laval, Canada, (in preparation).

Samson, E., Marchand, J. (1998). Numerical solution for the extended Nernst-Planck model, submitted for publication to Journal of Colloid and Interface Science, 22 p.

Samson, E., Marchand, J., Beaudoin, J.J. (1998). Describing ion diffusion in cement-based materials using the homogeneization procedure, submitted for publication to Cement and Concrete Research, 12 p.

Samson, E., Marchand, J., Robert, J.L., Bournazel, J.P. (1998). Modeling the mechanisms of ion diffusion in porous media. submitted for publication to International Journal for Numerical Methods in Engineering, 25 p.

Silberbush, M., Sorek, S., Yakirevich, A., (1993). K^+ uptake by root systems grown under salinity (part I) – A mathematical model, Transport in Porous Media, Vol. 11, pp. 101-116.

Xi, Y, Bazant, Z.P., Jennings, H.M., (1994a). Moisture diffusion in cementitious materials: Adsorption isotherms, Advanced Cement-Based Materials, Vol. 1, N° 6, pp. 248-257.

Xi, Y, Bazant, Z.P., Molina, L., Jennings, H.M., (1994b). Moisture diffusion in cementitious materials: Moisture capacity and diffusivity, Advanced Cement-Based Materials, Vol. 1, N° 6, pp. 258-266.

Zienkiewicz, O.C., Taylor, R.L., (1989). The finite element method, McGraw-Hill, fourth edition, USA.

Diffusivity-porosity relation for cement paste

E.J. Garboczi and D.P. Bentz
226/B350, Building Materials Division
National Institute of Standards and Technology
Gaithersburg, Maryland 20899 USA

ABSTRACT

The development of a ionic diffusivity-porosity relation for cement paste, soundly based on microstructure, along with future developments in this area, are discussed. The importance of such relations for sulfate attack and other mechanisms of deterioration in concrete, as well as for models of the chloride diffusivity of concrete, are outlined.

INTRODUCTION

The dependence of ionic diffusivity on cement paste pore space characteristics is an important quantity in modelling and understanding ionic transport through concrete. An equation for water-saturated cement paste, based on microstructural modelling, was developed several years ago [1]. This equation has since been used as the heart of a multi-scale model for concrete diffusivity [2,3], and as a key component in a sulfate attack model described in the paper by J. Marchand et al. [4] in this proceedings. The present paper serves as a supplement to Marchand's paper, and discusses the development of this equation, and present work being done to refine and extend this equation.

DEVELOPMENT OF DIFFUSIVITY-POROSITY EQUATION

The equation to be discussed is the following:

$$D/D_o = H(\phi - 0.18)\ 1.8\ (\phi - 0.18)^2 + 0.07\ \phi^2 + 0.001 \qquad (1)$$

where D is the water-saturated bulk or effective ionic diffusivity measured by a barrier experiment, in which a slab of cement paste is used as the barrier through which ions can diffuse. D_o is the diffusivity of the ion in question in unconfined water, ϕ is the capillary porosity, and H(x) ($H(x > 0) = 1$, $H(x \leq 0) = 0$) is the Heaviside function. The three terms will be discussed in detail below, once the microstructural model used to generate this equation is introduced.

The microstructure model used to generate this equation is no longer used by us in that form, although it has been incorporated into improved models [5,6]. It was a model for C_3S hydration, where the C_3S particles formed specified amounts of CH and C-S-H [1,7]. No other phases were considered. It was a cellular automaton model, as are the present models. In this model, material first dissolved from the cement grains to form diffusing species. These species diffused through the pore space, and reacted and formed C-S-H material and CH crystals. Only after every diffusing species was reacted would a new dissolution cycle begin.

After a specified amount of hydration was carried out, as determined by the degree of hydration or by the capillary porosity reached, the resulting microstructure was saved in a computer file and read into a 3-D conjugate gradient program that would solve for the effective diffusivity, D/D_o, of the structure. The capillary porosity was given a diffusivity of D_o. The C-S-H phase was considered as a solid phase that had small but finite diffusivity, which was chosen so as to roughly match experimental data. The ultimate value of the C-S-H diffusivity was taken to be 0.0025 D_o. Since only the ratio D/D_o was being determined, the results were of course independent of the value of D_o. Many computations were carried out for w/c ratios ranging from 0.4 to 0.6, and degree of hydrations ranging up to 1. These data were all collected together on one graph, and showed that a universal curve could be obtained, depending only on capillary porosity, and not w/c ratio. Fitting this curve was done in steps, analyzing each part of the curve and its dependence on microstructural parameters. The three terms in eq. (1) were the result.

The first term, $H(\phi - 0.18)\ 1.8\ (\phi - 0.18)^2$, came from percolation considerations. This model was found to have a critical porosity of about 18%, below which the capillary porosity became disconnected [8]. If the C-S-H diffusivity was turned off, which can be easily done in this model, then it was found that the overall diffusivity also went to zero at this percolation threshold in a power law of the difference of the porosity from 0.18, with a power of 2, in agreement with percolation theory [9]. The prefactor of 1.8 was found by fitting this data. This term was then included in the fit of the overall data, with the Heaviside function added in order to turn off this percolation term when the capillary porosity was disconnected.

The third term was determined by considering fully hydrated systems, where the main diffusion was through the C-S-H phase. Using effective medium theory [1], it was determined that there was only a small dependence on w/c ratio at late stages of hydration, so a best-fit constant of 0.001 was used, which of course depends on the value chosen for the C-S-H diffusivity, 0.0025.

The second term, 0.07 ϕ^2, was chosen to connect the percolation behavior, which tends to dominate in early hydration, with the nearly constant diffusivity, which dominates the late hydration behavior. The form of this term was chosen to reflect an Archie's law-type form [1]. The value of the exponent and the prefactor were chosen from a fit to the master curve generated by simulation.

Further simulations of the C-S-H nanostructure have given values of diffusivity relative to D_o around the value of 0.0025, lending support to this choice [10], and comparisons with careful experiments on white cements (used because of similarities to pure C_3S cements) showed that eq. (1) was indeed reasonable. This equation has since been used for predicting portland cement paste diffusivity, with reasonable results [2,4], although in principle, the simulations that supplied the data for eq. (1) were only for C_3S cement pastes.

AREAS OF IMPROVEMENT NEEDED

Over the last six years, changes in computer memory and speed, and new knowledge, have forced us to begin re-evaluating eq. (1). These reasons are discussed below.

The original microstructure model based its volume stoichiometry for the C_3S-water reaction on work by Young and Hansen [12]. It has since been realized, by those investigators and by us, that the volume of C-S-H produced per volume of C_3S reacted, 1.7, did not properly take into account chemical shrinkage. A more correct value of this number is 1.52 [5]. Also, the original model had the dissolution and diffusion/reaction cycles occurring separately. The latest version of the cement paste hydration model has these cycles occurring simultaneously, more in accord with the real experimental situation.

In Refs. [1] and [8], a cement particle size distribution was used that we have come to realize is quite unrealistic. This was four sizes of particles, with equal numbers of all four sizes. It is much more realistic to have approximately equal masses of different size particles, so that the number of small particles is enormously larger than the number of large particles. Recent results have shown that the capillary percolation threshold depends somewhat on the particle size distribution used, ranging from about 0.18, for the unrealistic particle size distribution described above, to 0.22 for more realistic size distributions [13,14]. The exact value of the capillary porosity percolation threshold will affect the form of eq. (1) to some extent.

Experimental evidence has shown that there are some differences, on the order of a factor of two, between white cement and portland cement pastes for the value of D/D_o, especially at late degrees of hydration [15]. The microstructural model can now do portland cement hydration, so this difference needs to be checked carefully. Also, at late degrees of hydration, the value of D/D_o obtained in the model is dominated by the value of diffusivity assigned to the C-S-H phase. This value needs to be re-evaluated as well in light of the experimental evidence and the changes in the model.

All the percolation and diffusivity work that has been done with the old microstructural model, on which eq. (1) is based, used 100^3 size systems, where each pixel had an edge length of one micrometer. Therefore the smallest capillary pore was of necessity one micrometer. This choice reflected the computing power of the computers available to us at that time. Computers have changed enormously over the past 6-8 years, in memory and computing speed, and so we are now able to test the percolation and diffusivity results obtained from the model at different choices of resolution, probably up to 1000^3 size models or a smallest pore size of 0.1 micrometers.

Discussion

All of the improvement needs listed in the last section are being carried out at present. We would like to emphasize that we do not expect huge qualitative changes in eq. (1). In fact, the early hydration parts of eq. (1) will probably change very little. It is in the late hydration stages that improvements are needed and hopefully will be seen. At the very least, the new version of eq. (1) obtained will be on a sounder scientific footing than the old version.

Qualitatively, the present form of eq. (1) is to be expected. This is because the percolation behavior of the capillary porosity must dominate the early hydration diffusivity. The competition between the well-percolated but low diffusivity C-S-H sets the scale for the intermediate hydration stage, and the increasing C-S-H phase and decreasing, disconnected capillary porosity phase must determine the late hydration diffusivity. We hope to correct and improve the quantitative details of these stages represented in eq. (1).

Finally, a possibility should be mentioned that is at present beyond the capabilities of our model, but may have to be dealt with in the future. That is the possibility of two kinds of C-S-H, present in different amounts and locations at different hydration times [16]. This would of course affect the diffusivity. The difficulty would be in determining the kind of C-S-H a particular pixel should be, as the diffusivities to be assigned would probably be very different. At present, since further specific information is not available, all the C-S-H in the microstructural model is given the same diffusivity. However, the model is easily capable of handling several different kinds of C-S-H, due to its cellular automaton, digital-image-based structure.

One main use of eq. (1) is as a key step in predicting the difusivity of concrete in a multi-scale model [2,3]. In Marchand's sulfate attack model, changes in the porosity of a cement paste due to consumption or growth of various compounds change the overall diffusivity of the microstructure via eq. (1). CH leaching has a similar effect on diffusivity via changes in porosity amount and topology [17,18]. The proposed re-evaluation of eq. (1) almost certainly will not qualitatively change these results, but should introduce some quantitative changes, on the order of a factor of two, at the later stages of hydration. We have also carried out some work on modifying eq. (1) for partially-saturated cement pastes, but this work is only preliminary for now [19].

REFERENCES

*Note: Refs.1,2,3,6,8,10,11,and 17 are also available on-line at http://ciks.cbt.nist.gov/garboczi/

[1] E.J. Garboczi and D.P. Bentz, "Computer simulation of the diffusivity of cement-based materials," J. Mater. Sci. **27**, 2083-2092 (1992).

[2] D.P. Bentz, E.J. Garboczi, and E.S. Lagergren, "Multi-scale microstructural modelling of concrete diffusivity: Identification of significant variables," ASTM Cement and Aggregates **20**, 129-139 (1998).

[3] E.J. Garboczi and D.P. Bentz, "Multi-scale analytical/numerical theory of the diffusivity of concrete," Adv. Cem. Based Mater. **8**, 77-88 (1998).

[4] J. Marchand, E. Samson, and Y. Maltais, "Modeling microstructural alterations of concrete subjected to external sulfate attack," in these proceedings.

[5] D.P. Bentz, "Three-dimensional computer simulation of portland cement hydration and microstructure development," J. Amer. Ceram. Soc. **80**, 3-21 (1997).

[6] D.P. Bentz, "Guide to using CEMHYD3D: A three-dimensional cement hydration and microstructure development modelling package," NISTIR 5977 (1997).

[7] D.P. Bentz and E.J. Garboczi, "Guide to using HYDRA3D: A three-dimensional digital-image-based cement microstructural model," NISTIR 4746 (1992).

[8] D.P. Bentz and E.J. Garboczi, "Percolation of phases in a three-dimensional

cement paste microstructural model," Cem. Conc. Res. **21**, 325-344 (1991).
[9] D. Stauffer and A. Aharony, *Introduction to Percolation Theory* (Taylor and Francis, London, 1992).
[10] D.P. Bentz, D.A. Quenard, V. Baroghel-Bouny, E.J. Garboczi, and H.M. Jennings, "Modelling drying shrinkage of cement paste and mortar: Part I. Structural models from angstroms to millimeters," Mater. Struc. **28**, 450-458 (1995).
[11] B.J. Christensen, T.O. Mason, H.M. Jennings, D.P. Bentz, and E.J. Garboczi, "Experimental and computer simulation results for the electrical conductivity of portland cement paste," in *Advanced Cementitious Systems: Mechanisms and Properties, Vol. 245*, edited by F.P. Glasser, G.J. McCarthy, J.F. Young, T.O. Mason, and P.L. Pratt (Materials Research Society, Pittsburgh, 1992), pp. 259-264.
[12] J.F. Young and W. Hansen, in *Microstructural Development During Hydration of Cement*, edited by L.J. Struble and P.W. Brown (Materials Research Society, Pittsburgh, 1987).
[13] D.P. Bentz and C.J. Haecker, "An argument for using coarse cements in high performance concretes," Cem. Conc. Res., in press (1999).
[14] E.J. Garboczi and D.P. Bentz, "Re-examination of capillary porosity percolation in a cement paste microstructural model," to be submitted to Cem. Conc. Res. (1998).
[15] B.J. Christensen, Ph.D. thesis, Northwestern University (1992).
[16] C.M. Neubauer, T.B. Bergstrom, K. Sujata, Y. Xi, E.J. Garboczi, and H.M. Jennings, "Drying shrinkage of cement paste as measured in an ESEM and comparison to microstructural models," J. Mater. Sci. **32**, 6415-6427 (1997).
[17] D.P. Bentz and E.J. Garboczi, "Modelling the leaching of calcium hydroxide from cement paste: Effects on pore space perclation and diffusivity," Mater. Struc. **25**, 523-533 (1992).
[18] S. Rémond, "Evolution de la microstructure des bétons contenant des déchets au cours de la lixiviation," Ph.D. thesis, L'Ecole Normale Supérieure de Cachan, 1998.
[19] N. Martys, "Diffusion in partially saturated porous materials," Mater. Struc., in press (1999).

DAMAGE ASSESSMENT AND SERVICE LIFE PREDICTION OF CONCRETE SUBJECT TO SULFATE ATTACK

J.W. Ju and L.S. Weng
Department of Civil and Environmental Engineering
University of California, Los Angeles
Los Angeles, CA 90095-1593, U.S.A.

S. Mindess and A.J. Boyd
Department of Civil Engineering
University of British Columbia
Vancouver, B.C., Canada V6T 1Z2

ABSTRACT

Under external sulfate attack, Portland cement concrete suffers not only the formation of new sulfate compounds such as ettringite and gypsum, but also the deleterious consequences of the dissolution of calcium hydroxide and the decalcification of the C-S-H phase. As a result, porosity and permeability increase. Under combined sulfate attack and mechanical loading, distributed microcracks and microvoids increase significantly, leading to reduced concrete durability and substantial losses in elastic stiffness and tensile strength. Both nondestructive and tensile testing are employed in this study. The detrimental consequences of sulfate attack on the quality and strength of concrete (even with Type V cement) are demonstrated for concrete with high water/cement ratios.

1. INTRODUCTION

Under aggressive external sulfate attack, Portland cement concrete is known not only to undergo the formation of new sulfate compounds such as ettringite and gypsum, but also to suffer the deleterious consequences of the dissolution of calcium hydroxide $Ca(OH)_2$ and the decalcification of the C-S-H phase. We refer to Mindess and Young (1981), Bonen and Cohen (1992a, 1992b), Gallop and Taylor (1992, 1994, 1995, 1996), Mehta and Monteiro (1993), Clifton and Pommersheim (1994), Ferraris et al. (1995), Carde et al. (1996), and Neville (1996) for a literature review. As a result, disruptive expansion and microcracking occur, and porosity and permeability increase, particularly for concretes with high water/cement ratios; see, e.g., Hearn et al. (1994), and Khatri et al. (1997). Under combined sulfate attack and mechanical loading, distributed microcracks and microvoids increase significantly, leading to reduced concrete durability and substantial reduction in its elastic stiffness and tensile strength, principally due to the loss of cohesion in the hydrated cement paste and of adhesion between it and the aggregates (cf. Ouyang et al., 1988; Ouyang, 1989; Gerard et al., 1993; Neville, 1996). Therefore, it is of great interest to assess the extent of elastic degradation and irreversible damage caused by sulfate attack in concrete components and systems, and to perform remaining service life predictions; see, e.g., Cohen and Mather (1991), Piasta and Schneider (1992), and Mehta (1997). Both nondestructive testing and tensile testing are employed in this study. The detrimental consequences of sulfate attack on the quality and strength of concrete (even with Type V cement) are demonstrated for concretes with high water/cement ratios.

Methods for predicting the remaining service life of concrete have been proposed in the literature; see, for example, Clifton (1991, 1993), and Clifton and Pommersheim (1992). On the other hand, according to the Uniform Building Code (1985, 1988), concrete to be exposed to sulfate-containing solutions or soils shall conform to the requirements of UBC Table 26-A-6, which is based on the ACI 201 document (Reading et al, 1977). Specifically, in severe sulfate exposure (1,500 to 10,000 ppm of SO_4 in water), the maximum allowable water/cement ratio is 0.45 with Type V cement.

First, we employ the Schmidt rebound hammer test to assess the relative quality and degradation for concrete foundations in-situ. We subsequently employ a high-resolution ultrasonic pulse velocity technique to estimate the effective elastic moduli of damaged concrete samples taken from *actual* concrete structures (such as slabs and foundations). The reduction in effective Young's moduli and normalized layered profile are calculated and plotted cm-by-cm to represent the extent of damage in an actual concrete core or sample from the top to the bottom. Effects of moisture content, aggregate settlement, and pulse frequency are also investigated for degraded concrete specimens subject to sulfate attack. Direct tension tests are performed to assess concrete damage in tension. Based on original concrete 28-day compressive strengths and current tensile strengths remaining service life predictions are also performed to assess the durability reduction and failure time in concrete components. Furthermore, according to our comprehensive finite element damage modeling and simulations (based on Ju and Taylor, 1988; Ju, 1989a, 1989b; Ju et al., 1989), the uniaxial compression test is *not* indicative or suitable for layer-damaged concrete cylinders. This is due in part to the frictional lateral confinement effect between the steel platen of the testing frame and the top and bottom surfaces of a concrete cylinder. Furthermore, compressive loading tends to *close* cracks in concrete perpendicular to the loading direction, thus masking the effects of layered damage on the true strength of concrete.

2. SCHMIDT REBOUND HAMMER TESTING

The Schmidt rebound hammer was developed in Switzerland in 1948 by Ernst Schmidt (1950), and has since become by far the most widely used method for the nondestructive *in situ* testing of concrete. The device is a surface hardness tester. Its construction and operation have been described in detail by Malhotra (1991) and by Mindess and Young (1981). The Schmidt rebound hammer measures the rebound distance of a hammer mass which is fired, by the energy stored in a spring, against a steel plunger held firmly in contact with the concrete surface. The rebound distance is recorded as a "rebound number", running from "10" to "100" on an arbitrary scale. During the impact event, a large compressive stress wave and a complex series of reflected waves are produced both in the steel plunger and in the concrete, as shown by Akashi and Amasaki (1984). While no theoretical relationship between the stress wave (or the rebound number) and the concrete strength has been established, there is an empirical correlation between the compressive strength and the rebound number: a higher number implies a stronger concrete, at least near the surface. According to BS 1881: Part 202: 1986, the rebound distance is affected by only about the outermost 30 mm of the concrete being tested.

It is well known that the rebound number is affected by a number of factors, including the surface finish of the concrete, the type of formwork against which the concrete was cast, the age of the concrete, the size and rigidity of the concrete member, the type of coarse aggregate, the moisture conditions of the concrete, and the depth of carbonation. These effects are by now reasonably well understood (Malhotra, 1991; Mindess and Young, 1981) and will not be discussed further here. It cannot be emphasized too strongly that the rebound hammer is ***not*** an appropriate test method for determining the real *in situ* strength of the concrete. Even using the detailed procedures for correlating rebound numbers with the measured strengths of cores outlined in ACI 228.1R-89 (1994), the compressive strength of the concrete can then only be estimated at best to within about +/-

20%. Moreover, such correlation data are rarely available for field (as opposed to laboratory) studies. Thus, the rebound hammer is best used to determine *differences* in the concrete being examined and to survey, rapidly and inexpensively, large areas of nominally similar concretes in the structure that is being investigated. As stated in ASTM C805-94 (1994), "This test method may be used to assess the in-place uniformity of concrete, to delineate regions in a structure of poor quality or deteriorated concrete, and to estimate in-place strength development". In addition, if differences in rebound number are found, the test provides no information as to the reasons for these differences, which must be determined by other means.

However, despite these limitations of the rebound test, in the hands of a skilled operator the rebound hammer can provide valuable information about the relative qualities of the concretes in a structure, and can be used both to help guide further testing, and to corroborate the results of other tests.

2.1. Analysis of concrete degradation in foundation footings

The Schmidt rebound hammer has recently been used to help identify areas in which concrete degradation in foundation footings had occurred. Other investigations of this concrete (which will not be reported here) had suggested the likelihood of external sulfate attack, since the soils and groundwater at the site contained significant quantities of sodium and magnesium sulfates. Unfortunately, there was insufficient information available regarding the precise mix design and raw materials that were used in the original construction to permit concrete specimens to be cast to replicate the original concrete, and thus to provide even a rough correlation between the rebound numbers and concrete strengths. Moreover, for the reasons stated above, compression tests on drilled cores were considered unreliable in this situation for use in establishing a rebound number vs. strength relationship. Thus, in this particular investigation, the rebound hammer was used primarily to see whether the rebound numbers could be used to determine whether there were systematic variations in the subject concretes.

To this end, pits were dug in order to expose the full depths of the footings (which were typically about 60 cm deep) at a large number of locations within the housing developments being investigated. At each location, rebound numbers were taken (following the ASTM C805-94 procedures) in a vertical line: (1) just *above* grade where the concrete had not been in direct contact with the soil or groundwater; (2) a few cm *below* grade; and (3) as near to the bottom of the footing as was practicable. Comparisons were then made amongst the three sets of readings at each location, to see whether there was a systematic difference in the rebound numbers. As may be seen from the representative data summarized in **Table 1**, a consistent pattern did emerge. At each of the more than 100 locations examined, the rebound numbers above grade were always the highest, and those nearest the bottom of the footings were always the lowest; intermediate values were found for readings taken just below grade. Even when the rebound numbers below grade were adjusted to take into account the fact that they were generally taken on wet concrete, compared to those taken on dry concrete above grade (which was done by increasing the wet concrete readings by 5 units; see Zoldners, 1957), the differences in readings were still significant. That is, at each location, where the concrete can be assumed to have been cast at the same time and cured in the same manner, there is a consistent decrease in rebound number with increasing depth. This is indicative of a degradation mechanism in which the damage is working its way up from the bottom of the footing. This is consistent with the chemical and scanning electron microscopy investigations which showed that sulfate attack was indeed occurring, and with the other test data presented below.

As noted earlier, the rebound numbers reported here cannot be used to determine, even approximately, the "true" strength of the concrete. Indeed, as may be seen from **Table 1**, while the sets of rebound numbers all show the same pattern, their absolute values vary considerably, from 40 to 55 above grade, and from 17 to 33 below grade. [Using the conversion tables supplied with the Schmidt hammer, this would translate to *apparent* "compressive strengths" from 34.4 MPa to 58.6 MPa above grade, and from 5 MPa to 24.2 MPa below grade]. It is unlikely that the concrete strengths are as widely variable as this, since similar mix designs were used throughout the project. The differences in the absolute values of the rebound numbers at different locations are more likely due to local differences in mix composition, in curing, or in the degree of carbonation. However, this does not affect the conclusion that the concrete below grade, in contact with sulfate-bearing soils and groundwater, has degraded compared to the concrete above grade.

It should be noted that these rebound numbers were all obtained at a single point in time. It would have been useful to try to repeat these readings after a time interval of one or two years, for an indication of the *rate* of concrete degradation. Unfortunately, it has not been possible to carry out this further study.

3. ULTRASONIC NONDESTRUCTIVE TESTING OF CONCRETE

The ultrasonic pulse velocity through concrete is well documented in ASTM C597-83 (1983) and RILEM NDT1 (1972). Our objective here is to detect the layer-by-layer (along the cm by cm grid) stiffness degradation (damage) in concrete specimens. That is, by using high spatial resolution, we are able to determine the layer-by-layer normalized profile of the pulse wave speed and Young's modulus variations in a concrete specimen by ultrasonic nondestructive testing. Furthermore, we can employ empirical formulae to relate the pulse velocities (or the dynamic Young's moduli) to the uniaxial compressive strengths, thus assessing the qualitative trend of (sulfate-attack induced) damage in concrete. We refer to Mindess and Young (1981), Sturrup et al. (1984), and Neville (1996) for some details.

3.1. Ultrasonic measurement setup and procedure

The placement of ultrasonic transducers upon the concrete specimens is shown in **Figure 1**, and the state-of-the-art ultrasonic NDT setup is shown in **Figure 2** (cf. Ju and Weng, 1998). The function generator is SRS Model DS-345; the power amplifier is Digital Wave Model UTA 3000; the broadband pre-amplifiers are DW Model PA 2040G/A; the filter trigger module is a DW model; and the computer data acquisition system includes a modern PC with DW A/D converter and LPWA software. In addition, small (1 cm in diameter), broadband, high fidelity, low frequency, contact type transducers (DW model B425 or B225) are used to measure the pulse wave velocities and attenuation of signals. The first arrival signals of ultrasonic bulk waves are very clear, pin-pointed by the mouse cursor, and automatically digitized. We perform swept sine-wave face-to-face sensor calibration before measuring each concrete specimen. For a cylindrical concrete core, we perform four *radial* ultrasonic measurements at each centimeter layer along the curved surfaces. Subsequently, we compute the mean, the standard deviation, and the coefficient of variation of the four radial measurements at each layer. The reduction in pulse velocities, effective Young's moduli, and *normalized* layered profiles are calculated along the cm-by-cm mesh to represent the extent of damage in a cored concrete sample from the top (or outside) to the bottom (or inside).

Since concrete is a dissipative medium, both the pulse wave velocity and wave attenuation (such as the peak-to-peak amplitude and frequency spectrum) are important. The pulse velocity and wave attenuation are affected by the microstructure and damage in a particular concrete specimen, such as the leaching of calcium hydroxide, the decalcification of C-S-H, the increase in crack and void densities and sizes, and the mixture type (e.g., the aggregate type, size and volume fraction), etc. It is therefore desirable to employ *low-frequency* ultrasonic transducers set at a low frequency (such as 50 KHz) in order to achieve low wave attenuation and high penetration power through damaged concrete specimens. **Figure 3** displays typical time domain signals in channel 1 (without concrete) and channel 2 (through concrete).

3.2. Pulse wave velocity measurements and normalized layered profiles for damaged concrete specimens

According to our study, the calculated values of the dynamic Young's moduli E_d are *not* very sensitive with respect to the Poisson's ratio ν, if ν is between 0.2 to 0.3. The Poisson's ratio of normal concrete is usually about 0.24. Effects of moisture content (Sturrup et al., 1984), large aggregate settlement, and pulse frequency upon measured pulse wave velocities are also investigated in this study for degraded concrete specimens subject to sulfate attack. **Figure 4** and **Figure 5** exhibit the layer-by-layer (1 cm by 1 cm mesh) *normalized* pulse velocity and Young's modulus profiles, respectively, for a representative damaged 4-in. diameter foundation core (Sample No. 4) under sulfate attack. It is clear that, near the bottom layers, the Young's moduli have degraded significantly (about 20%) relative to the top layer. Moreover, **Figure 6** shows the effects of moisture on the measured pulse velocities for a damaged 4-in. diameter garage *slab* core (Sample No. 6), under oven-dry, as-is (air-dry) and fully wet moisture conditions. We observe that the moisture effect accounts for about 5% to 10% differences in measured pulse velocities. However, in terms of the *normalized* profiles under different moisture conditions, the results are very similar; cf. **Figure 7**. The frequency effect (250 KHz vs. 50 KHz) causes about a 3% difference.

4. TENSILE TESTING OF CONCRETE

As previously stated, the uniaxial compression test is not suitable for detecting or representing damage and degradation in concrete under chemical attack. In addition, in an actual concrete structure or system, concrete components are more likely to deform and fail under flexural bending, instead of under direct compression. Therefore, it is more appropriate and indicative to assess the damage and degradation in concrete under sulfate attack by means of *tensile* testing. These include the direct tension test (RILEM CPC7, 1975; U.S. Bureau of Reclamation 4912-92, 1992; ASTM D2936-95, 1995), the splitting tension test (ASTM C496-

96, 1996), and the flexural tension test (ASTM C78-94, 1994). See also Kuzmar (1994) for sulfate attack on Portland cement paste via fracture mechanics concepts. For a cored concrete cylinder of circular cross-section, the direct tension test is more suitable for detecting and assessing the *layered* damage and degradation (along the longitudinal direction of the core) due to chemical alteration in the cement paste under sulfate attack.

4.1. Direct tension test

The guidelines in ASTM D2936-95 and RILEM CPC7 are followed here. We prepare concrete cores by trimming (using a diamond-edge blade wet saw) their bottom and/or the top parts to form smooth and parallel surfaces. A high-strength epoxy adhesive is employed to glue the top and bottom surfaces of the trimmed cores to the circular metal end plates. An accurate glue alignment fixture is used to guarantee the alignment of the concrete/ end plate assembly. Subsequently, the concrete/ end plate assembly is properly cured to achieve the desired high bonding strength. The metal end plates have precision-machined screw holes, which the tension grip fixture assembly can connect to and center easily and automatically. The direct tension testing system consists of a 22-Kips MTS loading frame, an Instron automated control panel, and a computer data acquisition system. See **Figure 8** for an illustration. The direct tension test is performed under computerized displacement control.

4.2. Analysis of test results

The direct tensile strengths measured from the direct tension testing need further corrections by using appropriate correction factors. These include the corrections for the size effects (Bazant et al., 1991; Tang et al., 1992; Rossi et al, 1994), the moisture effect, the maturation effect, and the "core" effect. The tensile fracture typically occurs suddenly during the direct tension test. Along the fracture surfaces, we observe mostly aggregate pull-out. The direct tensile strength is approximately 75% of the flexural tensile strength (the modulus of rupture).

5. SERVICE LIFE PREDICTIONS BASED ON TENSION TESTING

According to Shah et al. (1995), a structure is built to serve a certain function. If it does not in some way fulfill its intended purpose, then it has failed. In most civil engineering structures, failures may be classified as serviceability or strength. In the case of serviceability, the designer, owner and user are generally concerned with deflections, appearance and environmental impact, which must be satisfied for the desired economic life of the structure. Overriding these serviceability concerns usually is the question of safety, or the strength of the structure. The term "strength" implies some kind of limit (usually related to level of internal stresses) beyond which additional loads will cause material and structural distress (Shah et al., 1995). On the other hand, following the work of Clifton (1991, 1993) and Clifton and Pommersheim (1992), we assume a simplified *linear* damage accumulation law for sulfate attack induced damage. That is, conservatively, we assume a *constant* rate of sulfate induced damage per year during the service life of concrete foundation. Usually, the accumulation of sulfate-induced damage is nonlinear and accelerated in subsequent years.

The tensile cracking (strength) criterion can be established by comparing the current tensile strength f_t (after corrections) with the critical (design) tensile "cracking" stress f_{crack}. When f_t is less than f_{crack} at a location x, the concrete material at x initiates a tensile crack (fracture).

5.1. Reinforced concrete slabs and foundation footings

According to the design strength of concrete in the UBC (1985, 1988), the strength reduction factor ϕ is commonly taken as 0.9 for a concrete member or cross-section under flexure. The "original" (undamaged) modulus of rupture is denoted by f_{cr}. It is logical to write $f_{crack} = 0.9 \times f_{cr}$. From the UBC (1985, 1988), the modulus of rupture of concrete is commonly taken as $f_{cr} = 7.5\sqrt{f_c'}$ (psi). For example, if the minimum specified compressive concrete strength is f_c' = 2,000 psi, then we have the minimum modulus of rupture f_{cr} = 335 psi. The actual average f_c', however, is determined to be 2,730 psi at 28-day by the compressive testing. Therefore, the likely actual average f_{cr} is about 392 psi. The minimum critical (design) tensile "cracking" stress f_{crack} is thus taken as 302 psi. If the current corrected f_t is greater than f_{crack}, but less than the actual average f_{cr}, then the concrete has sustained some damage. Accordingly, we can compute the remaining service life (in number of years from the present) it will take for a concrete specimen to reach the minimum critical (design) tensile "cracking" stress f_{crack} under continued sulfate attack. The concrete foundations under investigation are assumed to contain Type V cement and high water/cement ratios around 0.65.

Again, as a conservative measure, we assume a *linear* damage accumulation law, which is likely to provide an *upper-bound* value in our remaining service life prediction. The formula for predicting the remaining service life is thus derived as:

$$remaining\ life = \frac{f_t - f_{crack}}{\Delta d \times f_{cr}}$$

where

$$\Delta d = \frac{1 - \frac{f_t}{f_{cr}}}{age\ (yr.)}$$

For example, if the corrected direct tensile strength is 245 psi (after 7 years of service in situ), then the equivalent modulus of rupture is about 325 psi. Therefore, according to the above formula, the remaining service life is about 2.4 years from the present (a likely upper bound value). Based on our testing and analysis of numerous concrete foundation and garage cores, the majority of concrete specimens in this investigation either has failed already or has little remaining service life due to in-situ sulfate attack.

6. CONCLUSION

Concrete is known to suffer the deleterious consequences of the dissolution of calcium hydroxide and the decalcification of the C-S-H phase under sulfate attack. Disruptive expansion and microcracking occur, and porosity and permeability increase, particularly for concrete with high water/cement ratios. Under combined sulfate attack and mechanical loading, distributed microcracks and microvoids increase significantly, leading to reduced concrete durability and substantial reduction in its elastic stiffness and tensile strength. Both nondestructive testing and tensile testing are employed in this study to assess the extent of elastic degradation and irreversible damage caused by sulfate attack in concrete components and systems. Furthermore, the remaining service life predictions are performed. The detrimental consequences of sulfate attack on the quality and strength of concrete (even with Type V cement) are demonstrated for concrete with a high water/cement ratio. It is noted that the degradation in tensile strength is much more severe than in compressive strength for damaged concrete under sulfate attack.

7. REFERENCES

[1] ACI 228.1R-89 (1994), "In-place methods for determination of strength of concrete", *ACI Manual of Concrete Practice – Part 2*, American Concrete Institute, Detroit, MI.

[2] Akashi, T. and Anasaki, S. (1984), "Study of the stress waves in the plunger of a rebound hammer at the time of impact," in V.M. Malhotra, ed., *In Situ/ Nondestructive Testing of Concrete,* ACI SP-82, American Concrete Institute, Detroit, MI., pp. 17-34.

[3] ASTM (1983), "Standard test method for pulse velocity through concrete", ASTM C597-83, *Annual Book of ASTM Standards*, pp. 286-288.

[4] ASTM (1994), "Standard test method for rebound number of hardened concrete," *Annual Book of ASTM Standards*, ASTM C805-94, Vol. 04.02, pp. 393- 395.

[5] ASTM (1994), "Standard test method for flexural strength of concrete (using simple beam with third-point loading", ASTM C78-94, *Annual Book of ASTM Standards*, pp. 31-33.

[6] ASTM (1995), "Standard test method for direct tensile strength of intact rock core specimens", ASTM D2936-95, *Annual Book of ASTM Standards*, pp. 272-274.

[7] ASTM (1996), "Standard test method for splitting tensile strength of cylindrical concrete specimens", ASTM C496-96, *Annual Book of ASTM Standards*, pp. 263-266.

[8] Bazant, Z.P., Kazemi, M.T., Hasegawa, T., and Mazars, J. (1991), "Size effect in Brazilian split-cylinder tests: measurement and fracture analysis", *ACI Materials Journal,* Vol. 88, No. 3, pp. 325-332.

[9] Bonen, D., and Cohen, M.D. (1992a), "Magnesium sulfate attack on Portland cement paste – I. Microstructural analysis", *Cement and Concrete Research*, Vol. 22, pp. 169-180.

[10] Bonen, D., and Cohen, M.D. (1992b), "Magnesium sulfate attack on Portland cement paste – II. Chemical and mineralogical analyses", *Cement and Concrete Research*, Vol. 22, pp. 707-718.

[11] BS (1986), "Recommendations for surface hardness testing by rebound hammer, 1881:Part 202:1986, British Standards Institute.

[12] Carde, C., Francois, R., and Torrenti, J.-M. (1996), "Leaching of both calcium hydroxide and C-S-H from cement paste: modeling the mechanical behavior", *Cement and Concrete Research*, Vol. 26, No. 8, pp. 1257-1268.

[13] Clifton, J.R. (1991), "Predicting the remaining service life of concrete", *NISTIR* 4712, National Institute of Standards and Technology.

[14] Clifton, J.R., and Pommersheim, J.M. (1992), "Methods for predicting remaining life of concrete in structures", *NISTIR* 4954, National Institute of Standards and Technology.

[15] Clifton, J.R. (1993), "Predicting the service life of concrete", *ACI Materials Journal*, Vol. 90, No. 6, pp. 611-617.

[16] Clifton, J.R., and Pommersheim, J.M. (1994), "Sulfate attack of cementitious materials: volumetric relations and expansions", *NISTIR* 5390, National Institute of Standards and Technology.
[17] Cohen, M.D., and Mather, B. (1991), "Sulfate attack on concrete – research needs", *ACI Materials Journal*, Vol. 88, No. 1, pp. 62-69.
[18] Ferraris, C.F., Clifton, J.R., Stutzman, P.E., and Garboczi, E.J. (1995), "Mechanisms of degradation of Portland cement-based systems by sulfate attack", Materials Research Society, 1995.
[19] Gerard, B., Breysse, D., and Lasne, M. (1993), "Coupling between cracking and permeability, a model for structure service life prediction", Proc. Int. Conference on Safe Management and Disposal of Nuclear Waste, 1993.
[20] Gollop, R.S., and Taylor, H.F.W. (1992), "Microstructural and microanalytical studies of sulfate attack. I. Ordinary Portland cement paste", *Cement and Concrete Research*, Vol. 22, pp. 1027-1038.
[21] Gollop, R.S., and Taylor, H.F.W. (1994), "Microstructural and microanalytical studies of sulfate attack. II. Sulfate-resisting Portland cement: Ferrite composition and hydration chemistry", *Cement and Concrete Research*, Vol. 24, No. 7, pp. 1347-1358.
[22] Gollop, R.S., and Taylor, H.F.W. (1995), "Microstructural and microanalytical studies of sulfate attack. III. Sulfate-resisting Portland cement: Reactions with sodium and magnesium sulfate solutions", *Cement and Concrete Research*, Vol. 25, No. 7, pp. 1581-1590.
[23] Gollop, R.S., and Taylor, H.F.W. (1996), "Microstructural and microanalytical studies of sulfate attack. IV. Reactions of a slag cement paste with sodium and magnesium sulfate solutions", *Cement and Concrete Research*, Vol. 26, No. 7, pp. 1013-1028.
[24] Hearn, N., Hooton, R.D., and Mills, R.H. (1994), "Pore structure and permeability", ASTM STP 169C, Chap. 25, pp. 240-262.
[25] Ju, J.W. and Taylor, R.L. (1988), "A perturbed Lagrangian formulation for the finite element solution of nonlinear frictional contact problems", *Journal de Mecanique Theorique et Appliquee*, Special Issue, Supplement No. 1 to Vol. 7, pp. 1-14.
[26] Ju, J.W. (1989a), "On energy-based coupled elastoplastic damage models at finite strains", *J. of Eng. Mech.*, ASCE, Vol. 115, No. 11, pp. 2507-2525.
[27] Ju, J.W. (1989b), "On energy-based coupled elastoplastic damage theories: constitutive modeling and computational aspects", *Int. J. Solids & Struct.*, Vol. 25, No. 7, pp. 803-833.
[28] Ju, J.W., Monteiro, P.J.M., and Rashed, A.I. (1989), "Continuum damage of cement paste and mortar as affected by porosity and sand concentration", *Journal of Engineering Mechanics*, ASCE, Vol. 115, No. 1, pp. 105-130.

[29] Ju, J.W., and Weng, L.S. (1998), "Damage assessment of concrete subjected to environ-mental attack using ultrasonic NDE", in Proceedings (Abstracts) of the 1998 QNDE Conference, Ed. By R.B. Thompson, July 19-24, 1998, Snowbird Resort, Utah.

[30] Khatri, R.P., Sirivivatnanon, V., and Yang, J.L. (1997), "Role of permeability in sulfate attack", *Cement and Concrete Research*, Vol. 27, No. 8, pp. 1179-1189.

[31] Kuzmar, A.S. (1994), "Sulfate attack on Portland cement paste via fracture mechanics concepts", Ph.D. Dissertation, Dept. of Civil and Environmental Engineering, Duke University.

[32] Malhotra, V.M. (1991), "Surface Hardness Methods," in V.M. Malhotra and N.J. Carino (eds.), *CRC Handbook on Nondestructive Testing of Concrete,* CRC Press, Boca Raton, Florida, pp. 1-17.

[33] Mehta, P.K. (1997), "Durability – critical issues for the future", *Concrete International*, July 1997, pp. 27-33.

[34] Mehta, P.K., and Monteiro, P.J.M. (1993), "Concrete: Microstructure, Properties and Materials", McGraw-Hill Co.

[35] Mindess, S., and Young, J.F. (1981), "Concrete", Prentice Hall Inc., New Jersey.

[36] Neville, A.M. (1996), "Properties of Concrete", John Wiley & Sons, Inc., 4th Edition.

[37] Ouyang, C., Nanni, A., and Chang, W.F. (1988), "Internal and external sources of sulfate ions in Portland cement mortar: two types of chemical attack", *Cement and Concrete Research*, Vol. 18, pp. 699-709.

[38] Ouyang, C. (1989), "A damage model for sulfate attack of cement mortars", *Cement, Concrete and Aggregates*, Vol. 11, No. 2, pp. 92-99.

[39] Piasta, W., and Schneider, U. (1992), "Deformations and elastic modulus of concrete under sustained compression and sulphate attack", *Cement and Concrete Research*, Vol. 22, pp. 149-158.

[40] Reading, T.J. et al. (1977), "Guide to durable concrete", ACI 201.2R-77, by ACI Committee 201, American Concrete Institute.

[41] RILEM (1972), "Testing of concrete by the ultrasonic pulse method", NDT1, TC7-NDT and TC43-CND, in "RILEM Technical Recommendations for the Testing and use of Construction Materials", pp. 73-82 E&FN Spon, London.

[42] RILEM (1975), "Direct tension of concrete specimens", CPC7, in "RILEM Technical Recommendations for the Testing and use of Construction Materials", pp. 23-24, E&FN Spon, London.

[43] Rossi, P., Wu, X., Le Maou, F., and Belloc, A. (1994), "Scale effect on concrete in tension", *Materials and Structures*, RILEM, Vol. 27, pp. 437-443.

[44] Scmidt, E. (1950), "Der Beton-Prufhammer [The Concrete Test Hammer]", *Schweizerische Bauzeitung* (Zurich), Vol. 68, No. 28, p. 378.
[45] Shah, S.P., Swartz, S.E., and Ouyang, C. (1995), "Fracture Mechanics of Concrete", John Wiley & Sons, Inc.
[46] Sturrup, V.R., Vecchio, F.J., and Caratin, H. (1984), "Pulse velocity as a measure of concrete compressive strength", in "In-Situ testing of Concrete", edited by V.M. Malhotra, ACI SP-82, pp. 201-227, Detroit.
[47] Tang, T., Shah, S.P., and Ouyang, C. (1992), "Fracture mechanics and size effect of concrete in tension", *Journal of Structural Engineering*, Vol. 118, No. 11, pp. 3169-3185.
[48] "Uniform Building Code" (1985), International Conference of Building Officials, 1985 Edition, Whittier, California.
[49] "Uniform Building Code" (1988), International Conference of Building Officials, 1988 Edition, Whittier, California.
[50] U.S. Bureau of Reclamation 4912-92 (1992), "Procedure for direct tensile strength, static modulus of elasticity, and Poisson's ratio of cylindrical concrete specimens in tension", in "*Concrete Manual*", Part 2, 9th Edition, pp 726-731, Denver, Colorado.
[51] Zoldners, N.G. (1957), "Calibration and Use of Impact Test Hammer," *ACI Journal Proceedings*, Vol. 54, No. 2, pp. 161-165.

Table 1. Representative rebound hammer test results

	Rebound Number (Arbitrary Scale)									
Location	1	2	3	4	5	6	7	8	9	10
Just Above Grade	54	40	54	47	47	51	48	52	45	55
Just Below Grade	45	33	38	39	31	40	38	39	38	39
Bottom Of Footing	28	22	33	32	25	27	17	29	24	25

Figure 1.

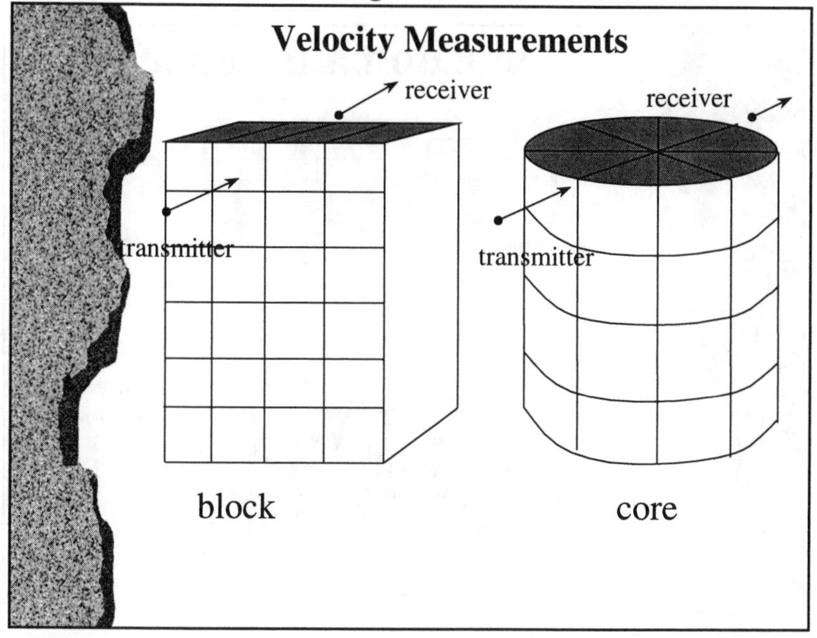
Velocity Measurements
receiver
receiver
transmitter
transmitter
block
core

Figure 2.

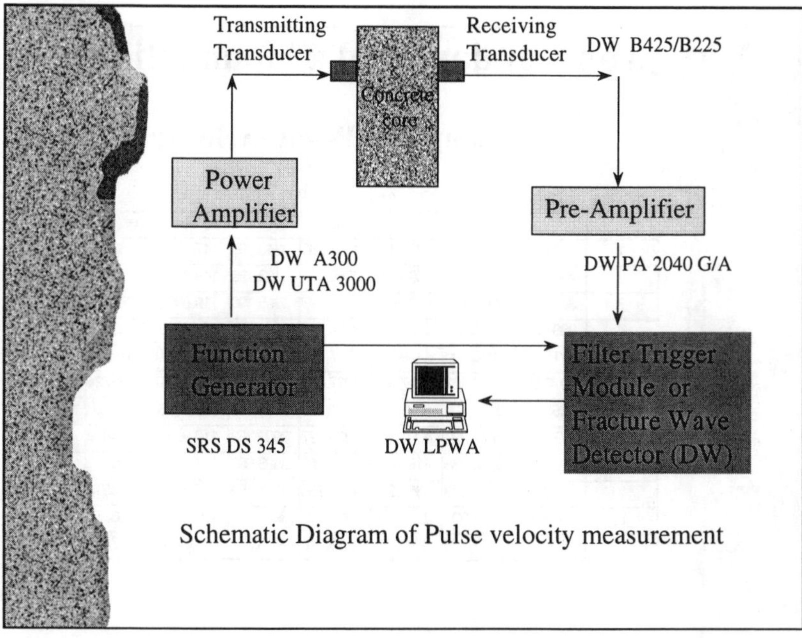
Transmitting
Transducer
Concrete
core
Receiving
Transducer
DW B425/B225
Power
Amplifier
Pre-Amplifier
DW A300
DW UTA 3000
DW PA 2040 G/A
Function
Generator
Filter Trigger
Module or
Fracture Wave
Detector (DW)
SRS DS 345
DW LPWA
Schematic Diagram of Pulse velocity measurement

Figure 3.

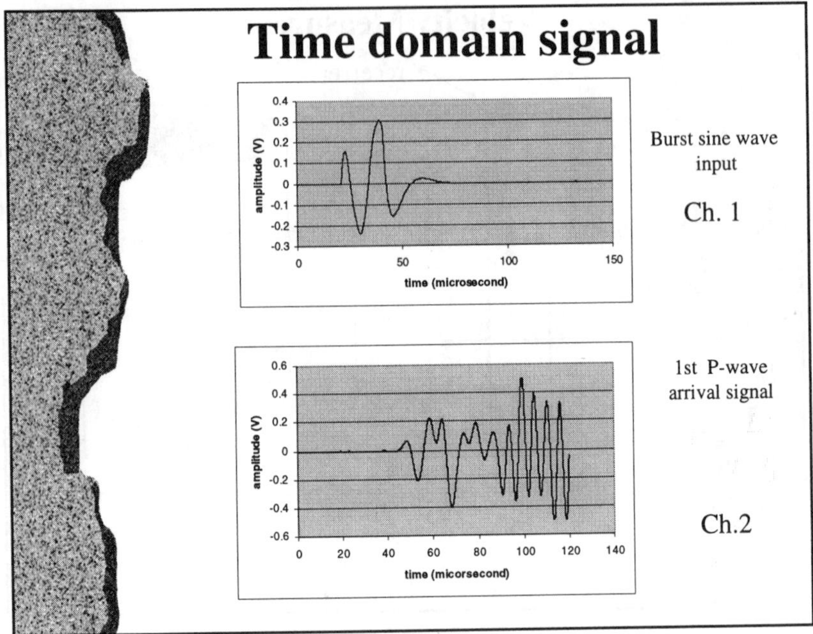

Figure 4.

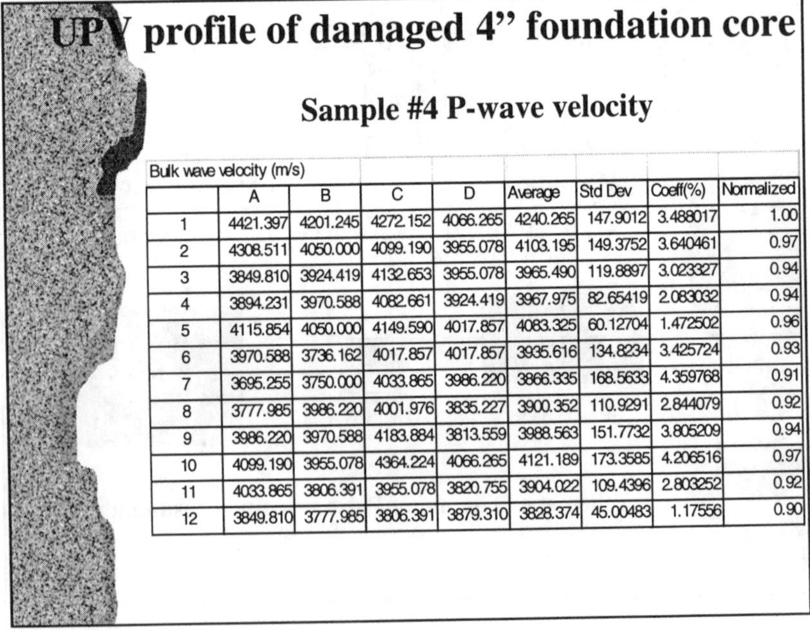

UPV profile of damaged 4" foundation core
Sample #4 P-wave velocity

Bulk wave velocity (m/s)

	A	B	C	D	Average	Std Dev	Coeff(%)	Normalized
1	4421.397	4201.245	4272.152	4066.265	4240.265	147.9012	3.488017	1.00
2	4308.511	4050.000	4099.190	3955.078	4103.195	149.3752	3.640461	0.97
3	3849.810	3924.419	4132.653	3955.078	3965.490	119.8897	3.023327	0.94
4	3894.231	3970.588	4082.661	3924.419	3967.975	82.65419	2.083032	0.94
5	4115.854	4050.000	4149.590	4017.857	4083.325	60.12704	1.472502	0.96
6	3970.588	3736.162	4017.857	4017.857	3935.616	134.8234	3.425724	0.93
7	3695.255	3750.000	4033.865	3986.220	3866.335	168.5633	4.359768	0.91
8	3777.985	3986.220	4001.976	3835.227	3900.352	110.9291	2.844079	0.92
9	3986.220	3970.588	4183.884	3813.559	3988.563	151.7732	3.805209	0.94
10	4099.190	3955.078	4364.224	4066.265	4121.189	173.3585	4.206516	0.97
11	4033.865	3806.391	3955.078	3820.755	3904.022	109.4396	2.803252	0.92
12	3849.810	3777.985	3806.391	3879.310	3828.374	45.00483	1.17556	0.90

Figure 5.

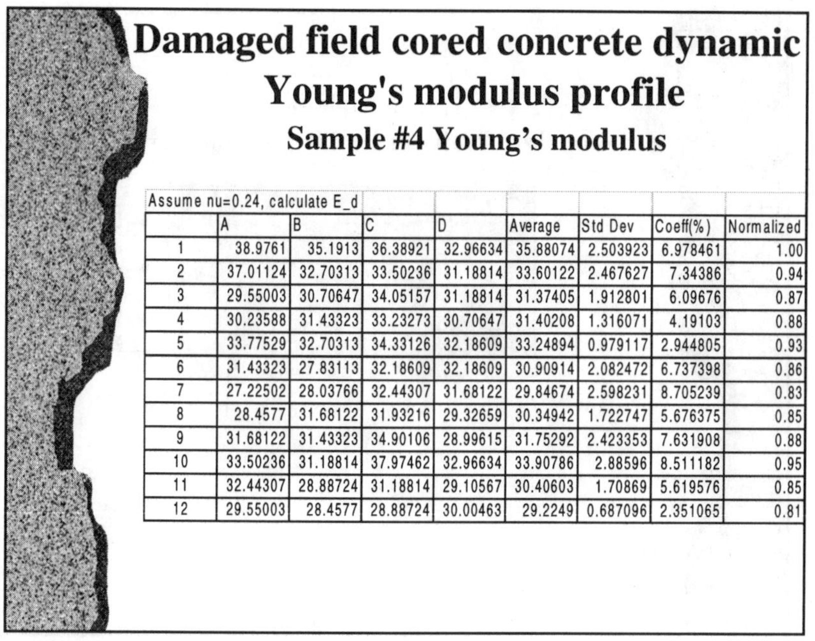

Assume nu=0.24, calculate E_d								
	A	B	C	D	Average	Std Dev	Coeff(%)	Normalized
1	38.9761	35.1913	36.38921	32.96634	35.88074	2.503923	6.978461	1.00
2	37.01124	32.70313	33.50236	31.18814	33.60122	2.467627	7.34386	0.94
3	29.55003	30.70647	34.05157	31.18814	31.37405	1.912801	6.09676	0.87
4	30.23588	31.43323	33.23273	30.70647	31.40208	1.316071	4.19103	0.88
5	33.77529	32.70313	34.33126	32.18609	33.24894	0.979117	2.944805	0.93
6	31.43323	27.83113	32.18609	32.18609	30.90914	2.082472	6.737398	0.86
7	27.22502	28.03766	32.44307	31.68122	29.84674	2.598231	8.705239	0.83
8	28.4577	31.68122	31.93216	29.32659	30.34942	1.722747	5.676375	0.85
9	31.68122	31.43323	34.90106	28.99615	31.75292	2.423353	7.631908	0.88
10	33.50236	31.18814	37.97462	32.96634	33.90786	2.88596	8.511182	0.95
11	32.44307	28.88724	31.18814	29.10567	30.40603	1.70869	5.619576	0.85
12	29.55003	28.4577	28.88724	30.00463	29.2249	0.687096	2.351065	0.81

Figure 6.

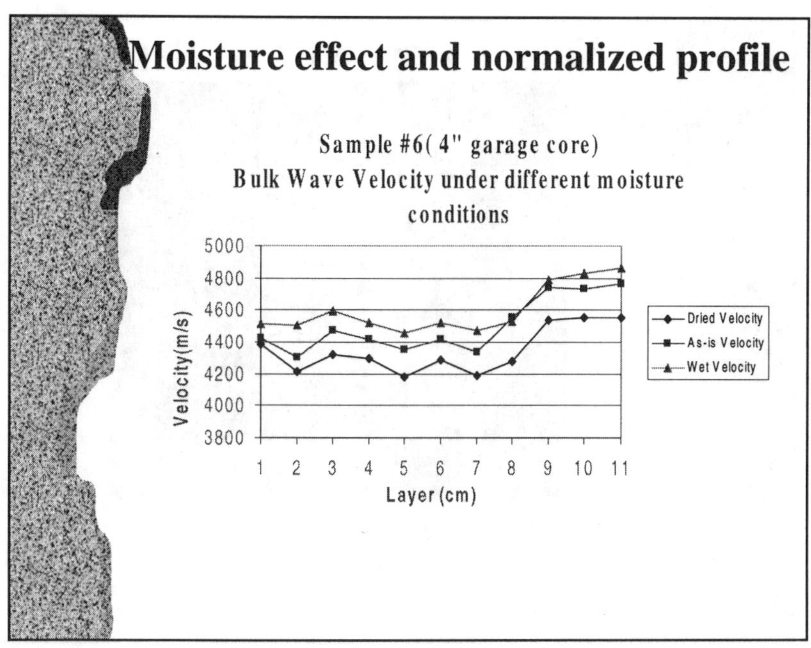

Figure 7.

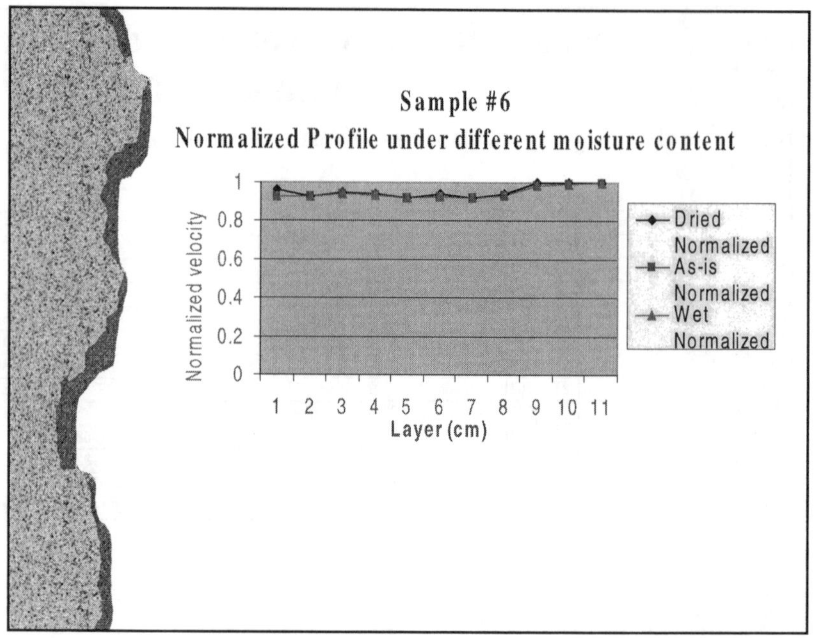

Figure 8.

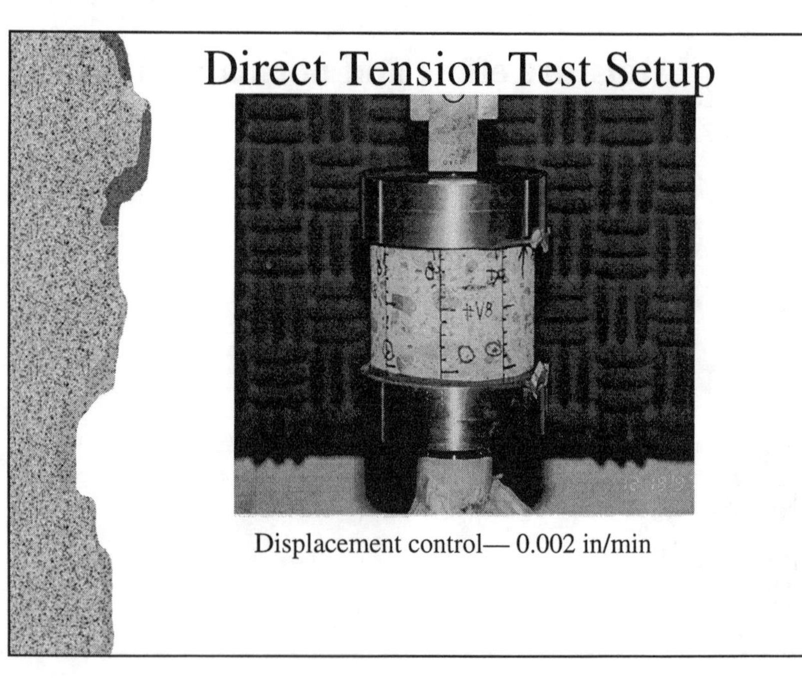

Direct Tension Test Setup
Displacement control— 0.002 in/min

INFLUENCE OF CALCIUM HYDROXIDE DISSOLUTION ON THE ENGINEERING PROPERTIES OF CEMENT-BASED MATERIALS

J. Marchand[1-2], J.J. Beaudoin[3] and M. Pigeon[1]

(1) Centre de Recherche Interuniversitaire sur le Béton
Université Laval, Sainte-Foy, Canada, G1K 7P4

(2) SIMCO Technologies inc.
1400, boul. du Parc Technologique, Québec, Canada, G1P 4R7

(3) Institute for Research in Construction
National Research Council, Ottawa, Canada

Abstract

In addition to the formation of sulfate-bearing phases, the penetration of sulfate ions in cement-based materials also contributes to the dissolution of hydrated compounds such as calcium hydroxide. The consequences of calcium hydroxide dissolution on the engineering properties of cement systems are briefly reviewed and discussed. The survey clearly indicates that the dissolution process markedly contributes to an increase in the porosity of the hydrated cement paste. This increase in porosity results in a significant reduction of the material mechanical properties. It also contributes to an increase in the value of parameters that affect the material transport properties.

Introduction

It has been clearly demonstrated that the penetration of sulfate ions can significantly alter the microstructure of cement-based materials. The detrimental consequences of sulfate attack on the concrete microstructure are not solely limited to the formation of new sulfate-bearing phases. Sulfate attack is also known to involve dissolution of $Ca(OH)_2$ and the decalcification of C-S-H [1-3].

It is now well established that the decalcification of C-S-H results in a major reduction of the concrete mechanical strength [1, 4]. The end product of the dissolution reaction is a silica-rich phase (often called silica « gel ») with little, i any, binding properties [4, 5]. Numerous investigations have also clearly demonstrated that the dissolution of $Ca(OH)_2$ contributes significantly to the alteration of the mechanical properties of concrete. A brief discussion of this phenomenon is presented in the following discussion.

LITERATURE SURVEY

Physical properties of $Ca(OH)_2$

$Ca(OH)_2$, along with C-S-H, are the end products of the reaction of alite and belite with water. The abundance of $Ca(OH)_2$ in the hydrated cement paste varie with the degree of hydration of the cement, and it can reach approximately 26 % of the total volume of a mature paste. Contrary to C-S-H that is an ill-crystallized phase, $Ca(OH)_2$ is present predominantly in the form of well-defined crystals in the hydrated cement paste. The size of these crystals tends to vary significantly from a few to several microns in diameter.

Data obtained by Beaudoin [6] on compacted samples indicate that the elastic modulus and the flexural strength of $Ca(OH)_2$ are in the same range as those o the cement paste. From a micromechanical point of view, the fact that $Ca(OH)_2$ crystals have nearly the same elastic modulus as that of the bulk cement paste will tend to reduce stress concentrations at the interfacial zone between the $Ca(OH)_2$ and C-S-H. These results are summarized in Table I and in Figure 1.

Table I. Mechanical properties of C-S-H and $Ca(OH)_2$ (from ref. 6)

Phase	Porosity (%)	Elastic Modulus (GPa)	Flexural Strength (MPa)
$Ca(OH)_2$	9 to 30	23 to 14	30 to 17
Cement paste	17 to 30	16 to 2	14 to 6

Influence of $Ca(OH)_2$ dissolution on the solid porosity and volume stability

$Ca(OH)_2$ is also characterized by its relatively high solubility (with respect to other hydrated cement paste phases). Studies by Feldman and Ramachandran [8, 9] have clearly indicated that the dissolution of $Ca(OH)_2$ results in a marked reduction of the hydrated cement paste density and a local increase of the solid porosity (see Figure 2). The data appearing in the figure represent changes to the solid density of the hydration products in cement paste as determined by helium pycnometry. Density values are corrected for the volume of helium that diffuses with time into the C-S-H structure.

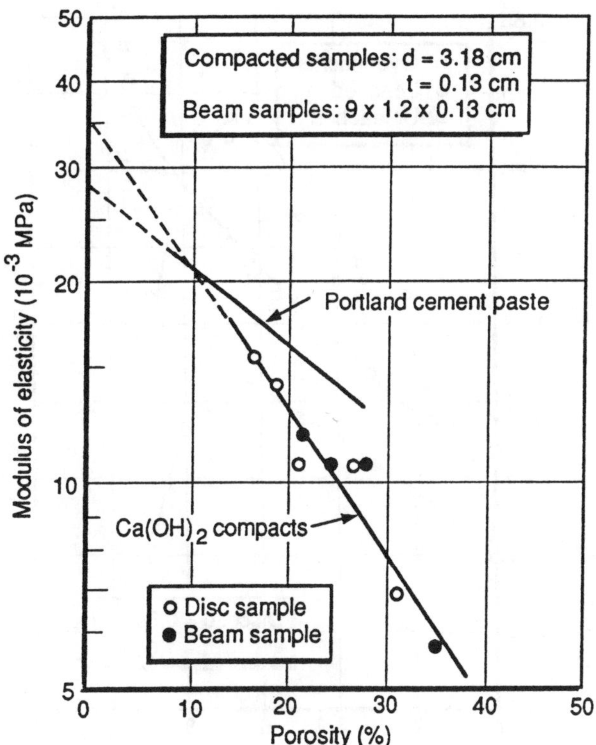

Figure 1 – Modulus of elasticity of $Ca(OH)_2$ compacts and portland cement paste vs porosity (adapted from ref. 6)

Density values decreases after a maximum mass loss for the leached and unleached cases due to the collapse of the interlayer structure restricting the flow of helium. No restriction access is observed for the overleached specimens and the density increases monotically over the entire range of mass loss. The leached specimens were prepared in lime solutions of selected molar concentrations to avoid perturbation to the C-S-H solid itself. The overleached specimens result in extraction of lime from the C-S-H solid. It is clear from the density determinations that overleaching produces "holes" in the structure that may have implications with respect to the volume stability of the solid.

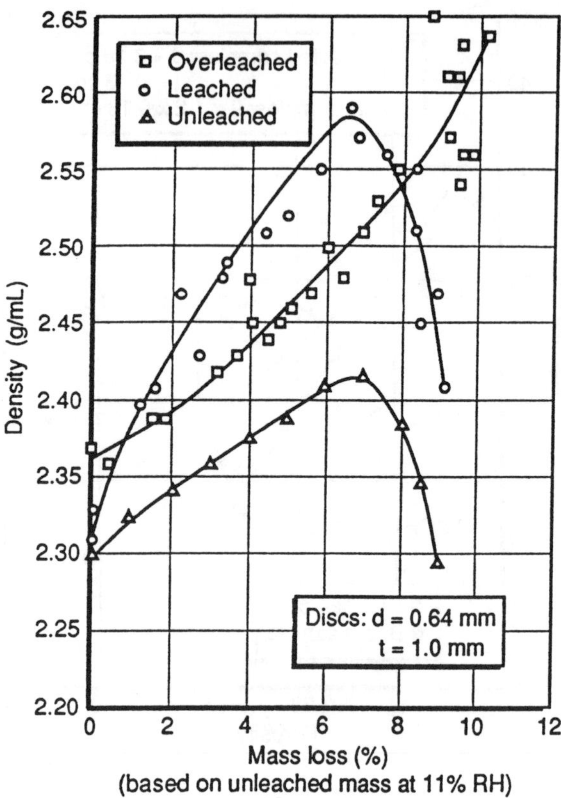

Figure 2 – Density of $Ca(OH)_2$-depleted cement paste (adapted from ref. 8)

Length change results (for a 0.40 water/cement ratio paste hydrated for 14 years) during the leaching process are plotted in Figure 3. Expansion occurs during the first 10 days followed by contraction. Most of the free lime is removed at 26 days after which the rate of removal decreases. It appears that lime associated with the C-S-H sheets is removed after 26 days.

The expansion during the first 10 days may be due to strain energy release occurring as the porosity increases due to $Ca(OH)_2$ crystal dissolution. This is a strong indication that $Ca(OH)_2$ has a structural role in the engineering behavior of hydrated cement paste. It is also clear that time-dependent volume stability is a function of removal of $Ca(OH)_2$ from the paste and C-S-H structure.

These results are in good agreement with the observations of Litvan [10, 11] who has reported significant length changes for porous solids immersed in aggressive solutions. The volume change induced by the dissolution of $Ca(OH)_2$ may lead to local microcracking of the paste.

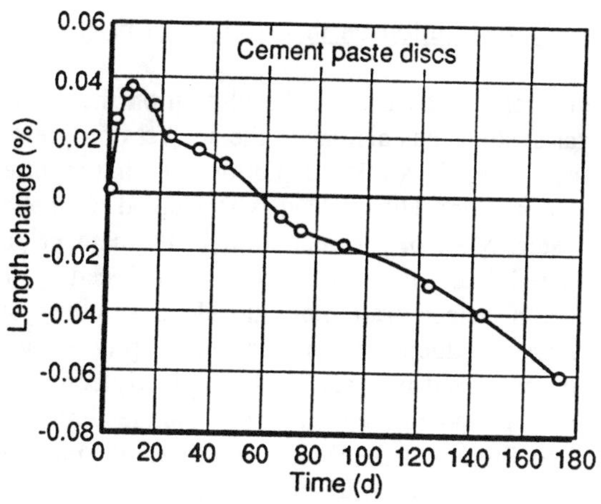

Figure 3 – Length change of $Ca(OH)_2$-depleted cement pastes (adapted from ref. 9)

Influence of $Ca(OH)_2$ dissolution on the diffusion properties of cement-based materials

Some authors have tried to assess the effect of $Ca(OH)_2$ dissolution on the diffusion properties of hydrated cement pastes and mortars. Using a cellular automaton-type digital-image-based model, Bentz and Garboczi [12] demonstrated that the dissolution process has a direct influence on the pore system connectivity and could contribute to markedly increase the diffusion properties of the cement paste. Their conclusions were recently supported by various experimental studies where the transport behavior of paste and mortar samples was found to be markedly modified by the dissolution process [13-15].

From the standpoint of the service-life of concrete, this increase in the value of parameters related to the diffusion characteristics is particularly important. As emphasized by Gérard et al. [16], most chemical degradation processes are diffusion controlled and any local increase of the concrete transport properties may contribute to significantly reduce the durability of the structure.

Influence of $Ca(OH)_2$ dissolution on the mechanical properties of cement-based materials

The influence of calcium hydroxide dissolution on the mechanical properties of paste and mortar systems has also been the subject of a great deal of research. The detrimental influence of $Ca(OH)_2$ depletion was first reported by Regourd et al. [17-18] and Feldman and Huang [19] who studied the behavior of concrete immersed in seawater. Most of the reduction in the cement paste compressive strength was attributed to the increase in porosity induced by the dissolution of $Ca(OH)_2$ resulting from the penetration of chloride ions in the system.

The very detrimental influence of the dissolution process on the compressive strength of cement pastes and mortars has been clearly confirmed by Carde and co-workers [20-22] who studied the behavior of samples immersed in ammonium nitrate solutions (see Figures 4-5). Buffer solutions of ammonium nitrate have been found to accelerate the dissolution of calcium hydroxide without forming any new (nitrate or ammonium bearing) products [20-22]. These results have later been confirmed by Gérard who studied the effect of calcium hydroxide depletion on the mechanical behavior of cement paste by micro-hardness testing [15].

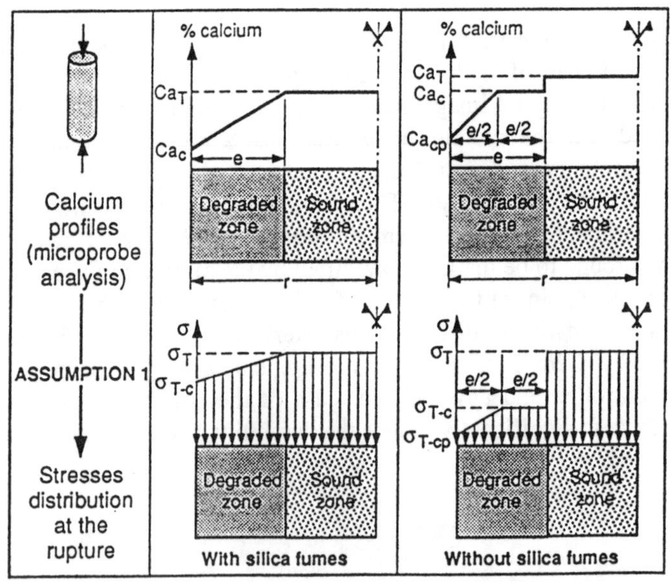

Figure 4 – Influence of $Ca(OH)_2$ leaching on the compressive strength of cement paste samples (adapted from ref. 21)

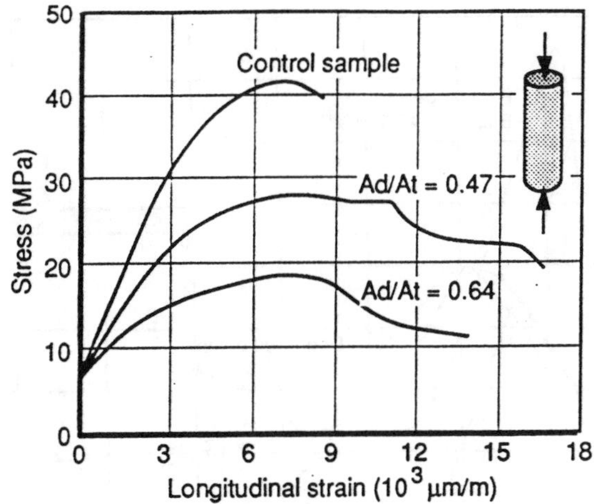

Figure 5 – Influence of $Ca(OH)_2$ leaching on the stress/strain curve of cement mortar samples (the ratio Ad/At increases with the depth of $Ca(OH)_2$ leaching – adapted from ref. 22)

The dissolution of calcium hydroxide has recently been found to markedly reduce the flexural strength of mortar [23]. Test results obtained by Schneider and co-workers indicate that the microstructural alterations induced by sulfate attack (ettringite formation, calcium hydroxide dissolution, ...) also contribute to a decrease in the mechanical properties of normal-strength and high-performance mortar and concrete mixtures [24-26]. Typical results are given in Figure 6. In all cases, the dissolution process was found to reduce the peak strain and the peak stress of the material. The substantial reduction of the flexural strength of concrete could be particularly important for slabs exposed to sulfate attack for which the degradation process starts from the top and bottom surfaces where the stress concentrations are the most important.

The importance of the $Ca(OH)_2$ dissolution process on the mechanical behavior of concrete is now the subject of major concern for civil engineers. The influence of the process on the design of concrete structures has been extensively studied. Numerical simulations clearly indicate that the dissolution process largely contributes to increase the cracking susceptibility of concrete subjected to flexural loading [27-29].

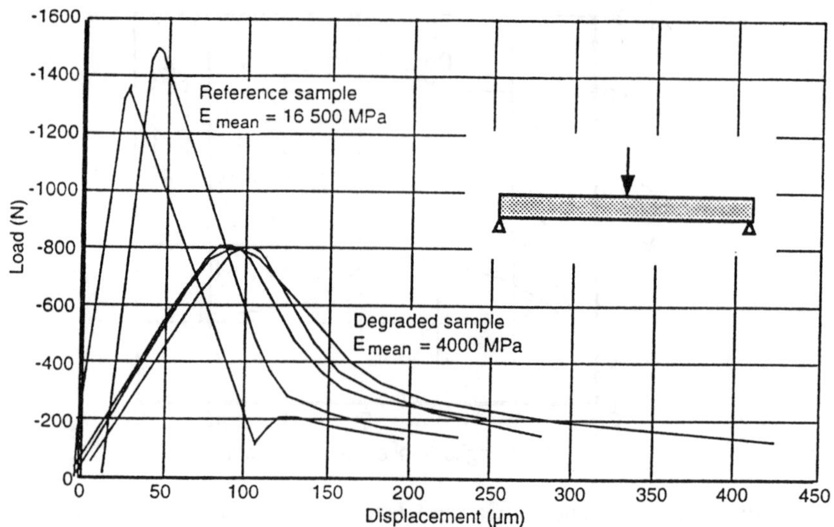

Figure 6 – Influence of $Ca(OH)_2$ leaching on the flexural strength of cement mortar samples (depth of $Ca(OH)_2$ leaching = 10 mm – adapted from ref. 23)

Concluding remarks

Calcium hydroxide dissolution is known to markedly increase the porosity of cement systems. The dissolution process may even lead to significant volume changes.

Calcium hydroxide dissolution has been found to increase the diffusion properties of cement-based materials. From the standpoint of the service life of concrete, this increase in the diffusion properties is particularly important since most sulfate degradation processes are diffusion controlled.

Many recent investigations have clearly established the very detrimental influence of calcium hydroxide dissolution on the mechanical properties of cement-based materials.

References

[1]Neville, A, "Properties of concrete", *John Wiley & Son*, 4th Edition, 844 p. (1995)

[2]Gollop, R.S., Taylor, H.F.W., "Microstructural and microanalytical studies of sulphate attack - Part 1: Ordinary Portland cement", *Cement and Concrete Research*, **22**, 1027-1038 (1992).

[3]Bonen, D., Cohen, M.D., "Magnesium sulfate attack on Portland cement paste - Part II: Chemical and mineralogical analyses", *Cement and Concrete Research*, **22,** 707-718 (1992).

[4]Adenot, F., "Durability of Concrete: Characterization and modeling of the physical and chemical degradation processes of cement", *Ph.D. Thesis*, Orleans University, France, 238 p. (in French) (1992).

[5]Faucon, P., "Durability of concrete : Physico-chemistry of the degradation induced by pure water", *Ph.D. Thesis*, Cergy-Pontoise University, France, 260 p. (in French) (1997).

[6]Beaudoin, J.J., "Comparison of mechanical properties of compacted calcium hydroxide and portland cement paste systems", *Cement and Concrete Research*, **22**, 707-718 (1983).

[7]Adenot, F., Buil, M., "Modelling the corrosion of the cement paste by deionized water", *Cement and Concrete Research*, **22**, 489-496 (1992).

[8]Feldman, R.F., Ramachandran, V.S., "Microstructure of calcium hydroxide depleted cement paste – 1 : Density and helium flow measurements", *Cement and Concrete Research*, **12**, 179-186 (1982).

[9]Feldman, R.F., Ramachandran, V.S., "Length changes in calcium hydroxide-depleted cement pastes", *Il Cemento*, 87-96 (1989).

[10]Litvan, G.G., "Volume instability of porous solids – Part 1", *7th Congress on Cement Chemistry*, Paris, **VII**, 46-50 (1980).

[11]Litvan, G.G., "Volume instability of porous solids – Part 2 : Dissolution of porous silica in sodium hydroxide", *Journal of Materials Science*, **19**, 2473-2480 (1984).

[12]Bentz, D.P., Garboczi, E.J., "Modelling the leaching of calcium hydroxide from cement paste: Effects on pore space percolation and diffusivity", *Materials and Structures*, **25**, 523-533 (1992).

[13]Delagrave, A., Gérard, B., Marchand, J., "Modeling calcium leaching mechanisms in hydrated cement pastes", in *Mechanisms of Chemical Degradation of Cement-Based Systems*, E & FN Spon, 38-47 (1997).

[14]Bourdette, B., "Durability of Mortars", *Ph.D. Thesis*, INSA-UPS, Toulouse, France, 203 p. (in French) (1994).

[15]Gérard, B., "Influence of the couplings chemistry-mechanics-mass transfer on the long-term durability of concrete nuclear waste storage facilities", *Ph. D. Thesis*, ENS-Cachan/Université Laval, Canada, 289 p. (in French) (1996).

[16]Gérard, B., Didry, O., Marchand, J., Breysse, D., Hornain, H., "Modelling the long-term durability of concrete for nuclear waste disposals", in *Mechanisms of Chemical Degradation of Cement-Based Systems*, E & FN Spon, 331-340 (1997).

[17]Regourd, M., Hornain, H., Mortureux, B., "Microstructure of Concrete in Aggressive Environments", *ASTM Special Technical Publication 691*, Edited by P.J. Sereda and G.G. Litvan, 253-268 (1978).

[18]Regourd, M., "Effects of sea water on cement systems", *Annales de l'Institut Technique du Bâtiment et des Travaux Publics*, N° 329, Paris, France, 87-102 (in French) (1975).

[19]Feldman, R.F., Huang, C-Y, "Resistance of mortars containing silica fume to attack by a solution containing chlorides", *Cement and Concrete Research*, **15**, 411-420 (1985).

[20]Carde, C., François, R., Ollivier, J.P., "Microstructural changes and mechanical effects due to the leaching of calcium hydroxide from cement paste", in *Mechanisms of Chemical Degradation of Cement-Based Systems*, E & FN Spon, 30-37 (1997).

[21]Carde, C., François, R., "Effect of the leaching of calcium hydroxide from cement paste on the mechanical and physical properties", *Cement and Concrete Research*, **27**, 539-550 (1997).

[22]Carde, C., François, R., "Effect of ITZ leaching on durability of cement-based materials", *Cement and Concrete Research*, **27**, 971-978 (1997).

[23]Molez, L., "Couplings chemistry-mechanics-mass transfer in concrete: Multi-scale characteristics and localization", *D.E.A. Dissertation*, ENS-Cachan, France, 47 p. (1997).

[24]Schneider, U., Piasta, J., Nagele, E., Piasta, W., "Stress corrosion of cementitious materials in sulphate solutions", *Materials and Structures*, **23**, 110-115 (1990).

[25]Schneider, U., Piasta, W., "The behaviour of concrete under Na_2SO_4 solution attack and sustained compression or bending", *Magazine of Concrete Research*, **43**, 281-289 (1991).

[26]Piasta, W., Schneider, U., "Deformations and elastic modulus of concrete under sustained compression and sulphate attack", *Cement and Concrete Research*, **22**, 149-158 (1992).

[27]Molez, L., Gérard, B., Pijaudier-Cabot, G., "Interactions between damage and calcium leaching in concrete : towards a calibration and validation procedure", *Engineering Mechanics – A Force for the 21st Century*, H. Murakami Ed., ASCE Publications, 5 p. (1998).

[28]Pijaudier-Cabot, G., Gérard, B., Burlion, N., Molez, L., "Localization of damage in quasi-brittle materials and influence of chemically activated damage", in *Materials Instabilities in Solids*, R. de Borst and E. van der Giessen Eds., J. Wiley and Sons Ltd., 15 p. (1998).

[29]Carde, C., François, R., "Aging damage model of concrete behavior during the leaching process", *Materials and Structures*, **30**, 465-472 (1997).

SULFATE ATTACK IN MARINE ENVIRONMENT

P.K. Mehta
University of California
Berkeley, CA., USA

ABSTRACT

Seawater typically contains 2700 mg/liter sulfate. In spite of the high sulfate content, concretes made with high – C_3A portland cement do not exhibit expansion and cracking that is normally associated with ettringite formation. This paper contains a brief historical account of field experiences with test mortars and concretes subjected to long-term exposure to marine environment. A review of chemical attacks that can occur when a cement paste is exposed to seawater is also included.

COMPOSITION OF SEAWATER

Most seawaters contain approximately 3.5 percent soluble salts and, with a few exceptions, are fairly uniform in chemical composition. Typical concentration of major ions present in seawater is as follows:

Na^+	11 000 mg/liter
Mg^{2+}	1 400
Cl^-	20 000
SO_4^{2-}	2 700

Additionally, some dissolved gases such as CO_2 are present. Depending on the content of free CO_2, the pH of seawater can vary from 7.0 to 8.4.

EXPERIENCE WITH MORTARS AND CONCRETE EXPOSED TO SEAWATER

According to Lea[1], Vicat was the first to propose that the chemical attack of seawater on lime mortar was mainly a result of interaction between the uncombined $Ca(OH)_2$ in the mortar and magnesium sulfate present in seawater. Vicat began his meticulous experimental studies in 1812 when, according to him, there was a 'chaos of opinion' on the subject. The first results of his work were published in 1818, but the complete work entitled, 'Research on the Causes of Physical Destruction of Hydraulic Mortars by Seawater,' was published in French in 1857. For his study Vicat received two prizes of 2000 French francs each, which had been offered by the Société d'Encouragement pour l'Industrie Nationale (Society for Encouragement of National Industry) for the best work on durability of cement mortars for marine construction. Vicat made this profound observation:

> On being submitted to examination, the deteriorated parts (of the mortar) exhibit much less lime than the others; what is deficient then, has been dissolved and carried off; it was in excess in the composite cement mixture. Nature, we see, labors to arrive at exact proportions, and to attain them corrects the errors of the hand that had formulated the original proportions. Thus the effects which we have just described, and in the case alluded to, become more marked the further we deviate from these exact proportions.

In 1880, an investigation by Prazier into the causes of concrete deterioration at Aberdeen Harbor in Scotland provided a confirmation of Vicat's findings. Prazier concluded that hydraulic cements used in concrete exposed to seawater lost lime and absorbed magnesium from seawater by an ion-exchange reaction that caused the degradation of concrete. In 1924, based on a comprehensive review of the previous 100 years of worldwide experience with concrete in seawater, Atwood and Johnson[2] essentially reconfirmed what Vicat had discovered more than one hundred years before. The oldest cementing material cited by the authors – an AD 60 lime-pozzolan mortar – when compared to five recent mortars (1886-89), showed that after the action of seawater (i.e. after removal of a part of the original lime present) the residual lime content in all six specimens was similar.

More recent investigations of deteriorated marine structures have shown that with permeable concrete, in addition to the magnesium ion attack, the carbonic acid attack (from dissolved CO_2 in seawater) is also an important factor. According to Feld[3] in 1955, after 21 years of service the concrete piles and caps of the trestle bends of the James River Bridge in Newport News, Virginia, required a $1.4 million repair involving 70% of the 2500 piles. Similarly, after 25 years of service, in 1957, approximately 750 pre-cast piles near Ocean City, New Jersey, had to be repaired because of severe loss of mass; some of the piles had been reduced from their original 550 mm diameter to 300 mm. In both cases the disintegration and loss of material was primarily attributed to abnormally high concentrations of dissolved CO_2 in seawater; the pH of the seawater was found to be close to 7 instead of 8.2 to 8.4, which is the pH value of normal seawater.

Mehta and Haynes[4] described a field investigation involving 18 unreinforced concrete test blocks (1.8 by 1.8 by 1.1 m) which were partially submerged in 1905 in seawater at San Pedro harbor near Los Angeles, California. Examined in 1972, after 67 years of exposure to seawater, it was found that the dense concrete blocks (1:2:4 parts cement, sand, and gravel, respectively) were in excellent condition although some of them had been made with high-C_3A portland cements. On the other hand, lean concrete blocks (1:3:6) had lost some mass and had a soft surface covered with marine growth. Mineralogical analyses of the deteriorated concrete showed significant amounts of magnesium hydroxide, gypsum, ettringite/thaumasite, aragonite (calcium carbonate), and hydrocalumite (carboaluminate hydrate). The original products of portland cement hydration, namely calcium silicate hydrate and calcium hydroxide, had decomposed as a result of magnesium sulfate and CO_2 attack. From long-time field tests in southern

France, Regourd[5] reported similar findings. After 66 years of exposure to seawater, the test mortar cubes made with 600 kg/m^3 cement were in good condition even though they had been made with a high-C_3A (14.9%) portland cement. Those containing 300 kg/m^3 cement were destroyed; the deteriorated cement paste contained aragonite, brucite, ettringite, thaumasite, and magnesium silicate hydrate.

Fluss and Gorman[6] surveyed the condition of 46 year old reinforced concrete piles of the San Francisco ferry building constructed in 1912. The concrete mixture had a high cement content (400 kg/m^3) and was made with high – C_3A portland cement (14-17% C_3A). Most of the piles were found in good condition; only some were found cracked due to corrosion of the reinforcing steel. According to the authors, microcracking of concrete from excessive deflection and poor workmanship must have preceded the reinforcement corrosion. Similar findings were reported from the surveys of hundreds of reinforced concrete structures in Denmark by Idorn[7], and from the Norwegian seaboard by Gjorv[8]. From field experience with reinforced concrete structures in marine environment, it was concluded by Mehta[9] that concrete deterioration occurs due to a number of intertwined physical and chemical causes whose significance is controlled by the microclimate at the structure with respect to the tidal zone. This is illustrated in Fig. 1.

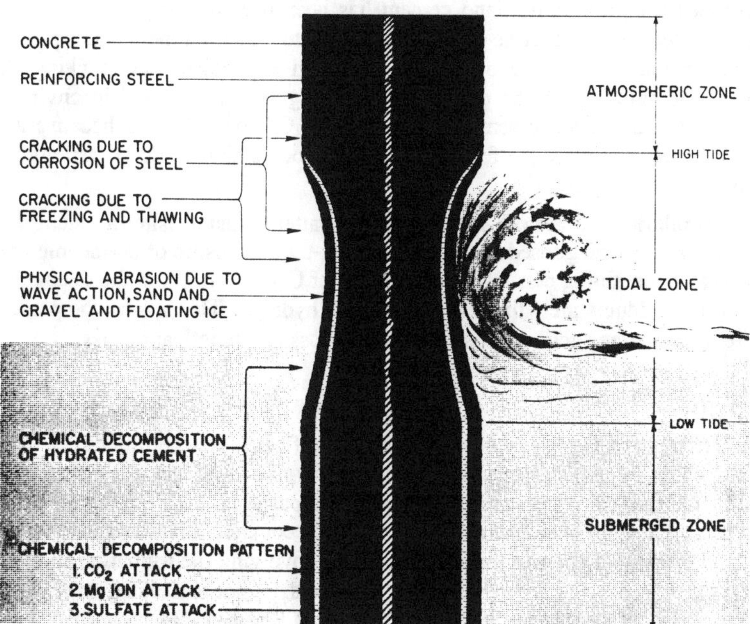

Fig. 1. Physical and chemical processes responsible for deterioration of a reinforced concrete element exposed to seawater.

CHEMICAL ATTACKS ON CEMENT PASTE EXPOSED TO MARINE ENVIRONMENT

Aragonite, not calcite, is the crystalline form of $CaCO_3$ that is stable in seawater. With the low concentrations of CO_2 normally present in seawater, the consequences of the CO_2 attack on the hydration products of portland cement are beneficial because the formation of insoluble aragonite reduces the permeability at the concrete surface, which is the first line of defense against seawater penetration.

Magnesium salts from seawater can also enter into chemical reactions with $Ca(OH)_2$ present in hydrated cement paste, resulting in the formation of brucite -- another highly insoluble mineral that further reduces the permeability of concrete.

However, if a concrete is highly permeable, then a long-term exposure to seawater is known to cause "Mg^{2+} ion attack". This attack involves ion exchange between the magnesium ions of seawater and calcium silicate hydrate present in the hydrated portland cement paste, resulting in the formation of a non-cementitious mineral, namely magnesium silicate hydrate ($4MgO.SiO_2.8H_2O$), which makes the concrete quite weak.

What about sulfate attack? With groundwater, sulfate ion concentrations higher than 1500 mg/liter are considered as 'severe' from the standpoint of potential for sulfate attack on the constituents of hardened portland cement paste. Interestingly, in spite of the undesirably high (2700 mg/liter) sulfate content of seawater, it is a common observation that even when a high-C_3A portland cement has been used and significant amounts of ettringite have formed in concrete as a result of sulfate attack on the cement paste, there is no deterioration of concrete by sulfate-generated expansion and cracking. It was proposed by the author[9] that the expansion of ettringite is suppressed in environments where OH^- ions have been essentially replaced by Cl^- ions. This is because alkaline environment is necessary for the formation of a microcrystalline type of ettringite which expands by water adsorption.

Due to similarities in their X-ray diffraction patterns, thaumasite, a calcium silico-carbonate, is often misdiagnosed as ettringite. Long-term exposure of permeable concrete to cold seawater containing normal concentrations of CO_2 can result in the decomposition of cementitious products including calcium silicate hydrates. This is why the presence of thaumasite in concrete is generally associated with loss of strength and mass.

CONCLUSION

In addition to CO_2, seawater contains significant concentrations of sodium, magnesium, chloride, and sulfate ions. Due to the formation of dense films of aragonite and brucite on the concrete surface, the cement paste in high-quality concretes is generally protected from chemical attacks. Permeable concrete mixtures made with high-C_3A portland cements do permit the sulfate reaction involving the formation of ettringite. However, ettringite formation in chloride environments is not associated with expansion and cracking. Recent experience in the Arabian Gulf Region[10] has reconfirmed that, in the case of waters containing excessive chloride and sulfate concentrations, it is the chloride-induced reinforcing steel corrosion and not sulfate attack on the cement paste which is the main cause of deterioration of concrete structures. This is the reason that cases of premature deterioration of concrete made with sulfate-resisting portland cements have occurred in the Gulf environment,. The issue of sulfate attack in marine environment offers a good example why a holistic rather than a reductionistic approach to concrete durability problems is essential.

REFERENCES:

[1] F.M. Lea, *The Chemistry of Cement and Concrete*, 3rd Edition, Chemical Publishing Company, New York, (1971).

[2] W.G. Atwood and A.A. Johnson, "The Disintegration of Cement in Seawater," *Trans. ASCE*, 87, pp. 204 – 230, (1924).

[3] Feld, J., *Construction Failures*, John Wiley, New York, (1968).

[4] P.K. Mehta and H. Haynes, "Durability of Concrete in Seawater," *J. ASCE Structural Division*, 101 (ST8), pp. 1679 – 1686, (1975).

[5] M. Regourd, "The Action of Seawater on Cements," *Annales de l'Institute Technique du Batiment et des Travaux Publics*, No. 329, pp. 86 – 102, (1975).

[6] P.J. Fluss and S.S. Gorman, "Discussion of Wakeman et al's Paper," *J. ACI*, Proc. 54, pp. 1309 – 1346, (1958).

[7] G.M. Idorn, *Durability of Concrete Structures in Denmark*, Technical University of Denmark, (1967).

[8] O.E. Gjorv, *Durability of Reinforced Concrete Wharves in Norwegian Harbors*, Ingenior Plaget A/S Oslo, (1968).

[9] P.K. Mehta, *Concrete in the Marine Environment*, Elsevier Applied Science, London, (1991).

[10] Z.G. Matta, "Concrete Practices in the Arabian Gulf," *ACI Concrete International*, Vol. 20, No. 7, pp. 55 – 52, (1998).

SULPHATE ATTACK IN A MARINE ENVIRONMENT

M.D.A. Thomas, R.F. Bleszynski and C.E. Scott
Department of Civil Engineering, University of Toronto,
35 St George St, Toronto, Ontario M5S 1A4, Canada.

ABSTRACT

This paper reports the findings of an examination of 10-year-old marine-exposed concrete samples of various strength grade and composition. Plain Portland cement concretes were found to lose strength during exposure and this was attributed to a slow progressive softening of the cement paste from the surface inwards. Examination by scanning electron microscope showed that the surface layers were characterized by a decalcification of the CSH with aragonite, magnesium silicate and thaumasite being the primary reaction products. Ettringite was only found beneath the softened surface zone, whereas gypsum was found both within and below the surface layers. Concrete containing fly ash showed a lower degree of strength loss and the surface was generally hard and intact in specimens with either 30 or 50% fly ash. There was no evidence of a brucite layer forming on the surface on any of the specimens examined.

INTRODUCTION

The chemical attack of concrete in a marine environment has been well documented by Mehta (1). The main deleterious effects result from reactions between CO_2 dissolved in the seawater and magnesium salts. Both components can attack the calcium-silicate hydrates (CSH) in addition to the calcium aluminates and lime compounds present in Portland cement concrete. Mehta (1) describes the decomposition of the CSH by the following reactions (Eqn. 1 & 2):

$$3CO_2 + 3CaO \cdot 2SiO_2 \cdot 3H_2O \rightarrow 3CaCO_3 + 2SiO_2 \cdot 2H_2O$$

$$MgSO_4 + Ca(OH)_2 + 3CaO \cdot 2SiO_2 \cdot 3H_2O \rightarrow 4MgO \cdot SiO_2 \cdot 8H_2O + CaSO_4 \cdot 2H_2O$$

The loss of CSH renders the concrete soft and weak, and may lead to loss of material especially in moving water. The sulphates associated with the magnesium and other cations present in seawater also attack the lime and calcium aluminates to form gypsum and ettringite. Sulphate attack in seawater does not usually lead to expansion and cracking of the concrete even though large quantities of ettringite may be produced, as ettringite is generally considered to be non-expansive in the presence of chloride ions.

Despite these destructive processes low-permeability concrete is highly resistant to chemical attack by carbonates and magnesium salts in seawater. Indeed, the initial products of attack, aragonite ($CaCO_3$) and brucite ($Mg(OH)_2$), may actually form dense protective skins on the surface of good quality concrete exposed to seawater. In the case of reinforced concrete, there is a much greater risk of chloride-induced corrosion of the steel reinforcement than chemical attack of the surrounding concrete. Indeed, the original purpose of the present study was to examine the effects of fly ash and curing on chloride penetration and steel corrosion in marine-exposed concrete. Results from these studies have been presented in a series of papers (2-6). However, results from these studies showed a marked strength reduction in Portland cement (OPC) concretes after 10 years exposure and the surface of these specimens was found to be soft and easily removed. Fly ash concretes were in much better condition with little or no visible deterioration in specimens with 30 or 50% fly ash. Although none of the specimens were air-entrained the deterioration of the OPC concretes was considered to be inconsistent with the action of freezing and thawing. Consequently, it was decided to perform chemical and petrographic examination of the specimens to determine the cause of damage and elucidate the role played by fly ash in mitigating deterioration. At the time of writing this paper only preliminary data, mainly from the examination of the 25-MPa and 35-MPa OPC concretes, were available. More detailed analysis of the concrete, especially those containing fly ash, will be presented at a later date.

EXPERIMENTAL METHODS

The concrete specimens examined in this study were 100-mm cubes and 100x100x300-mm prisms produced from three series of concrete mixes (with design strengths of 25, 35 and 45 MPa) using a single source of ordinary Portland cement, various levels of a low-lime fly ash (0 to 50% by mass of cementitious material) and siliceous coarse and fine aggregate. Details of the cementitious materials and mix proportions are given in Tables 1 and 2.

Table 1 Chemical Composition of Cementitious Materials

	SiO_2	Al_2O_3	Fe_2O_3	CaO	MgO	Na_2O	K_2O	SO_3	C_3A	LOI
OPC	20.6	5.07	3.10	64.5	1.53	0.15	0.73	2.53	8.00	-
Fly Ash	48.2	26.7	11.6	1.71	1.62	0.65	3.18	0.83	-	4.34

Table 2 Concrete Mix Proportions

Strength Grade	Fly Ash (%)	Mix Proportions (kg/m^3)		W/CM	Strength at 28 days (MPa)
		OPC	Fly Ash		
C25	0	250	-	0.68	32.5
	15	226	40	0.61	33.0
	30	202	87	0.54	34.5
	50	162	162	0.44	33.0
C35	0	300	-	0.57	41.5
	15	271	48	0.51	44.5
	30	242	104	0.45	45.5
	50	196	196	0.37	41.5
C45	0	350	-	0.49	50.0
	15	314	55	0.44	50.0
	30	280	120	0.39	53.0
	50	226	226	0.32	48.0

The specimens were left in the moulds under damp sacking and plastic at 20°C for 24 hours following casting after which time they were air-stored in the laboratory at 20°C and 65% RH. At 28 days concrete specimens were placed in the tidal zone of the BRE marine exposure site on the Thames Estuary. The average annual temperature of the sea water at this site is 10°C and its composition is compared with typical Atlantic Ocean water in Table 3.

Table 3 Composition of Seawater

Ions Analyzed	Composition (ppm)	
	BRE Site	Atlantic
SO_4	2600	2540
Cl	18200	17800
Ca	400	410
Mg	1200	1500
Na	9740	9950
K	400	330

Compressive strength tests were carried out on 100-mm cube specimens after 1, 2 and 10 years exposure. Powder samples were taken by drilling (in 5-mm depth increments) in eight locations on the four side faces of prisms; the samples were combined to provide a representative samples of each depth increment. Powder samples were ground and analyzed by XRF for chloride and calcium, and by LECO for carbon and sulphur. The cementitious content of the sample was calculated using the known *CaO* content of the cement and fly ash (i.e. non-calcareous aggregate was used) and the chloride then expressed as a percentage of the cementitious content. Powder samples milled in 1-mm depth increments from cube specimens were analyzed by differential thermal analysis (DTA) and x-ray diffraction (XRD) to determine the nature of interaction, if any, between the seawater and the cement hydrates. Furthermore, exposed surface layers, and internal fracture and polished surfaces were examined by scanning electron microscopy (SEM) equipped with energy-dispersive x-ray (and wavelength) analysis (EDS) to further determine the extent and nature of any chemical interaction.

DISCUSSION OF RESULTS

Results from compressive strength tests on 35-MPa concrete are shown in Fig. 1. During the first two years exposure all the concretes gained strength. However, between 2 to 10 years the control concrete lost more than 30% of its compressive strength. The extent of the strength loss was somewhat less for concrete with 15% fly ash and little change in strength was observed for concrete with 30% fly ash. Concrete with 50% fly ash continued to gain strength during this period, the 10-year strength being 11% higher than the 2-year strength. Similar behaviour was observed for the 25-MPa concretes.

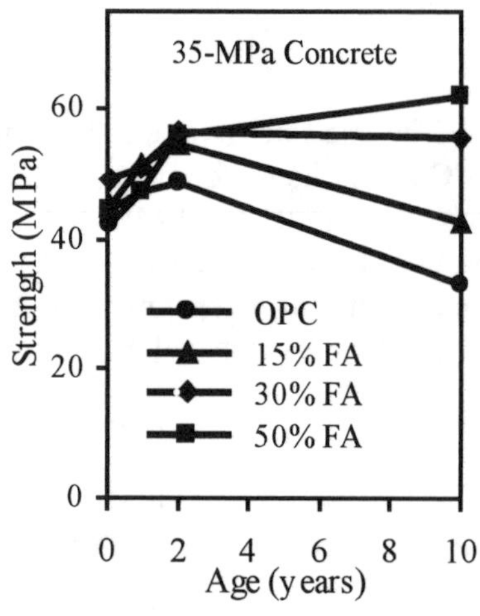

Fig. 1 Strength Results

Although, after 10 years the reinforced concrete specimens showed various amounts of cracking and rust staining consistent with corrosion of the embedded steel reinforcement, the visual damage on the non-reinforced specimens was confined to slight erosion at the corners of cubes from mixes without fly ash. There was little evidence of microcracking or surface scaling often associated with freeze-thaw damage in this environment. However, the surface layer of the OPC concrete cubes was soft and could be easily removed by gentle abrasion.

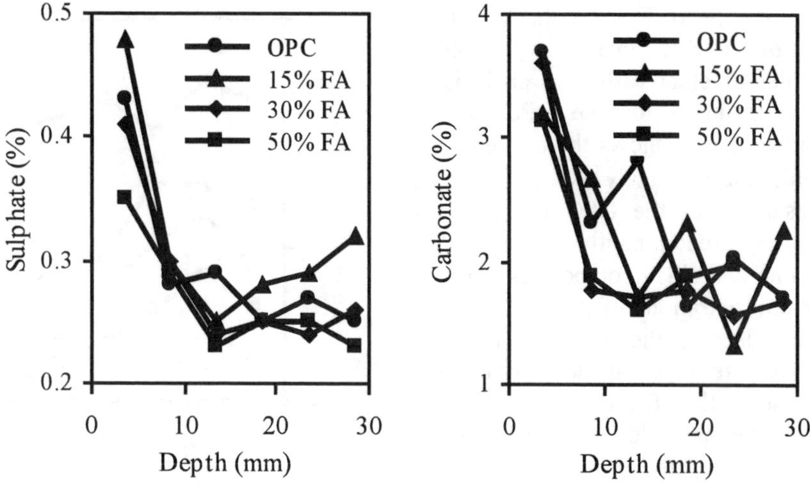

Fig. 2 Chemical Profiles in 10-year-old 25-MPa Concretes

The results from LECO analyses carried out on powder samples taken at varying depth increments by drilling are presented in Figure 2 for the 25-MPa concretes. The results are expressed as SO_3 and CO_2 contents as a percentage of the mass of concrete sample. Only the surface-most drilling sample with a depth interval of 1-6 mm contained enhanced levels of SO_3 and CO_2, indicating that no significant penetration of these compounds had occurred below a depth of 6 mm even after 10 years tidal exposure. Little consistent difference is observed between these profiles due to the incorporation of fly ash. Figure 3 shows chloride profiles for the same concretes (expressed as a percentage of the mass of cementitious material). Chlorides clearly penetrate much deeper into the concretes, especially the OPC concrete. The increased penetrate may reflect a higher diffusivity of chlorides in concrete compared with sulphates and carbonates, but is also likely due to the relatively small amount of chemical interaction between chlorides and the products of cement hydration compared to the other compounds.

Powder samples from the surface layers of the concrete specimens were also analyzed by x-ray diffraction (XRD). Small amounts of gypsum and limited quantities of ettringite were detected in most samples, but unique to the OPC samples was the presence of the mineral thaumasite ($CaSiO_3.CaSO_4.CaCO_3.15H_2O$) as shown in Figure 4. This sample also showed the presence of the mineral aragonite ($CaCO_3$) although this coexisted with the common form of calcium carbonate, calcite.

Figs. 5 to 9 show back-scattered electron (BSE) images of polished samples together with EDS analysis and x-ray dot-maps for the OPC concrete samples. Fig. 5 shows the existence of a 'skin', approximately 30 μm thick, on the surface of the OPC concrete. The skin is predominantly comprised of alumina and silica compounds with very low concentrations of calcium. There is no evidence of the formation of a Mg-rich brucite layer at the surface. It is interesting that the outer skin contains very little chloride compared with the underlying layers.

Figure 6 shows details of the concrete layer immediately below this skin. The calcium silicate hydrates have been altered and there is no evidence of

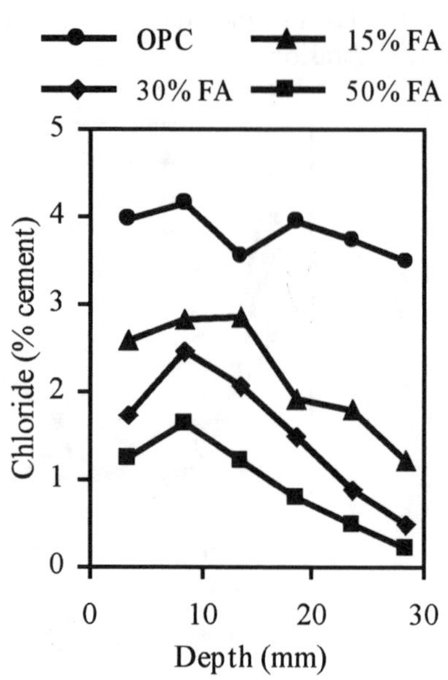

Fig. 3 Chloride Profiles in 10-year-old 25-MPa Concretes

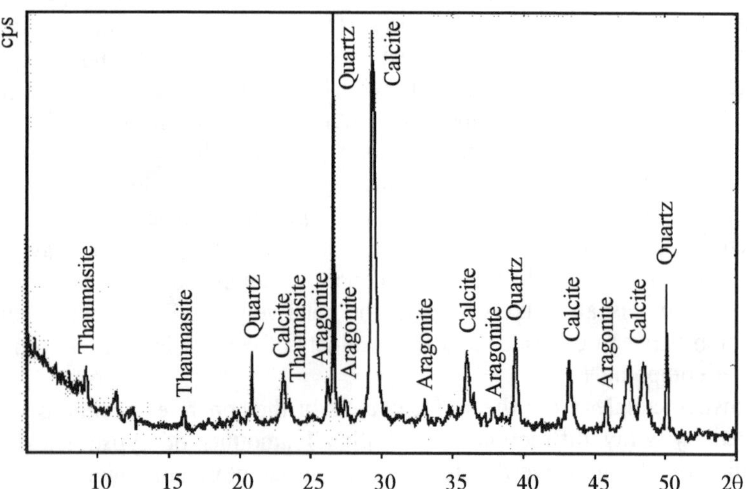

Fig. 4 XRD Results from Surface Layer of 25-MPa OPC Concrete

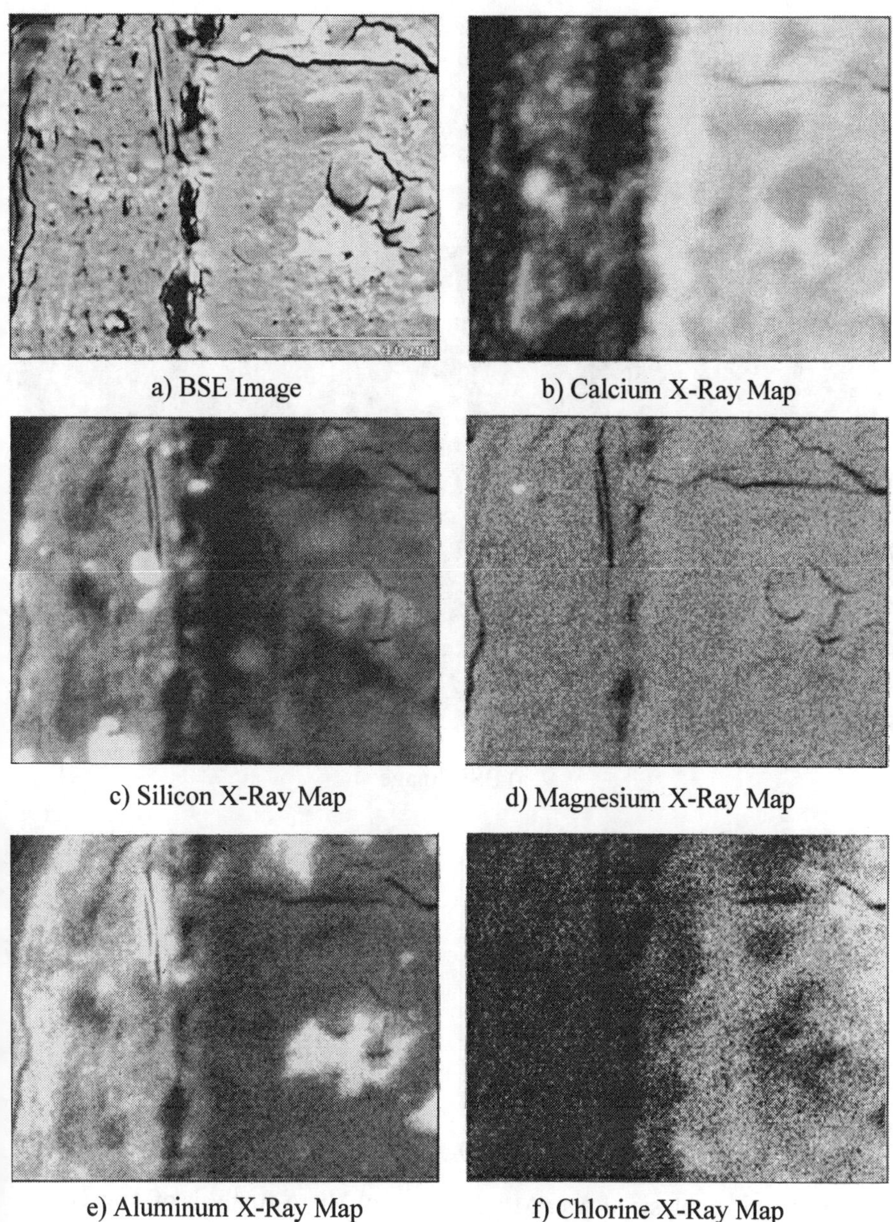

a) BSE Image

b) Calcium X-Ray Map

c) Silicon X-Ray Map

d) Magnesium X-Ray Map

e) Aluminum X-Ray Map

f) Chlorine X-Ray Map

Fig. 5 Surface Skin (approx. 30 microns thick) in OPC Concrete

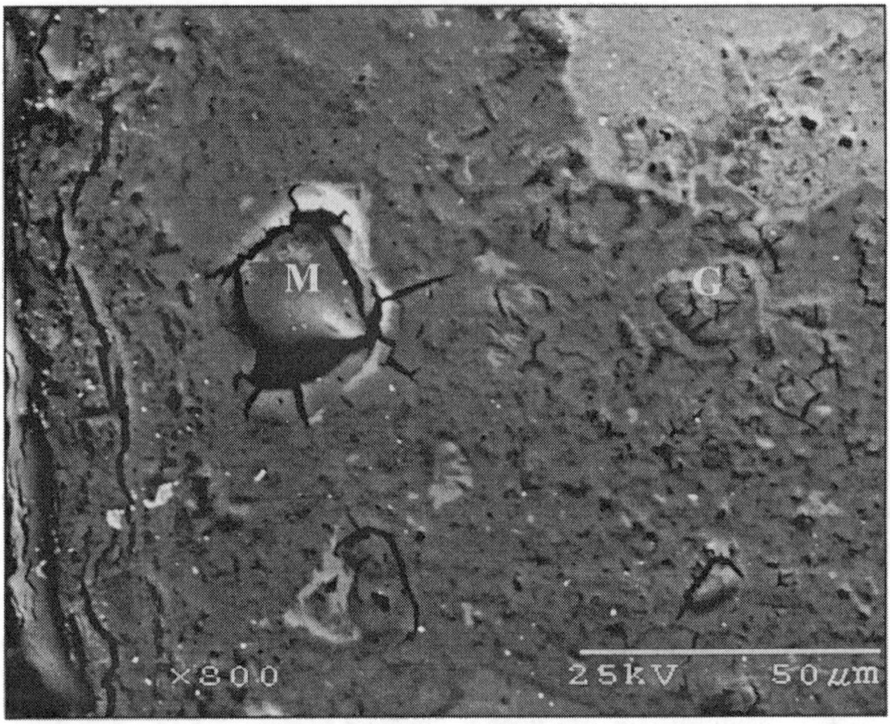

a) BSE Image

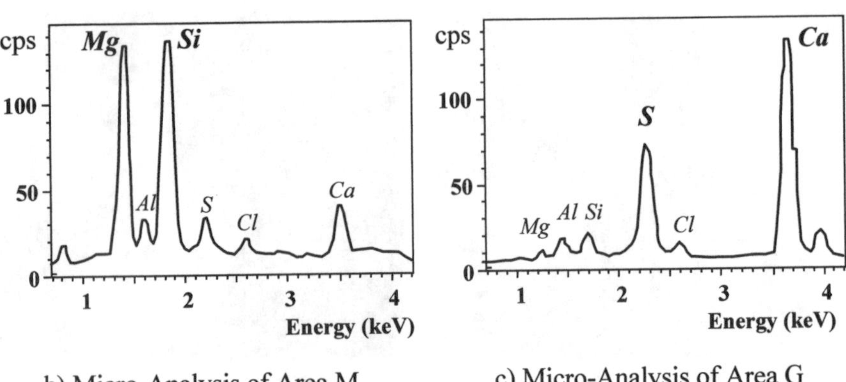

b) Micro-Analysis of Area M

c) Micro-Analysis of Area G

Fig. 6 Formation of Magnesium Silicate and Gypsum in OPC Concrete

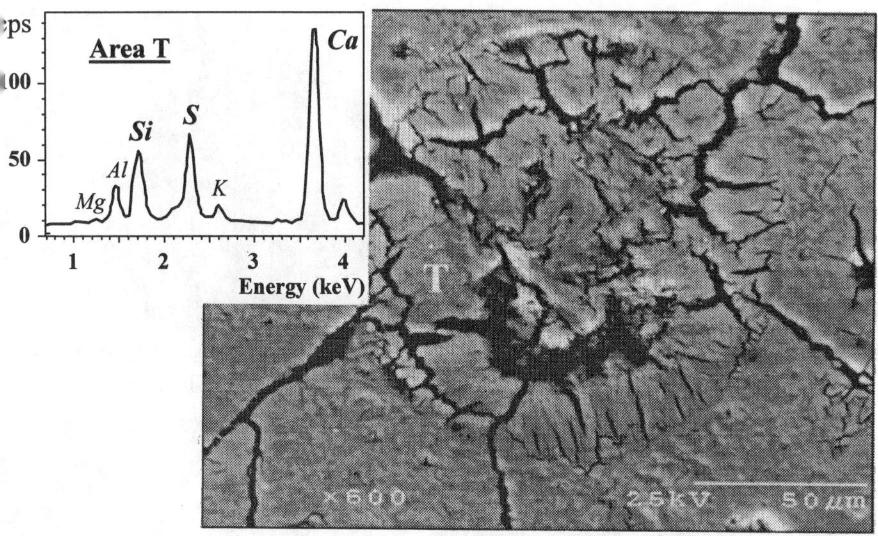

Fig. 7 Thaumasite in Void in OPC Concrete

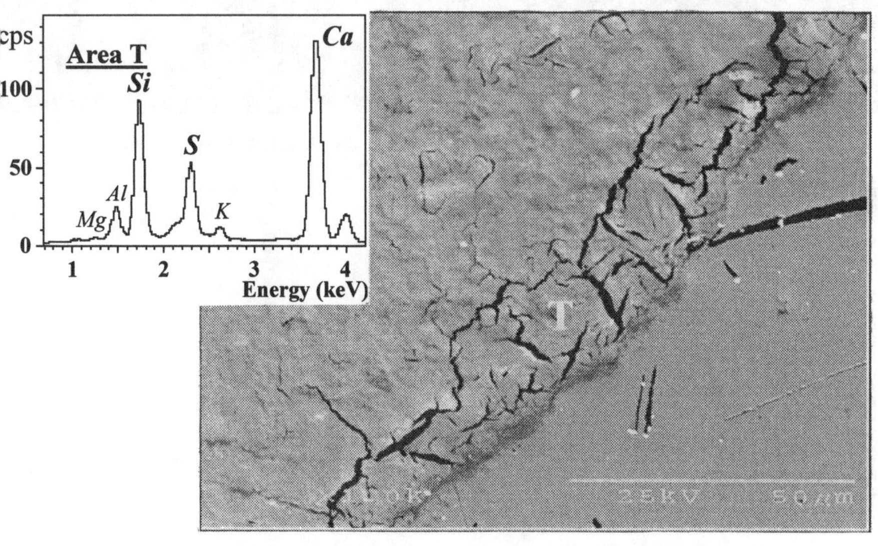

Fig. 8 Thaumasite in at Cement-Aggregate Interface in OPC Concrete

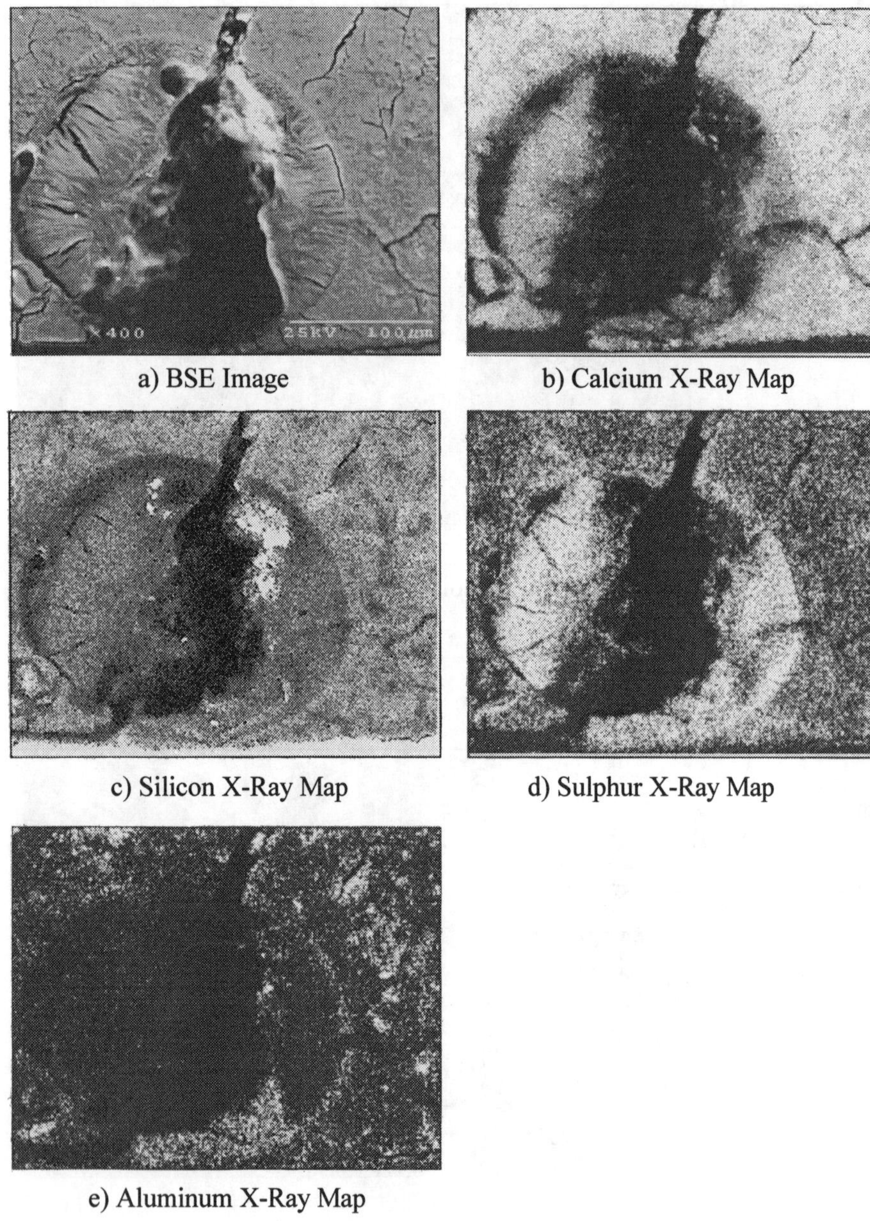

a) BSE Image

b) Calcium X-Ray Map

c) Silicon X-Ray Map

d) Sulphur X-Ray Map

e) Aluminum X-Ray Map

Fig. 9 Thaumasite in Void in the Surface Layer (0 - 5 mm) of OPC Concrete

residual unhydrated clinker particles, which are normally detected in mature concrete. This zone corresponds to the soft, weak layer of material and its composition is typical of seawater attack, with calcium being leached from the CSH leaving behind silica gel, magnesium silicates and gypsum (see Eqn. 1 and 2). These features are often present in concrete that has been attacked by magnesium sulphate (6-8). It is interesting though that little ettringite was found in the attacked layer in the present study in contrast to previous studies; this is probably a consequence of the carbonation of the surface layers. The surface layer was also characterized by the presence of significant quantities of the mineral thaumasite both in voids (Fig. 7 and 9) and at the periphery of some aggregate particles (Fig. 8). Thaumasite has been found in marine-exposed concrete by other workers (9), sometimes leading to severe deterioration (10).

A common feature of thaumasite formation, whether in seawater or sulphate-contaminated groundwater, is that the concrete contains a source of carbonate ions, such as limestone aggregate (10, 11). This is not the case in the presence study since both the coarse and fine aggregates are siliceous materials with a negligible content of carbonate.

In this study, it is possible that both the seawater and atmospheric CO_2 provided the source of carbonate required for thaumasite formation. Figures 1 and 2 indicate that significant levels of sulphate and carbonate ions penetrated the outer 5-mm thick layer of concrete during the 10-year exposure. The interaction of these ions with the cement hydrates produce, among other compounds, $CaCO_3 + 2SiO_2.H_2O + CaSO_4.2H_2O$. Since thaumasite is essentially comprised of these products, then the combined action of sulphates and carbonates has the potential to form thaumasite ($CaSiO_3.CaSO_4.CaCO_3.15H_2O$). However, thaumasite was not detected in specimens containing 30% or 50% fly ash in spite of the presence of similar concentrations of sulphate and carbonate as those found in the OPC concrete (see Figures 1 and 2). It is not quite clear how fly ash inhibits the formation of thaumasite in the presence of sulphate and carbonate ions, although it is possibly due to the pozzolanic reaction. The presence of aragonite (or calcite), silica gel and gypsum (from Eqns. 1 and 2) is not sufficient to form thaumasite without an additional source of lime, which may be provided by calcium hydroxide; for example Eqn. 3:

$$2C\overline{C} + S_2H + 2C\overline{S}H_2 + 2CH + 23H \longrightarrow 2\left(CS \cdot C\overline{S} \cdot C\overline{C} \cdot 15H\right)$$

The need for additional lime might also explain why thaumasite was detected at the cement-aggregate interface where increased levels of calcium hydroxide are generally found.

Although the chemical attack appeared to be confined to the surface zone of the 25-MPa and 35-MPa OPC concretes, this may be sufficient to explain most of the strength reduction observed between about 2 and 10 years. For example, it would take the loss of an 8-mm layer around a 100-mm cube to reduce the effective cross-section by approximately 30%. In addition, the deteriorated condition of the ends of the specimen in contact with the platens of testing machine would almost certainly result in non-uniform loading of the test specimen (e.g. stress concentrations) and reduced measured strengths. The consequences of this depth of deterioration would be less significant for larger elements. Furthermore, it should be noted that concrete designed for marine exposure by today's standards is likely to be of superior quality compared with the 25-MPa and 35-MPa concrete reported here. Unfortunately results from the 45-MPa concretes were not available at the time of writing. Thus the poor performance of the OPC concretes reported here may not be significant for properly designed concrete structures exposed to similar conditions. However, the results from this study do emphasize the benefits that may be attained through the use of fly ash in concrete exposed to aggressive environments.

CONCLUSIONS

Unreinforced OPC concretes with design strengths of 25 and 35 MPa were shown to deteriorate after 10 years exposure to a tidal marine environment. The deterioration took the form of surface softening due to chemical attack by the seawater, which was further manifested as a significant reduction in the strength of 100-mm cubes. The combined effect of sulphates (including $MgSO_4$) and carbonates resulted in a breakdown of the CSH to form aragonite (and calcite), magnesium silicates, gypsum and thaumasite. Little ettringite was found in the deteriorated surface.

The strength of the concrete increased with the level of fly ash addition and little damage was observed in concrete with 30% or 50% fly ash. The consumption of lime by pozzolanic reaction with the fly ash may be sufficient to explain the absence of deterioration due to the presence of thaumasite in these concretes and thus their superior performance in marine exposure.

REFERENCES

1. Mehta, P.K. Concrete in the Marine Environment, Elsevier Applied Science, London, 1991.
2. Thomas MDA. "Marine performance of pfa concrete." Magazine of Concrete Research, Vol. 43, No. 156, 1991, pp 171-185.
3. Thomas, MDA and Matthews, JD. "Chloride penetration and reinforcement corrosion in marine-exposed fly ash concretes." Third CANMET/ACI International Conference on Concrete in a Marine Environment, (Ed. V.M. Malhotra), ACI SP-163, American Concrete Institute, Detroit, August, 1996, pp.317-338.

4. Bentz, EC, Evans, CM and Thomas, MDA. 1996. "Chloride diffusion modelling for marine exposed concrete." Corrosion of Reinforcement in Concrete Construction, (Ed. CL Page, PB Bamforth and JW Figg), Royal Society of Chemistry, Cambridge, 1996, pp. 136-145.
5. Thomas, MDA. "Chloride thresholds in marine concrete." Cement and Concrete Research, Vol. 26, No. 4, 1996, pp. 513-519.
6. Bonen, D. and Cohen, M.D. “Magnesium sulphate attack on portland cement paste. I. Microstructural analysis.” Cement and Concrete Research, Vol. 22, No. 1, 1992, pp. 169-180.
7. Bonen, D. and Cohen, M.D. “Magnesium sulphate attack on portland cement paste. II. Chemical and mineralogical analyses.” Cement and Concrete Research, Vol. 22, No. 4, 1992, pp. 707-718.
8. Gollop, R.S. and Taylor, H.F.W. “Microstructural and microanalytical studies of sulfate attack. I. Ordinary portland cement pastes.” Cement and Concrete Research, Vol. 22, No. 6, 1992, pp. 1027-1038.
9. Mehta, P.K. and Haynes, H. “Durability of concrete in seawater.” Journal of the American Society of Civil Engineers Structural Division, 101 (ST8), 1975, pp. 1679-1686.
10. Bickley, J.A., Hemmings, R.A., Hooton, R.D. and Balinski, J. “Thaumasite related deterioration of concrete structures.” Concrete Technology: Past, Present and Future (Ed. P.K. Mehta), ACI SP-144, American Concrete Institute, Detroit, 1994, pp. 159-175.
11. Crammond, N.J. and Halliwell, M.A. “The thaumasite form of sulfate attack in concrete containing a source of carbonate ions – a microstructural overview.” Advances in Concrete Technology (Ed. V.M. Malhotra), ACI SP-154, American Concrete Institute, Detroit, 1994, pp. 357-380.

FIELD STUDIES OF SULFATE ATTACK ON CONCRETE

John Figg
John Figg and Associates
5 Andrewes Croft
Great Linford
Milton Keynes
MK14 5HP
UK

ABSTRACT

High sulfate ground conditions are found in many places, but widespread damage has largely been prevented by established precautions, although the mechanism of reaction and damage is still not completely understood. A wider range of options for combating adverse field conditions is now available. Nevertheless, new variants of sulfation degradation continue to arise and expensive cases of damage to buildings have occured in recent years. Some of these cases are discussed in this present paper.

INTRODUCTION

Attack on concrete foundations by sulfates derived from the ground was, some 60 to 70 years ago, a serious problem in those parts of the world where the ground naturally contains sulfate salts. Disintegration of concrete maritime works was also attributed, in part, to reaction between the sulfate in seawater and the hydrated cement paste in the concrete.

By the advent of World War II theses difficulties appeared to have been overcome and in the post-war period codified recommendations for buried concrete were established in which the use of maximum levels of C_3A in cement and maximum values for the water/cement ratio were set. The use of pozzolanic additives and slag mixes has also been found beneficial.

The objective of these precautionary measures has been to reduce the amount of reactive aluminates in the concrete whilst also making the concrete more impermeable, hence hindering the access of aggressive sulfate solutions.

The precautions have been extended from natural soils to high-sulfate ground conditions caused by industrial pollution. Further modifications have also allowed

for the simultaneous presence of sulfates and acidic conditions, as well as the influence of static or flowing below-ground waters.

The then state of knowledge of sulfate reactions, including field studies is essentially recorded in the Proceedings of the Thorberger Thorvaldson symposium held in Toronto in 1967 [1]. Contributors to this meeting were asked to emphasise results from field performance and from these it is clear that damage to structures depended greatly on whether sulfate-bearing solutions could penetrate the concrete by capillary action coupled with evaporation from and exposed surface. Damage could be especially marked where evaporation was possible at an exposed edge or corner of a concrete element.

The proceedings of the Thorvaldson symposium record that in high-sulfate environments the best defence against sulfate attack is provision of concrete of low permeability, coupled with use of adequate and durable waterproofing. Only then can the advantageous properties of sulphate-resisting cement be realised. Air entrainment is also advocated for enhanced sulfate resistance. The importance of proper curing is emphasised.

Both steam curing and deliberate carbonation of precast concrete elements are also identified as confering improved resistance to sulfate attack.

The symposium does not specifically deal with the mechanisms and chemical reactions of sulfate solutions, but it is interesting to note that in many instances the type of deterioration occurring is of the concrete disintegrating to a mushy paste; the expansive form of sulfate attack appears to be rarer. This suggests that in the low temperature soil conditions of Canada the Thaumasite mode of reaction may be the most prevalent circumstance (*see below*).

The occurrence, diagnosis and prevention of adverse sulfate (and chloride) reactions, with particular reference to UK conditions has subsequently been reviewed and discussed [2, 3]. No significant problems with adverse sulfation have been reported where recommended precautions have been adopted.

However, like a virus which mutates to wreak havoc in an unprepared population, new variants of sulfate reactions have been encountered and these show the necessity for constant awareness and a proper appreciation of the chemistry, volumetric changes and engineering consequences of sulfation of concrete.

FIELD EXPERIENCES

Combustion Products

Fuels used in industry, especially electricity generation, may contain sulfur which after combustion result in the formation of sulfur oxides (SO_2 and SO_3). In circumstances where condensation can occur these gases can cause severe damage to concretes due to the combination of acid and sulfate attack.

Typically the uppermost parts of chimneys are at risk and very often nowadays this part of the flue will be constructed with engineering bricks and chemically-resistant mortar. Concrete elements immediately downstream of a chimney may be attacked. One such incident involved a multi-storey building where the structural frame was exposed at rooftop level. The concrete had performed well for two decades, when the boiler installation was modified by fitting a new and more thermally efficient one. The temperature of the exhaust gases from the boiler was now much lower and condensation on the concrete frame was now a regular occurrence. After a few years the concrete, made with normal Portland cement, exhibited cracking and erosion of the cement paste matrix, a clear indicator of solution sulfation.

With this sort of condensation/chemical attack the actual mode of distress is largely a matter of the concrete quality. Structural elements, such as the reinforced concrete frame mentioned above, will be relatively impermeable and hence, at least initially, suffer the "exposed aggregate" or "acidic" type of attack. Only with lower quality, more permeable concretes, or after prolonged contact will the sulfate solution be able to penetrate the material to cause expansion and cracking damage.

Contaminated Fill

A number of cases have been reported of sulfation damage to concrete floor slabs due to use of sulfate-containing fill materials. In coal mining districts the huge piles of colliery spoil "slag tips" were at one time a characteristic sight and there has always been a temptation as well as a financial incentive to make use of the cheap material as a bulk fill or aggregate source. (Since the Aberfan disaster in Wales on 21 October 1966 [4] strenuous efforts have been made to remove and/or landscape the tips and in some circumstances this has increased the desire to utilise the industrial waste as a construction material).

Dwellings in the Midlands and North of England have experienced heave or "doming" of concrete ground floors where colliery spoil has been used as fill beneath the concrete. If rising damp conveys sulfate salts the concrete of the floor undergoes expansive sulfate attack. The brick cavity walls act as a constraint to the expansion of the concrete and the only relief is for the concrete to bulge upwards. In severe cases this can result in sticking of doors and development of large cracks in walls.

This should not happen if an effective damp-proof course is installed, but in the past it has been permissable for the damp-proof membrane to be on top of the concrete (to protect moisture susceptible floor finishes) and usually this will allow some damp penetration of the concrete.

Colliery spoil is liable to contain a significant proportion of sulfur, as sulfate, and this material should always be tested before use as a fill where water is available for leaching to occur.

In Scotland difficulties have arisen from the use of burnt colliery spoil, which locally is called *Blaes*. Combustion of spoil tips often occurs and some tips have burned for years. The burning process may result in oxidation of sulfur compounds to give enhanced sulfate concentration. The high-sulfate *Blaes* has been the cause of several instances of damage to ground-bearing concrete slabs, including in a major hospital in Glasgow.

It is, of course, significant that all instances of damage involve the presence of water and that incorrect construction has allowed the aqueous salts solutions to initiate the sulfate attack.

The *Mundic* Problem [5]

About as far as it is possible to get from Scotland in the UK is the County (actually Royal Duchy) of Cornwall. Here pyrite in the mining wastes from the, now abandoned, tin, lead and copper mines is known by the old Cornish word *Mundic*. Especially in the St Austell area the *Mundic* waste has been used as a concrete aggregate, in particular for the manufacture of concrete blocks.

The South West of England has a tradition of stone masonry construction and the masons skills have carried over to the use of concrete block masonry for small commercial and domestic buildings.

The *Mundic* blocks when used in rendered cavity wall construction slowly carbonate from the inside. In the outer leaf (waithe) the sulfide minerals oxidise to sulfate and in the presence of moisture result in expansive reaction with the cement minerals and disintegration of the blockwork walls.

The *Mundic* Problem has caused particular distress to individual householders as it is considered an inherent defect which results in the property being uninsurable and ineligible for a mortgage loan. Until recently the mortgage companies took the view that all houses in the area where problems had been encountered were a bad risk, but the Royal Institute of Chartered Surveyors has now established a diagnostic test based on the petrographic examination of block samples, which has enabled some 80% of properties to be restored to the market.

Pyrites Oxidation

Pyrites occurs in many gravels in the Southeast of England and also in other geological strata elsewhere in the Country. In gravels the concentration may be well under 1% but if the pyrites particle is incorporated in concrete and finishes up close to an outside surface, subsequent oxidation can cause pop-out spalling which appears to be a combination of sulfation expansion and rust formation. The end result is an unsightly blemish which is highly visible due to the pock-marked surface coupled with iron-staining of the concrete.

Where pyrites-induced damage has affected concrete slabs, the depressions can fill with water and in Winter result in further damage due to freeze-thaw action.

Tunnelling works, especially in compressed air conditions, can cause oxidation of pyrites in the sides of the excavation which can affect the setting/hardening of concrete cast in contact with the now sulfate-enhanced and acidic surface. The influence on overall durability is marginal, but it has been the cause of contractual arguments where concrete quality has been questioned.

Sewers and Drains

Sewage and industrial effluent frequently includes sulfur-containing material. In warm, anaerobic conditions sulfurphile bacteria can thrive with formation of hydrogen sulfide gas, which in turn, can later be oxidised in other parts of the sewage system by aerobic organisms to produce sulfur oxides. The resultant formation of sulfurous/sulfuric acids has been the origin of severe damage to concrete sewers and drainage installations.

South Africa experienced major problems of this type which are summarised in the remarkably comprehensive 1959 CSIR publication on Corrosion of Concrete Sewers [6]. This book sets out the results of both field and laboratory studies of the problem and gives recommendations for avoidance of future occurrences of this nature.

The meticulous field observations of damage to sewers in service are coupled with concurrent microbiological and chemical/physical tests to conclusively demonstrate the processes and mechanisms of this type of deterioration. Despite the warm temperatures experienced in sewer conditions, most degradation is reported to be of the eating-away or softening of the cement matrix, although some instances of sulfate-induced expansion are noted.

Some quotations are well worth repeating, e.g. if concrete pipes "—have a low permeability and are made of high quality concrete, and if the pipes are well-cured and carbonated before use, no trouble from sulphate expansion should be experienced" and "The resistance of concrete to sulphate solutions can be improved by reducing the penetration of the solution into the body of the concrete, and confining its action to the surface, so that only the surface of the concrete will deteriorate, the inner mass being affected only as the outer corroded material falls away" and again "The proportioning of the concrete constituents, the grading of the aggregate and the water:cement ratio will vitally affect its physical characteristics".

Key findings from this study were that significantly improved resistance to sulfate attack could be obtained by:

- decrease in water/cement ratio
- increase in cement content
- pre-exposure carbonation

improved performance could also be obtained by:

- use of resistant cement (pozzolanic cement, Portland cement of low C_3A content, calcium aluminate cement, blastfurnace slag cement)

The CSIR volume remains an important source of information on sulfate attack which diserves greater attention.

South Africa is by no means the only place where such microbiologically induced attack has occurred and other countries where problems have been reported are Egypt, Cyprus, Australia and the USA (California and Wisconsin)

More recently rather similar problems have been reported with sewers in Japan, especially Tokyo. Here the damage has been exacerbated by scouring of the ground around buried sewer pipes by escaping effluent to create large cavities, which eventually collapse causing major traffic problems, since most pipes are beneath roads. The unusually prevalent sulfate attack is attributed to the widespread use of soy sauce in the Japanese diet (it has a high sulfur content). Japanese researchers have developed the use of ground-penetrating radar to detect cavities at an early stage in order that defects can be rectified before damage occurs [7]. The formation of sulfate, prior to attack on the concrete, involves sulfate-reducing and sulfur-oxidising organisms.

The Thaumasite Type of Degradation

In the UK recently there has been great concern following discovery of damage to concretes, in particular bridge abutments, where the cement paste of the concrete has disintegrated to a mushy paste, with the principal identified deterioration product being thaumasite. This type of attack involves the calcium silicates of the cement as well as the calcium aluminates and is favoured by low temperature conditions. It also appears that limestone aggregates and, especially calcium carbonate addition in cement, can act as a source of carbon dioxide as well as the atmosphere.

Thaumasite and ettringite are end members of a continuous series of solid solutions having similar structure and morphology, and it has been known for over three decades that in many instances concrete deterioration products have been mis-diagnosed because of the difficulty in distinguishing between them [8].

One of the peculiarities of the thaumasite type of attack is that sulfate-resisting cements appear to be more vulnerable than normal Portland cements. It is also worrying because up to 5% limestone dust addition to cement is expressly permitted by the British Standard (BS 12) for Portland cement.

A number of case studies have been reported involving concretes suffering from thaumasite type sulfation [9-12]. However, despite rather emotive alarms in the press, it is clear that concrete of good quality will not suffer deleterious strength reduction from this form of sulfate attack [13].

PROSPECTIVE DURABILITY EXPERIMENTS

Sheerness Seawater Trials

Although the types of concrete investigated by the Sea Action Committee of the Institution of Civil Engineers [14] are now well outdated, the results of this seawater immersion and laboratory study, which commenced in 1929, are still valid.

In terms of overall durability, chemical (sulfate) attack on the concrete of the piles was less important than cracking caused by corrosion of reinforcement, but both degradation mechanisms were found to be dependent on the quality of the concrete i.e. impermeability of the cementitious matrix. Two principle factors influencing permeability were demonstrated to be the cement content of the concrete and the presence of an effective pozzolan. Where rusting of reinforcement was concerned, the thickness of the concrete cover was also a major factor.

This experiment commenced before sulfate-resisting cement came into common use, but the importance of achieving low permeability for maximising resistance of concrete to chemical attack, is well exemplified.

Northwick Park, Long-Term Sulfate Durability Tests

When ground condition surveys were made, preparatory to construction of a new hospital at Northwick Park, near Harrow, it was discovered that the London Clay contained an unusually high proportion of sulfate (about 0.35% w/w). The opportunity was taken to establish a long term durability experiment to evaluate the performance of buried concretes, in particular the type of high-slump concrete mix used for cast in-situ piles.

Although laboratory, and field studies, had demonstrated that concretes could be attacked by sulfates, the effects in terms of structural performance were still in contention. For skin-friction piles, which are commonly used in clays, two possible scenarios could be envisaged. Firstly, that expansive sulfate attack on the cylindrical periphery of the piles could actually increase the load-bearing capability of the piles, but perhaps cause upthrust for the supported building. Or, secondly, that the deteriorated annulus of the pile would allow the undamaged central, and smaller diameter core, to sink through the rubble, with consequent subsidence damage to the structure.

It was originally intended that the Northwick Park experiment would include loading tests on the in-situ piles, but at a late stage, this part of the programme was omitted for financial reasons. Nevertheless, in 1970 both in-situ and precast pile specimens were buried adjacent to the hospital site, along with a number of other types of cementitious materials. A basement structure was also build with walls both cast against the clay and with shuttered walls, later backfilled with the site clay.

Samples were excavated after 5 and 15 years, but the intended 25-year samples have yet to be removed.

The results after 15 years [15] show that only incompletely compacted concretes have recognisable sulfate attack. Some Portland cement mixes were weaker by up to 5 per cent, but most concrete had gained in strength during burial. The distance sulfate had penetrated into the concrete was not dependent on cement type, but the with sulfate-resisting cements the concentration gradient was less steep, indicating that less reaction had occurred. Carbonation by exposure to the atmosphere signifantly reduced sulfate penetration. The main finding to date is the importance of impermeability for combating sulfate attack.

CONCLUSIONS

Field studies, both accidental and prospective, show that concrete quality, especially low permeability, is the key requirement for resistance to sulfate attack. This property is a function of adequate cement quantity and a minimum water/cementitious material ratio. Pozzolanic addition can significantly reduce permeability and hence improve sulfate resistance. Low C_3A (sulfate-resisting) cement is less effective against the thaumasite-type of sulfation than for the ettringite-type reaction. Proper curing and pre-carbonation are effective durability enhancing actions.

REFERENCES

[1] "Performance of Concrete: Resistance of Concrete to Sulphate and Other Environmental Conditions", A Symposium in Honour of Thorbergur Thorvalson, American Concrete Institute,1967, E G Swenson (Technical Editor), Canadian Building Series No.2, University of Toronto Press, 1968, 243+viii pp.

[2] John Figg, "Chloride and Sulphate Attack on Concrete", *Chemistry and Industry (London)*, 17 October 1983, pp 770-775.

[3] John Figg, "Salt, Sulphate, Rust and Other Chemical Effects", Proceedings, Ben C Gerwick Symposium on International Experience with Durability of

Concrete in Marine Environment, University of California, Berkeley, 16-17 January 1989, K C Mehta (Editor), pp 49-69.

[4] The Aberfan Disaster, http://www.nuff.ox.ac.uk/politics/aberfan/home.htm.

[5] Alan Bromley and Kelvin Pettifer, "Sulphide-Related Degradation of Concrete in Southwest England ("*THE MUNDIC PROBLEM*"), Building Research Establishment, Garston, 1997, ISBN 1 86081 137 X, 174+ix pp.

[6] "Corrosion of Concrete Sewers", South African Council for Scientific and Industrial Research, Series DR 12, 1959, 236 pp.

[7] H Tomita. H Tada, T Nanbu, K Chou, T Nakamura and T McGregor, "The Nature and Detection of Void-Induced Pavement Failures", *Transportation Research Record*, 1995, No. 1505, pp 9-16.

[8] B Erlin and D Stark, "Identification and Occurrence of Thaumasite in Concrete", *Highway Research Record*, 1966, No. 113, pp 108-113.

[9] J Bensted, "Thaumasite - A Deterioration Product of Hardened Cement Structures", *Il Cemento*, 1988, Vol 1, pp 3-10.

[10] N J Crammond and P J Nixon, "Deterioration of Concrete Foundation Piles as a Result of Thaumasite Formation", Proceedings, 6th Conference on the Durability of Building Materials, Japan, 1993, Vol 1, pp 295-305.

[11] J A Bickley, R T Hemmings, R D Hooton and J Balinski, "Thaumasite Related Deterioration of Concrete Structures", Proceedings, Conference on Concrete Technology: Past, Present and Future, American Concrete Institute, 1994, Special Publication SP 144-8, pp 159-175.

[12] N J Crammond and M A Halliwell, "The Thaumasite Form of Sulphate Attack in Concretes Containing a Source of Carbonate Ions", Proceedings, 2nd Symposium on Advances in Concrete Technology, 1995, American Concrete Institute, Special Publication SP 154-19, pp 357-380.

[13] John Bensted, "Scientific Background to Thaumasite Formation in Concrete", Proceedings, Seminar on Recognising Thaumasite: The Diagnosis of an Unusual Form of Sulphate Attack, University of Hertfordshire, Fielder Centre, 9 September 1998, 12 pp.

[14] F M Lea and C M Watkins, "The Durability of Reinforced Concrete in Sea Water", National Building Studies, Research Paper No. 30, *London, HMSO*, 1960, 42+vi pp.

[15] W H Harrison, "Sulphate Resistance of Buried Concrete: Third Report on a long term investigation at Northwick Park and on Similar Concretes in Sulphate Solutions at BRE", Building Research Establishment, Garston, 1992, Publication BR 164, 69pp.

LONG TERM TEST OF CONCRETE RESISTANCE AGAINST SULPHATE ATTACK

Björn Lagerblad, Swedish Cement and Concrete Research Institute (CBI)
S-10044 Stockholm Sweden

ABSTRACT

The most common way to test external sulphate attack is to measure the expansion of a mortar prism in a sulphate solution. The results presented in this paper come from concrete prisms stored in sodium sulphate solutions for an extended period of 5 years. The data gives more precise information on the expansion mechanism. The expansion is controlled by the content of C_3A in the cement, the diffusivity of sulphate ions into the concrete and the shape and size of the concrete body. The expansion is in two steps where a first small expansion, mainly controlled by the contents of C_3A, is followed by a breakpoint and an accelerated expansion. This is explained by a model where the expansion takes place in a sulphate-enriched shell with restrains from the interior. When the interior breaks due to the strain, the ingress of sulphate ions accelerates the expansion. Addition of silica fume and a low water/binder ratio hampers the ingress of sulphate ions but do not stop the expansive reactions. The geometric constraint makes it possible to get exfoliation of the concrete without measuring expansion on a slender mortar prism. Granulated blast furnace slag mainly dilute the contents of C_3A.

INTRODUCTION

It is well known that sulphate solutions attack concrete in several ways. The sulphate attack can lead to strength loss, expansion, spalling of surface layer and ultimately disintegration (1). This paper will mainly treat the expansion, which can give rise to spalling. In principle there are two major reactions that destroys the concrete, the reactions with the portlandite and the reactions with the aluminate components of the paste. The portlandite reactions mainly effect the surface, which breaks down while the aluminate reaction lead to an expansion, which eventually cracks the concrete. The major expansion is thought to be linked to reactions where monosulphate or other aluminum components in the paste reacts with sulphate ions to form ettringite which gives a volume increase and expansion. The mechanism behind the expansion is complicated and debated. The expansion is, however, linked to the contents of C_3A in the cement and sulphate resistant cements are stipulated to have a low content of this component. The mechanisms of expansion and the role of ettringite are discussed by Brown and Taylor (2). The expansion and cracking will, in turn, open up the concrete for an increased attack, which eventually breaks down the cement paste from the surface.

This paper will treat the expansion of concrete prisms in sodium sulphate solution, as this is a standard way to give expansion, and discuss the expansion mechanism. The common way to test sulphate resistance is according to ASTM C 1012-89. In this test long slender mortar prisms (250 x 25 x 25 mm) are subjected to a 0.35 molar sodium sulphate solution and the sulphate resistance is measured as expansion. Different standard for evaluating the susceptibility of cement-based materials to external sulphate attack is discussed by Clifton et al (3). The material properties of massive concrete are, however, different from that of a slender mortar prism. The larger mass will give physical constraints to the expansion and cracking pattern. In our test we have subjected concrete prisms (300 x 75 x 75 mm) to the same sulphate solution for more than 5 years in contrast to ASTM C 1012-89 where a test period of 6 month is stipulated. The concrete prism test will be less accelerated and will thus give another type of information. In the experiment series we have tested different water/ binder ratios, different binder amounts and different binder combinations including silica fume and granulated blast furnace slag. The tests with concrete prisms reveal more details than the mortar prisms.

This research was originally part of a larger research program on chemical resistans of high strength concrete. As the test gave inconclusive data during this research program the test period was prolonged. Besides the sulphate resistance the same concretes have been tested for acid attack and attacks in different agricultural environments. They were also tested in 1 molar NaOH solution for a couple of years. None of the mixes gave any alkalisilica expansion.

MATERIALS AND EXPERIMENTAL DATA

In the test series we have casted concrete prisms with a water/binder ratio (w/b) of 0.55, 0.45 and 0.35. We have used two different types of Portland Cement, one ordinary and one sulphate resistant. Some mixtures also include granulated blast furnace slag and silica fume. In the mixes we have used an alkaliresistant granitoid aggregates with a D-max of 18 mm.

In most of the mixes the amount of paste was standardised to 300 litres except in two series were the amount of paste was 250 and 350 litres. A compilation of all the mixes is in table 2. The concrete was cast in prismatic forms with an inner dimension of 75 x 75 x 300 mm. The casting was done in September, when the local climate is around 10-15 °C and fairly damp. A concrete factory, SABEMA AB, cast the concrete, in an industrial standard way. The concrete was remoulded after 24 hour after which they were stored outdoors for one month before delivery to CBI. At CBI they were stored indoors for two month before the test started. At the end of each prism a measuring pin was inserted. Two prisms of each mix was used and measured. They gave the same expansion and after four years we continued with only one prism.

Table I - Cement composition according to type analysis.
OPC = ordinary Portland cement. SR = Sulphate resistant Portland cement

Oxides	Slite Std (OPC)	Degerhamn Std (SR)
CaO	62.6	64.5
SiO2	19.8	22.2
Al2O3	4.3	3.5
Fe2O3	2.2	4.7
K2O	1.3	0.6
Na2O	0.3	0.1
MgO	3.4	1.0
SO3	3.2	2.0
LOI	2.5	0.7
Free Lime	1.3	0.8
Clinker (Bogue)		
C3S	58	51
C2S	13	25
C3A	8	1.2
C4AF	7	14

Table II - Concrete mixtures. Amount in weight %. Paste in litres per m^3 concrete.
SF = Silica Fume, BS = Blast Furnace Slag. OPC and SR see above.

Cement	OPC	OPC	SR	SR	OPC	OPC	OPC	OPC
Additive	SF %	BS %.	SF	BS	SF	BS	SF	BS
Paste	300 lit	300	300	300	250	250	350	350
W/B								
0.55	0	0	0	0				
0.55	8	0						
0.55	0	0						
0.55	0	50						
0.55	8	50						
0.45	0	0	0	0	0	0	0	0
0.45	8	0						
0.45	16	0	16	0	16	0	16	0
0.45	0	50						
0.45	8	50	8	50	8	50	8	50
0.35	0	0	0	0				
0.35	8	0						
0.35	16	0	16	0				
0.35	0	50						
0.35	8	50	8	50				

The composition of the liquid was the same as in ASTM C 1012-89 a 0.35 molar sodium sulphate solution. The solution was renewed once a year. Sulphuric acid was used to keep the pH below 10. The prisms were stored standing in containers with a pump that circulated the liquid. The uppermost parts of the prisms were above the liquid surface to enhance sulphate ingress. The measuring was done in a measuring frame with a standard and a digital measuring gauge with a precision of 0.001 mm. The precision of the measuring is around 0.05mm.

The dry prisms were measured and the experiment started 93-11-10. The first measuring was 93-12-29 and all the prisms had expanded. This was presumably mainly due to wetting and thus this value is used as starting point in all diagrams.

RESULTS

There are mainly three variables that are of interest, type of cement, type of additive and water/binder ratio. They will be treated consecutively.

Figure 1 shows the reactivity of the cements tested on mortar prisms according to ASTM C 1012-89 in a one-year test. Here we can notice as expected, from the contents of C_3A, that the OPC will give a distinct expansion. The sulphate resistant cement will also give some expansion but no acceleration and damage. The mortar prisms with low water/ binder ratio give a very small expansion. This indicates that a low W/B and silica fume hampers the sulphate expansion and concrete damage.

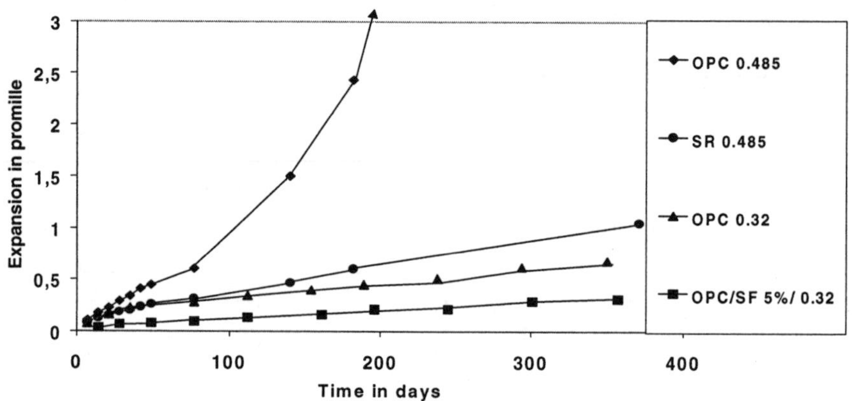

Figure 1 - Sulphate expansion of mortar prisms according to ASTM C 1012-89. Abbreviations as in Table II. The number after the binder type is the W/B.

Figure 2 shows the expansion of all the concrete prisms. As expected it takes much longer time for the expansion to occur. When the expansion of the prisms exceeds around 0.7 ‰ very small can be observed on the surface. A distinct map-cracking pattern can not be noticed until after the acceleration when the expansion exceeds 1 ‰. Cracks resulting from the sulphate expansion are less frequent and much thinner than cracks formed by the alkalisilica reaction at the same level of expansion. This is presumably due to that the expansion mechanism is different. When the paste expands as in the sulphate reactions the crack pattern will be different from that when the aggregate expands. Delayed aggregate expansion also gives a more distinct map-cracking pattern than the external sulphate attack. The most visible cracks in the external sulphate attack appear parallel to the surfaces and at the corners and can in some cases also be found on prisms that have not expanded significantly. These cracks are often filled with gypsum. The top part of the prisms that has been exposed to air shows more severe surface damage probably due to salt crystallisation processes.

During the first period, when the original research program was running (2 years), we did not get any significant expansion at all. The only expansion we noticed was a small expansion that mainly related to the amount of C_3A in the mix. It was not until after three years that the first prism expanded enough to be cracked. The results indicate that you first get a "small" expansion to a certain breakpoint after which the expansion accelerates. Thus there are three variables that must be understood. What is the effect of different mix variables on the small expansion?

1 What is the effect of different mix variables on the small expansion

2 At what level and time do the expansion accelerate and what is the cause of it?

3 How do silica fume and granulated blast furnace slag effect the sulphate reaction?

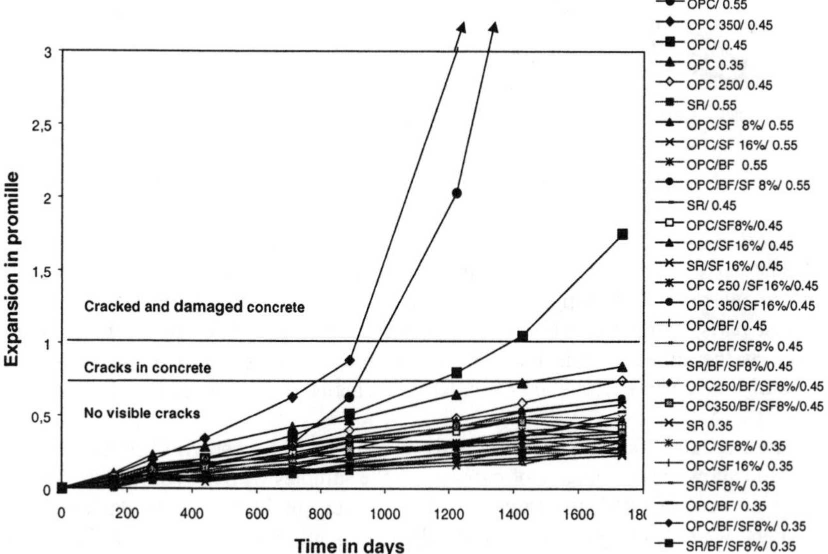

Figure 2 - Sulphate expansion of concrete prisms. All samples. Abbreviations as in Table II. The number at the end is the W/B ratio

Type and amount of cement

The expansion for the prisms casted with pure Portland cement is shown in Fig. 3 and 4. As we expected the sulphate resistant cement give a very small expansion.

During the first two years the expansion relates to the amount of OPC independent of the W/C (Fig. 3). The expansion presumably reflects the amount of C_3A. After 800 days the concrete with high W/C expands with much higher speed than the concrete with low W/C, although the concrete with low W/C contains more cement.

In Fig. 4 there are three mixes with the same W/C but with different amount of cement. The expansion here clearly shows that it is dependent on the amount of cement.

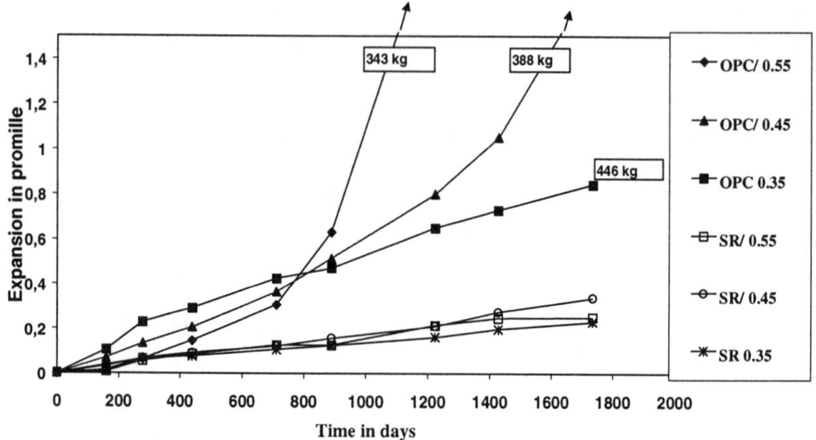

Figure 3 - Sulphate expansion with pure Portland cement and different W/C

Interpretation

The early expansion is related to the amount of C_3A and is independent of the water/cement ratio. Gypsum can be found in cracks parallel to the surface in the outermost parts of the concrete. The expansion can not be related to the formation of gypsum, as the sulphate resistant cement will give almost as much gypsum as OPC. Presumably the early expansion is related to ettringite formation as it is linked to the amount of C_3A.

After around 800 days the prisms with high W/C ratio and OPC starts to expand rapidly and show clear evidence of damage. In the samples with the same W/C but with different amount of OPC, the sample with most cement was severely damaged after 1200 days while the one with intermediate cement content at 1800 days starts accelerate and shows clear signs of cracking and damage. The concrete with least OPC is almost undamaged.

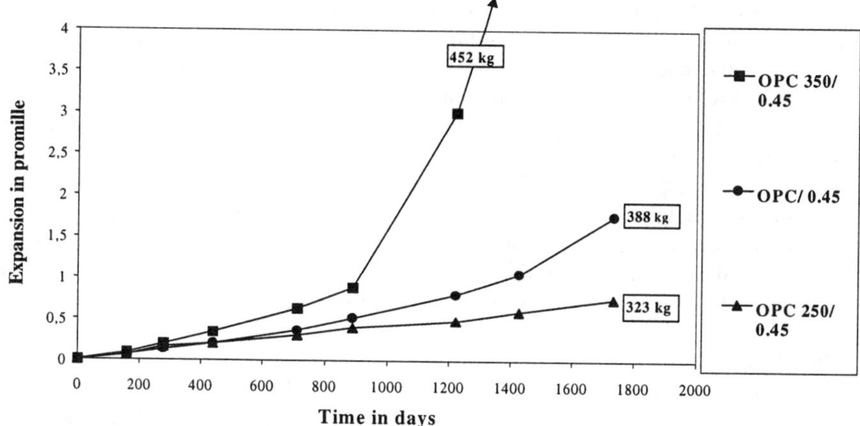

Figure 4 - Sulphate expansion with different amount of OPC and constant W/C (0.45)

There seems to be a break point when the concrete expansion starts to accelerate and becomes severely damaged. With the cement with a W/C ratio of 0,55 the breakpoint lies at around 0.3 ‰ while it with concrete with a W/C of 0.45 lies at 0.7 (with much cement) or higher. This may be due to the strength of the concrete. It must, however, be more complicated as the breakpoint for sample with more cement with the same W/C has a lower breakpoint. Actually the concrete with 388-kg OPC has slightly higher cube strength at 28 days and 6 month than the concrete with 452-kg OPC.

The results indicate that the sulphate ions can penetrate even relatively dense concrete. Energy dispersive analyses in SEM of hydrated paste have been performed on polished surfaces on a couple of samples. The interpretation of the data is difficult and more detailed work is needed. The outermost 2-4 mm is distinctly enriched in sulphate in almost all samples irrespective of binder type. Three samples have been analysed in more detail, the two with a W/C of 0.45 (323 and 452 kg of OPC) and the one with W/C 0.35 (446 kg of OPC). All three show high sulphate contents close t the surface. Gypsum can be recognised in cracks and around some large aggregate grains. A zone with slightly increased contents of sulphate follows that with high sulphate content. In this zone the sulphate content expressed as SO_3 has increased with about or slightly more than 1 %. It presumably represents a diffusion front with ettringite formation.

The sample with low cement content (Fig. 4) that did not accelerate has a slightly increased content of sulphate to a depth of around 20-mm while the sample with high cement contents that accelerated has an increased content all the way through the concrete. The concrete with W/C of 0.35 has higher contents of sulphate ions to a depth of 10 mm. The data shows, as expected, that the sulphate diffusion is controlled by the density of the paste. The volume of expanding concrete will be lower with lower W/C ratio. The data also suggest that the expansion also is related to the amount of paste and contents of C_3A. A higher cement content will give a larger expansive force. Thus the concrete with low W/C ratio gives a high initial expansion but slows down later as the volume of expanded concrete will be less. An interesting observation was that the paste with low W/C ratio contained more sulphate and aluminates in general than the other concretes. This is presumably due to that all the aluminates have hydrated while some of the alite and belite are still unhydrated. Thus the expansive capacity of the paste will be the same irrespectively of the W/C ratio of the mix. The expansive force in the concrete with a large amount of paste will be larger at the same penetration depth. When this force is large enough it will crack the concrete. A cracking will increase the diffusivity of the concrete and ingress of sulphate ions, which in turn will accelerate the expansion. This will be discussed further in the discussion.

Silica Fume

The mixes with silica fume are included in Fig. 5. All the prisms show a "small" expansion but do not show any tendency for acceleration. They have small cracks parallel to the surface, which indicates spalling. The expansion of all the prisms is fairly similar but one can notice some regularity. The expansion seems related to both the W/B and the amount of OPC.

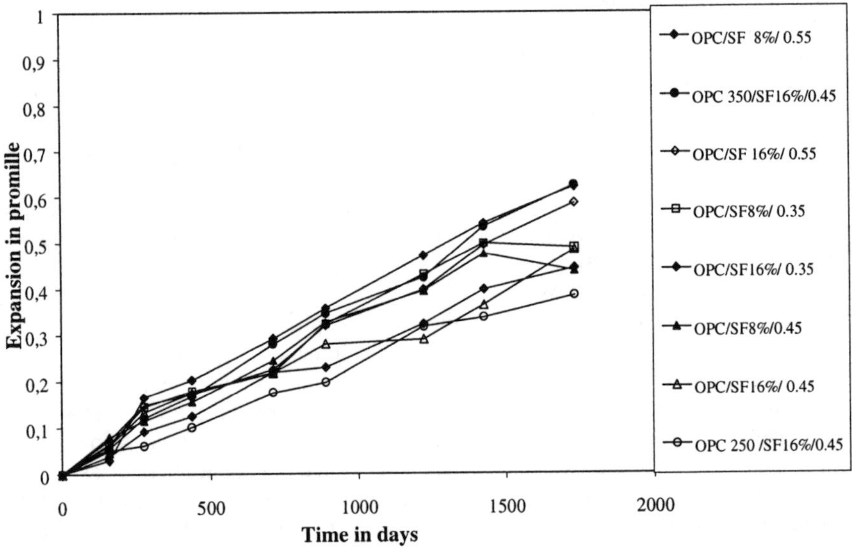

Figure 5 - Sulphate expansion with mixes including silica fume. The concretes contain different amounts of OPC, silica fume and have different W/B.

Interpretation

With silica fume we do not get any real damage of the concrete but it shows a "small" expansion that in part is related to the amount of cement. The expansion with silica fume is larger than with sulphate resistant cement or with granulated blast furnace slag. This indicates that the silica fume by itself do not stop the expansion but it is very slow and even after five years the prisms do not show any tendency towards accelerated expansion.

Chemical analyses have been performed on the sample OPC/SF 8 %/0.55 and OPC 350 SF 16/0.45 (FIG. 5). On the first a distinct sulphate increase could be observed the first 4 mm after which a slight increase (about 1 %) of sulphates could be detected to a depth of 9 mm. In the concrete with lower W/B and more silica fume the depth of sulphate increase was around 5 mm. This data indicates that the silica fume as such do not stop the expansion but hinders the ingress of sulphate ions. It is presumably only the outermost layer with higher sulphate contents, which expands.

Granulated blast furnace slag

All the mixes with granulated blast furnace slag (BF) are included in Fig. 6. Some of the mixes contains both slag and silica fume. The expansion of these prisms is between that of sulphate resistant cement and that of OPC with silica fume. None or the prisms show any cracking but the expansion presumably continuos. The surface cracks and spalling is less prominent on the prisms with slag. The mixes with most OPC seem to expand most and those that also contains silica fume least. A high W/B expand more than it should considering the relatively low content of OPC.

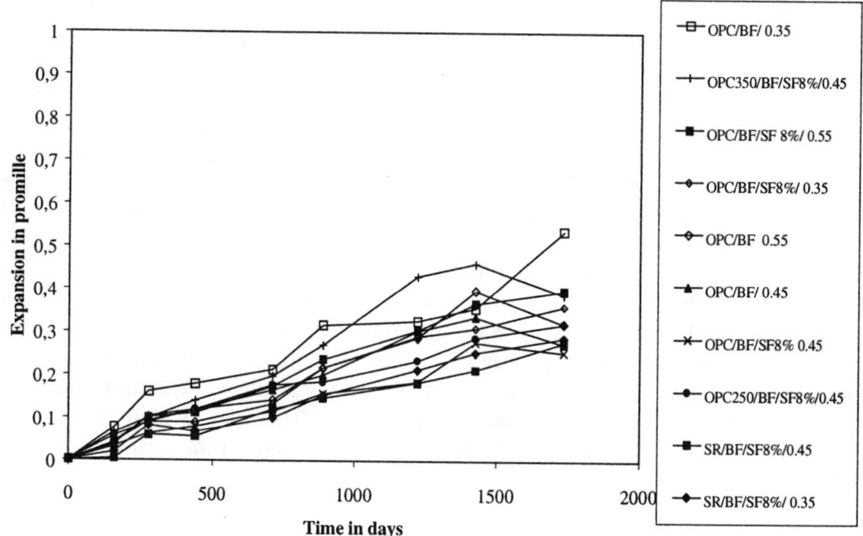

Figure 6 - Sulphate expansion with concrete mixes including granulated blast furnace slag. Abbreviations as in Table II.

Interpretation

The prisms with granulated blast furnace slag are not damaged but they expand more than those with sulphate resistant cement do. The prisms with both slag and sulphate resistant cement expand similarly to those with only sulphate resistant cement. If we only consider the "small expansion" the expansion is related both to a large amount of OPC and a high W/B ratio. Presumably the reaction with 50% replacement of OPC is enough to hinder severe expansion in a foreseeable future. The data indicate that slag dilutes the OPC but it does not hinder the expansive reaction as such.

DISCUSSION

The sulphate resistance of cement is normally tested with a mortar test according to ASTM C 1012-89. This test does give the really bad combinations but it is not a definite test of concrete combinations. The mortar prism test indicate that silica fume and slag makes the concrete sulphate resistant while the concrete prisms shows that they only slow down the process. The long term experiment with concrete prisms show the mechanism in greater detail.

The results show that we get an initial small expansion that at a certain level starts to accelerate and give a large expansion and severe deterioration of the concrete. Only a few of the prisms has expanded severely but some trends can be noticed. The small expansion seems to be controlled by a combination of amount of C_3A in the mix and the density of the paste (controlled by the W/B ratio).

Neither silica fume nor granulated blast furnace slag stops the "small" expansion. This indicates that they do not interfere with the expanding ettringite formation. Both the silica fume and slag dilute the cement and will thus give a smaller expansion. It also seems to be controlled by the chemistry and diffusivity of the paste. The chemical and petrographical analyses are still unfinished but they already give some important clues. The attack is from the surface and an increase in sulphate content can be detected to a certain depth that is related to W/B ratio and addition. A low W/B ratio and silica fume gives less depth of sulphate ingress. This suggests that the expansion only will effect the outer "shell" of the concrete and that the interior unaffected concrete will give a restrain. With time and deeper penetration of sulphate ions larger volumes of concrete will expand and the overall measured expansion will increase. Thus the effect of the expansion will be controlled by the geometry of the tested concrete body. The expansion of the sulphate enriched concrete will be in all direction and not only necessarily in the direction of the prism. This may be the reason for formation of cracks subsequently filled with gypsum and coarse grained ettringite parallel to the surface. A certain thickness of expanding sulphate-enriched concrete is necessary to overcome the restraint of the interior. This may explain why overall expansion can be noticed in some of the concrete prisms but not the prisms. With a ball the result would be spalling.

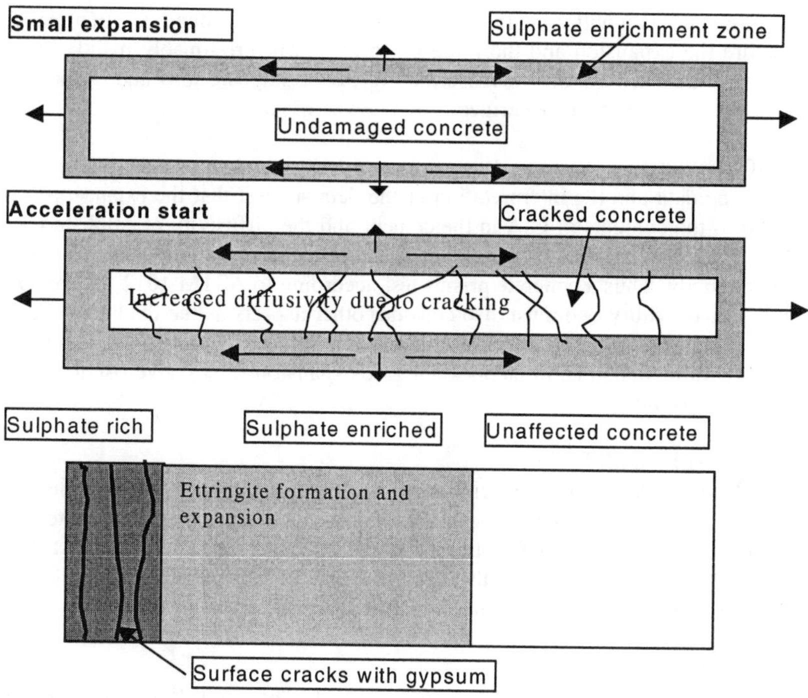

Figure 7 - Schematic sketch outline showing the mechanism of sulphate alterations, expansion and consequences for a concrete prism.

When the expansive force of the shell is larger than the restraint of the interior the interior will crack which in turn will increase the diffusivity and ingress of sulphate ions. This is presumably the reason for the accelerating large expansion. With a high W/C ratio the concrete is both more permeable and weaker thus the accelerating moment comes quicker. With constant W/C ratio and larger amount of cement the expanding force will be larger but the strength of the concrete almost the same which may explain the phenomena in fig. 4. With a low W/C ratio the concrete will be much stronger and can thus withstand the expansion of the shell much longer. Eventually, however, the expansion will crack the concrete and accelerate the overall expansion but it might also result in spalling. With sulphate resistant cement or with slag cement the expansive force governed by the amount of aluminates is to small to overcome the restraint of the interior. Silica fume as such do not seem to stop the expansive reaction but it does slow down the ingress of sulphates and makes the concrete stronger.

The data shows that the sulphate attack is slow with real concrete. Thus, in reality, most of the concrete mixes tested will not degrade in foreseeable time. A low W/C ratio or silica fume will delay the sulphate ingress and give such a strength to the concrete that it can withstand the sulphate attach for a very long time. In a real concrete structure the expansion and the force given by it will presumably mainly give rise to spalling which in turn will increase the ingress of sulphate ions and eventually break down the concrete from the surface.

CONCLUSIONS

The data and the interpretation of the data suggest that the expansion is directly related to the contents of C_3A in the cement and the diffusivity of sulphate ions in the paste. The consequences of the expansion are related to the geometry of the attached concrete body. Thus a concrete prisms test according to ASTM 1012-89 C only gives a guideline. In reality one must also consider other aspects as the thickness and strength of the concrete and the geometry of the concrete surface. Although a concrete mix with low W/C ratio and silica fume does not give expansion it may give rise to spalling and surface deterioration.

ACKNOWLEDGEMENT

I acknowledge the financial support provided by the Swedish Consortium for High-performance Concrete, The group consists of Cementa, Elkem material, Euroc Beton, Skanska, and the authorities BFR and NUTEC. The prolonged test period has been sponsored by MEROX. I also want to thank SABEMA and Kenneth Ottosson for the sample preparations and the personal at CBI for laboratory work.

REFERENCES

[1] H.W.F. Taylor, Cement chemistry, 2nd Ed., Thomas Telford, London (1997)

[2] P.E. Brown & H.W.F. Taylor, " The role of ettringite in external sulphate attack", Conference volume, Quebec Seminar on Sulphate Attack Mechanisms, Quebec city, Quebec, Canada, October 5& 6, 1998.

[3] J.R. Clifton, G. Frohnsdorff & C. Ferraris, " Standards for evaluating the susceptibility of cement-based materials to external sulphate attack, Conference volume, Quebec Seminar on Sulphate Attack Mechanisms, Quebec City, Quebec, Canada, October 5& 6, 1998.

STANDARDS FOR EVALUATING THE SUSCEPTIBILITY OF CEMENT-BASED MATERIALS TO EXTERNAL SULFATE ATTACK

James R. Clifton, Geoffrey Frohnsdorff and Chiara Ferraris
National Institute of Standards and Technology

ABSTRACT

In this paper, the current status of standards 1) for specifying cement-based materials to prevent deleterious reactions with external sulfates, and 2) for predicting the performance of concrete exposed to external sulfates, is reviewed. The applicability and reliability of these standards are discussed and deficiencies in them noted. Recommendations for the development of improved standard test methods are presented.

INTRODUCTION

The purpose of this paper is to discuss: 1) the current status of standards for controlling or predicting the durability or service life of portland cement-based materials exposed to aqueous sulfate solutions; and 2) improvements needed in the standards. Deficiencies in current standard tests include lengthy testing periods, the insensitivity of the measurement tools to the progression of sulfate attack, and uncertain relationship to field degradation mechanisms. Ways to erase these deficiencies are discussed based on the application of the methodology outlined in ASTM E 632, "Standard Practice for Developing Accelerated Tests to Aid Prediction of the Service Life of Building Components and Materials" [1]. To begin with, reliable specifications for selecting durable cements, and test methods for evaluating the sulfate resistance of cements and concrete made from them, should be based on knowledge of the reaction mechanisms. This conference has advanced our knowledge of the reaction mechanisms of external sulfate attack and hopefully will help to establish the agenda for future research leading to improved standards.

MANIFESTATION OF DEGRADATION

The nature of the manifestation of sulfate degradation depends on several factors including the exposure conditions that can range from continuous immersion to cycles of wetting and drying. In the following, rather simplistic descriptions of the manifestation of sulfate attack are related to the exposure conditions.

Continuous Immersion

If the concrete is essentially continuously immersed in a fixed volume of sulfate-containing water and sulfate attack occurs, either cracking or delamination of concrete surfaces or their softening, or all three, are usually observed. The affected area (reaction zone) follows the advancing sulfate interface. The processes result in a reduction of the concrete's mechanical properties, e.g., its load-carrying capacity [2]. Atkinson and Hearne [3] developed a model for predicting the rate of attack of concrete under continuous immersion in an sulfate solution, based on the following assumptions:

- sulfate ions from the environment penetrate into concrete, usually by diffusion,
- sulfate ions react expansively with certain aluminum-containing phases in the concrete,
- to simplify the problem, magnesium ions are not present (however, the effect of magnesium sulfate is predicted by the model) and the precipitation of gypsum does not cause significant expansion,
- the expansion of the surface layers results in stress buildup culminating in cracking and delamination (spalling) of the reacted material from the concrete surface,
- the rate of reaction is sufficiently rapid that there is a well-defined reaction front separating a layer in which the reaction is essentially complete from the inner volume in which it has not started.

The rate of degradation, R, is given by:

$$R = X_{spall} / T_{spall} = E\beta^2 c_s C_o D_i / \{\alpha_o \tau (1-\nu)\} \tag{1}$$

where X_{spall} is the thickness of a spalled layer,

T_{spall} is the time for a layer to spall,

E is Young's modulus,

β is the linear strain caused by one mole of sulfate ions, reacted in 1 m^3 of concrete,

c_s is the sulfate ion concentration in the bulk solution (which is assumed to remain constant),
C_o is the concentration of reacted sulfate ion from the solution present as ettringite,
D_i is the intrinsic diffusion coefficient of sulfate ions,
α_o is a roughness factor for the fracture path (usually assumed to be 1),
τ is the fracture surface energy of the concrete, and
ν is Poisson's ratio

This model predicts that the rate of degradation increases with increases in diffusivity, in the concentration of external sulfate ions in the solution, and in the concentration of reactive species (compounds) in the cement paste matrix. The results of experimental studies and modeling at NIST [4] are in general agreement with the concepts of the formation of a reaction zone that delaminates, exfoliates or spalls. A NIST model [5], developed based on the analysis of experimental data obtained using a constant pH test method, showed that the data was consistent with the occurrence of a two-stage process (Figure 1); with Stage 1 being diffusion-controlled, with little expansion occurring as voids in the concrete

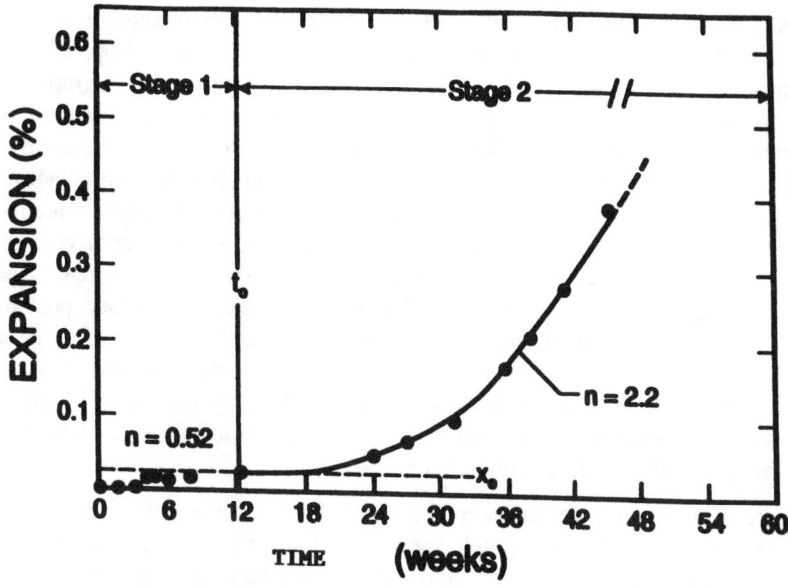

Figure 1. Schematic of sulfate attack model [5]. The value of n is the exponent of time, t.

are being filled with gypsum and ettringite; and, in Stage 2, continuing formation of ettringite causes expansion to accelerate with the resultant formation and propagation of cracks. Atkinson and Hearne [3] observed that only about one twentieth of the solid volume increase due to ettringite formation (also, for brucite) appears as bulk expansion. Therefore, the rate of degradation is not linearly related to the rate of ettringite formation. This is in agreement with a model by Clifton and Pommersheim [6] which predicted that, in the w/c range of normal concrete, sufficient void space is usually available to accommodate the maximum volume of ettringite formed from ASTM Types I and V portland cements. The above studies suggest that if the model is reliable, no testing should be needed for portland cements. If testing is to be accelerated so that the rate of spalling is increased, both the diffusivity of specimens and the exterior concentration of sulfates should be increased. The diffusivity can be increased by adjusting the mixture design, especially the water-to-binder ratio. Also, the measurement of expansion appears to be an insensitive parameter of ettringite formation.

Exposure to Cycles of Wetting/Drying.

Wetting and drying cycles with sulfate-containing water is another form of exposure which can be encountered when water levels rise and drop, e.g., by flooding and runoff of precipitation, and ocean tides. Also, cyclic migration of water into concrete can be induced by capillary flow and cyclic variation in atmospheric relative humidity. The result of such processes is the concentration of sulfates near the free concrete surfaces (at times denoted as "subsurfaces"). When soluble sulfate salts, as well as other soluble salts, are concentrated in sufficient amounts, disintegration of the surface layers can occur. The mechanism of the disintegration is controversial and both topochemical and precipitation/crystallization mechanisms have been proposed. Other possible degradation processes include reduction in the modulus of elasticity of concrete and thermal expansion/contraction incompatibilities of the affected and unaffected concrete [7]. Studies of wet and dry cycling by the U.S. Bureau of Reclamation [8] indicate that the accompanying degradation is more rapid than that associated with continuous immersion.

In a restricted sense, degradation of concrete by concentrating sulfate salts in the outer concrete surfaces through wetting and drying cycling is a physical process and other soluble salts may induce the same type of damage. However, if the wetting and drying processes are gradual then it is likely that both chemical and physical processes are involved in the degradation.

Partial Immersion with Evaporation

A particularly severe condition for sulfate attack is present when the lower portion of a concrete element is in contact with moist soil or ground water containing sulfates and the upper portion is exposed to a drying atmosphere. Damage occurs in areas exposed to the dry atmosphere in the form of spalling and scaling from both chemical and physical processes. Such problems have been observed [9, 10] in basements, foundations, and footings of residential dwellings, and service tunnels and dams. A standard test for concrete partially immersed in a sulfate solution or sulfate-laden soil, combined with drying conditions, was not found during the preparation of this paper.

SPECIFICATIONS FOR SULFATE RESISTANCE

Specifications for selecting cements and concrete resistant to external sulfate attack are numerous; the specifications have been developed by national organizations and local authorities such as highway departments. Rather than making a comprehensive survey of national and regional specifications, we will focus on current ASTM and European standards for concrete performance.

Because of the obvious correlation between the potential for ettringite formation and the amount of tricalcium aluminate (C_3A) in cements, specifications for sulfate resistance of cements usually stipulate limits on the maximum C_3A content. In ASTM C150 [11], Type V portland cement (sulfate resistant) is specified to have a maximum C_3A content of 5.0%, by mass, and Type II (moderately sulfate resistant) a maximum C_3A content of 8.0%, by mass. K. Mather [12] described the lengthy process in which ASTM Committee C-1 on Cement developed the above limits on C_3A contents. She commented on the need for a performance test, especially for evaluating the sulfate performance of blended cements. Through the work by Mather and other members of ASTM C-1, a performance test using mortars was adopted by ASTM and designated as ASTM C 1012 [13], which is applicable to both portland and blended cements. In this test method, the water-to-binder ratio is fixed and the mortar specimens are to attain an average compressive strength of 20 MPa before testing. Therefore, the performance of mortar specimens in the test is usually considered to be dependent primarily on the composition of the cement. The American Concrete Institute Guide to Durable Concrete [14] gives recommendations on durable concrete exposed to aggressive chemicals and salts which depends on severity of the exposure conditions. Based on the sulfate concentration in soil or water, recommendations are given on applicable cements and the maximum w/c ratio. The limit on w/c ratio indicates recognition of the influence of transport properties on the sulfate resistance of concrete.

European specifications for sulfate attack are undergoing change. Hobbs [15] has commented that in the forthcoming European Standard pr EN 206, "Concrete - Performance, Production and Conformity," durability of concrete will rely on prescriptive specification of minimum grade, minimum binder content, and maximum water-binder ratio for a series of defined environmental classes. He further comments that "it has not proved possible within the European Committee for standardization (CEN) to agree on common values for the specification parameters to cover the wide range of climates and wide range of cements in use in the EU Member States." Therefore, the standard will likely include indicated (default) values and individual Member States will be able to specify national values, when necessary. The book edited by Hobbs [15] gives minimum requirements for durable concrete considering corrosion, freeze-thaw attack, and chemical attack, which includes sulfate attack, in the U.K. Minimum requirements are given for concrete subject to exposure from sulfate in groundwater or in soil, that are necessary for service lives of 100 years. The recommendations cover the type of cement, cement content, and water-cement ratio. The concrete coarse aggregate should have a nominal maximum size of 20 mm and should conform to BS 882 or BS 1047 [15].

The effectiveness of mineral admixtures such as some fly ashes, silica fume, and granulated blast furnace slag in mitigating sulfate attack is well known [12,15,16] and is recognized in the ACI 201 Durability Guide and in pr EN 206. The type of cement, either portland cement or blended cement, and if needed, the type of mineral admixture to be added to the concrete, depends on the severity of the sulfate exposure. ASTM C 1157M is a recently established standard performance specification for blended cement [17]. It does not give any restrictions on the composition of a blended cement or its constituents. The sulfate performance of a blended cement is evaluated using ASTM C 1012.

In a review of sulfate attack on concrete, Metha [18] commented that the reliability of present specifications and test methods dealing with the resistance of concrete to sulfate attack are subject to question. He cited evidence that control of permeability of concrete is more important than control of the chemistry of the cement for sulfate-resistant concrete. This can be deduced from Eq. 1 since D_i varies by several orders of magnitude in concrete, whereas $[C_3A]$ only varies between about 1 and 12 percent.

TEST METHODS

Analysis of Standard Test Methods and Other Proposed Test Methods

Two ASTM test methods are available and other test methods have either been implemented or proposed by other organizations. Before reviewing the most widely used methods, the requirements for an acceptable test are explored. A reasonable approach for analyzing the reliability of test methods for determining sulfate resistance of concrete is to follow the generic procedures in developing accelerated tests described in ASTM E632 [1]. The major elements of ASTM E632 are applied to ASTM C452 [19] and C1012 [13] (both test methods are described in Section 4.2) in the following:

- PROBLEM DEFINITION - Define what the test should do and the degradation factors that should be included in the accelerated test.
  COMMENT - The purpose of the ASTM sulfate resistance test methods is well-defined - determine the performance of a cement paste matrix of concrete when exposed to sulfates at concentrations expected to be encountered in service.

- PRE-TESTING - Design and perform preliminary accelerated aging tests to demonstrate that rapid failures can be induced and measured and identify the main degradative factors.
  COMMENT - The work leading to the development of ASTM C452 and C1012 is reviewed by Mehta [18] and by Cohen and Mather [20]. Several ways to accelerate sulfate attack have been used or proposed [20, 21] including increasing reaction surface of test specimens (small specimens with large surface area); increasing the concentration of sulfate; increasing crystal pressure (i.e., continuous wetting-drying cycles); and raising the temperature of the solution. Acceleratory factors used in the ASTM tests are restricted to elevated concentrations of sulfates.

- TESTING - Design and perform predictive service life tests using the degradation factors of importance to determine the dependence of the rate of degradation on exposure conditions. Also, determine if the mechanisms of degradation in the accelerated test are representative of those observed in-service.
  COMMENT - This workshop should be a good starting point in determining if the conditions and procedures used in current test methods give rise to degradation mechanisms occurring in the field. If the field mechanisms are not being produced in the accelerated tests, then new test methods need to be developed.

- INTERPRETATION OF DATA - Develop mathematical models of degradation and compare rates of change in service life tests with those from in-service tests. Establish performance criteria for predictive service life tests. An alternative to actually predicting service life is to compare the relative behaviors of a number of materials based on the results of the same test.

  COMMENT - Failure criteria in the ASTM tests appear to be empirically based which are only useful in making relative durability predictions, i.e., Cement A is potentially more durable than Cement B. Cohen and Mather [20] suggested that a wide spectrum of properties should be measured including changes in visual appearance, in size and mass, and in mechanical properties.

In addition to the generic criteria of ASTM E632, Mehta [18] recommended that for an acceleratory test for sulfate attack to be meaningful: (a) it should give reliable results within a relatively short time (e.g., 4 weeks or less) if the test is to be of practical benefit to a cement manufacture or user; and (b) it should be applicable to both portland and blended cements. ASTM C452 is reasonably rapid, completed within 14 days, however, it is not considered to be applicable to blended cements [13, 18]. ASTM C1012 is applicable to both portland and blended cement; however, a cement meeting the performance criteria will require 6 months of testing.

ASTM C452

ASTM C452 [19] involves accelerating the expansion of mortar bars by adding gypsum to a portland cement prior to making the mortar bars and thus its rate is not controlled by the transport properties of the mortar. In the test procedure, sufficient gypsum is added to dry portland cement so that the mixture has a sulfur trioxide (SO_3) content of 7 percent, based on mass of the mixture. Then, mortar bars (25 x 25 x 285 mm) are made, cured in their molds for around 23 hours, demolded and their lengths measured, before they are immersed in water, and their 14-day expansions measured. This test is rapid and does differentiate between high-C_3A and low-C_3A portland cement. However, it is not considered to be applicable to blended cements as the reaction between gypsum and portland cement has been found to be essentially completed before the mineral admixture portion of the blended cement has appreciably reacted with the hydroxide ions liberated by the hydration of the portland cement fraction [22].

Contrary to recommendations of ASTM E632, even with portland cement, the test conditions of C452 do not simulate field exposure of concrete to sulfate, which involves the ingress of sulfates into concrete and the subsequent exfoliation of the attack layers, opening up free surfaces and accelerating the ingress of sulfate into the concrete. Therefore, as presently constituted, while C452 can be used to differentiate between the potential reactivity of portland cements, it cannot be used to differentiate between portland and blended cements (also, probably not between different blended cements), nor for predicting the sulfate resistance of concretes which depends on the rate of ingress of sulfates.

ASTM C1012

In the immersion test of ASTM C1012 [13], mortar bar specimens are immersed in a sulfate solution, 50 g/L, sodium sulfate, Na_2SO_4, (magnesium sulfate ($MgSO_4$) may also be used) and their expansions periodically measured for periods of 6 months, or less, if the expansion criterion is exceeded earlier. Expansion limits depend on the severity of anticipated exposure conditions; (a) for moderate sulfate exposure conditions, the maximum expansion is 0.10% after 6 months and (b) 0.05% after 6 months for cements for use in severe exposure condition. Prior to being immersed, mortars should achieve a compressive strength of 20 MPa.

ASTM C1012 clearly more closely simulates field exposure conditions than does ASTM C452, in that sulfates penetrate the mortar prior to reacting with calcium hydroxide ($Ca(OH)_2$) and the hydration products of C_3A. Also, in C1012 the cement is cured to an extent that most of C_3A is considered to have hydrated [22].

How well does C1012 simulate the field process? A possible significant problem is a change in the chemistry of test solutions. The test procedures stipulated that the beginning and replacement test solutions of Na_2SO_4 should have a pH of between 6 to 8. The test solution should be replaced when expansions are measured at specific periods. However, the pH rapidly rises approaching that of saturated calcium hydroxide, after a solution is replaced. In a report by Romanoff [23], the pH of 124 soils distributed throughout the United States averaged about 7 and ranged from 3.1 to 10.2. The pH of seawater is approximately 8 [24]. Reardon [25] modeled the effect of pH on phase changes of sulfoaluminate in concrete and found that ettringite decomposes at a pH<10.6 into gibbsite and gypsum. Along with a change in pH, the sulfate concentration decreases with time. Again following ASTM E632, the pH of the test solution should be close to the pH of the exposure, otherwise the degradation mechanisms could be different. Also, in an accelerated test, the sulfate concentration should either remain relatively constant or changes in concentration monitored so that reaction rates can be determined.

The model by Atkinson and Hearne [3] would predict that the results of C1012 should be sensitive to the diffusivity coefficient which is controlled by the curing of the mortar. Consistent with the modeling prediction, we have observed lengthy curing has significantly improved the sulfate resistance of test specimens. If curing is not well-controlled so that a specified diffusion or absorption is achieved, the test results will likely be too variable for a standard test method.

Constant-pH Test

Mehta and Gjorv [26] developed an immersion test in which the pH and the sulfate concentration were kept constant at a desired level, with the specimens being 10 mm cubes of high w/c paste. The idea of using a constant pH eliminating some of the problems associated with ASTM C1012. In their early experiments, Mehta and Gjorv kept the pH of the immersion solution at the original pH by manual titration with sulfuric acid (H_2SO_4); later the titration was automated by using a pH-stat [27]. Brown [28] further developed the test and investigated the effects of pH of the sulfate solution on strength and expansion. He found that control of the pH at 7 significantly increased the rate of sulfate attack, as measured by either strength loss or expansion, as compared to ASTM C 1012. The effect of controlling pH at 7 and simultaneously the sulfate concentration upon expansion is shown in Figure 2 [4].

An apparatus used by this paper's authors, similar to that used by Brown [28], is shown in Figure 3; it consists of a pH controller that measures the pH of the solution and, when the pH has decreased by a selected interval, the solenoid valve is opened to allow a dilute solution to sulfuric acid to flow at a low rate until a selected pH is attained. The solution is continuously circulated to mix each addition of sulfuric acid. In different series of experiments, we maintained constant sulfate concentrations and pH at 7, 9, and 11 units with an average deviation of $\pm$ 0.5 units.

The constant-pH appears to be a desirable feature of testing, since among other benefits, it allows simulation of field exposure conditions, e.g., the pH can be maintained at that of seawater, and if necessary a variable pH environment can be simulated. It can also accelerate the sulfate degradation process. However, its reproducibility has not been determined by inter-laboratory testing. The same comments made relative to curing in the analysis of C1012 are applicable to the constant-pH test.

Cyclic Wetting and Drying.

Even though degradation of concrete exposed to cycles of wetting (or soaking) and drying associated with a sulfate-containing water is likely to be more severe than continuous immersion [8], a standard wetting and drying test has not yet been developed. A soaking and drying sulfate test for high-strength concrete was described by de Almeida [29]. Cube specimens, 150 mm on edge, were immersed for 2 h. in a 16% Na_2SO_4 aqueous solution, followed by being dried at 105°C for 10 to 15 h. Then they were cooled to room temperature, their masses measured, and returned to the test solution for another test cycle. Performance was also measured by testing selected specimens in compression and by visual inspection. The results of the visual inspection were reported to be consistent with changes in mass and strength. In general, it was found that the porosity and capillary absorption controlled the performance of concrete more than the addition of mineral admixtures. Almeida suggested that, in rapid wetting and drying processes, the precipitation of soluble sulfate salts and accompanying phase changes control the degradation rather than the formation of calcium sulfate and ettringite.

In order to predict the service life of concrete exposed to sulfates, the Bureau of Reclamation [8] carried out a program lasting some 20 years in which the length change of 150 mm diameter, 300 mm long (6" x 12") concrete cylinders exposed to wetting and drying cycles were compared to length changes of companion concrete specimens exposed to continuous immersion. In both types of exposures a 2.1% Na_2SO_4 immersion solution was used. A wetting and drying

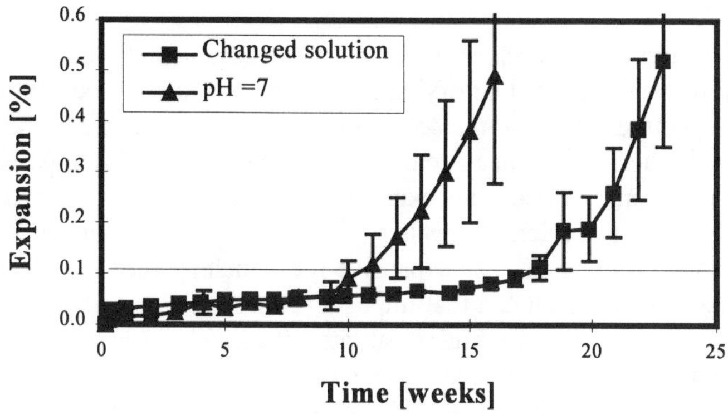

Figure 2. Comparison of expansion using ASTM C 1012, with 5% Na_2SO_4 solution changed each time a measurement was made, with a pH-controlled test at pH 7. [4].

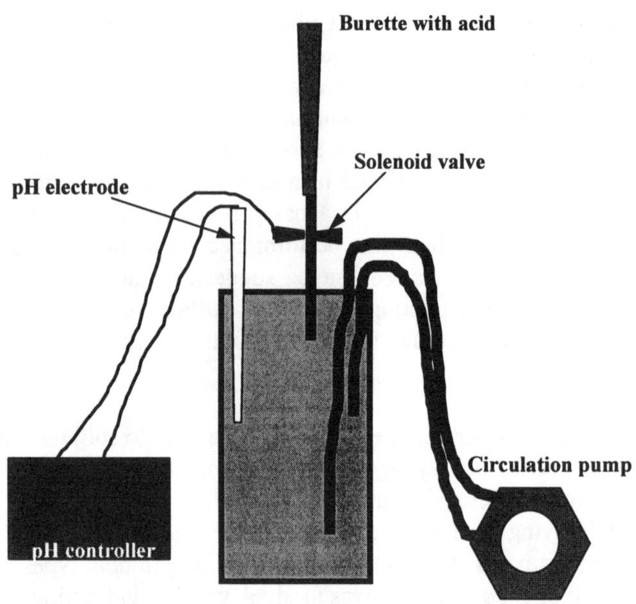

Figure 3. Schematic of pH-controlled test apparatus.

cycle consisted of immersing specimens in the 2.1% Na_2SO_4 solution for 16 hours at 21° to 27°C followed by forced air drying at 54°C for 8 hours. In the continuous immersion test, specimens were immersed at a temperature between 21° and 27°C and only removed when length measurements were made. From a comparison of the times for specimens to reach an expansion of 0.5% in both tests, it was estimated that 1 year of accelerated testing (i.e., wetting and drying) equaled 8 years of continuous immersion. The expansive mechanisms were not identified; however, the researchers concluded that imposing limits on C_3A and C_4AF were not the ultimate solution for preventing sulfate attack. Nevertheless, cements, portland or blended, with a maximum C_3A content of 6.5 % and a maximum C_4AF content of 12 % were recommended for making sulfate-resistant concrete.

Before we reject the wetting and drying test as not being applicable to sulfate attack in the field, more knowledge on the process should be obtained. None of the above studies were concerned with identifying the mechanisms of degradation by wetting and drying cycles, which poses the question "in these studies was the degradation process purely physical involving crystal pressure or were chemical processes also involved, e.g., formation of ettringite?" Depending on the time of drying, both physical and chemical processes may be involved, with the chemical processes likely dominating with slower drying rates.

Notched Beam Test.

The sulfate test methods described so far involve either the measurement of linear expansion or the compressive strength. As discussed previously, the model developed by Atkinson and Hearne [3] indicates that such expansion and strength measurements are insensitive indicators of the progress of sulfate attack. For example, the voids in specimens are nearly filled with ettringite before appreciable expansion is measured. Since sulfate attack involves a moving interface [3,5], only the degraded zone will contribute to expansion or to the loss of strength (actually the compressive strength first increases as voids in a specimen are being filled with reaction products or by the precipitation of penetrating soluble salts [28]). Hughes and Grounds [30] suggested that a more sensitive indicator would involve monitoring microstructural changes in the degraded zone and they explored the use of notch beam specimens. The notched beam test was found to give an earlier indication of sulfate attack than expansion measurements, while giving the same ranking of relative durability. A notch beam test should be sensitive to surface and subsurface discontinuities which arise from the delaminating of layers, from the outside surface inwards, during external sulfate attack. Similar to the results observed with other test methods, the maximum load for notched beam specimens was dependent on the period of curing, with blended cement specimens being especially sensitive.

DISCUSSION AND RECOMMENDATIONS

Based on our analysis, it appears that two rather distinct standards/specifications appear to be needed: 1) a rapid standard test for determining the potential reactivity of cements to sulfates; and 2) a standard method for the service life design of concrete in which sulfate attack is a one of the major degradation processes. Even if these needs are fulfilled, prescriptive specifications will likely continue to be used.

Test Methods for Determining the Potential Sulfate Reactivity of Cements

The present test ASTM C 1012 [13] for evaluating the sulfate resistance of mortars can require up to 6 months testing, which may not impede the introduction to the market place of new cements as numerous other tests would need to be performed. However, if the composition of a mineral admixture used to produce a blended cement is changed, the manufacturer may be required to subject the blended cement to the ASTM C 1012 test, to demonstrate performance equivalency. For example, for blended cement Types MS and HS, ASTM C1157 requires re-testing of sulfate resistance if analysis shows the change mass of any oxide of the blended cement or any constituent making up 10% or more of the cement is ≥ 3%; or if the oxide composition changes by ≥ 5% for constituents making up less than 10% of the blended cement. In such a case, 6 months of testing would almost certainly cause problems to a company striving to maintain its production. It follows that, for such cases, a new accelerated test should be developed, possibly based on the methodology outlined in ASTM 632 [1]. For example, combining the constant pH test, with a more sensitive monitor of sulfate attack than the measurement of expansion, appears to be a reasonable approach. Application of fracture mechanics techniques may offer a way to develop a more sensitive monitoring tool. Details such as the curing regime so that the test mortars have the same transport properties would need to be worked out. Other possible ways of accelerating the tests such as changing the dimensions (surface to volume ratio) and shape (tubular, cylindrical or spherical) of test specimens, as well as optimizing the mixture design should be explored. The possibility of modifying ASTM C452 [19] could be explored as it is more rapid and more closely evaluates the potential reactivity of cements than does ASTM C1012. One approach would be to combine gypsum with a partially hydrated cement paste and make compressed specimens that are exposed to a sulfate solution. By this approach, ASTM C452 may become applicable to all types of hydraulic cements. A similar test method used in some European countries, the Anstett test, has been described by Mather [12]. Hopefully, these suggestions will stimulate the development of new test methods that overcome the deficiencies recognized in the existing methods.

Service Life Design of Concrete.

One of the major concerns to an engineer/designer specifying a concrete that will be exposed to an aggressive sulfate environment, is its service life. Our present test methods and specifications for concrete only provide a qualitative basis for ranking relative durability with no quantitative assessment of service life. The model by Atkinson and Hearne [3] has been used in predicting the service life of underground concrete vaults [31] and appears to give predictions that relate to experience, which suggests that further improvements in the model should be pursued. A significant problem of using the model is the lack of validated sulfate diffusion data as such a standard test has not been developed [32]. A computer model has been developed [33] to predict the chloride ion diffusivity of concrete, which in principle could be used to predict the potential diffusivity of sulfate ions.

Prescriptive Standards versus Performance Standards.

A recommendation from the workshop on *Cement and Concrete Standards of the Future* [34] was that performance test methods, performance standards and performance prediction methodologies should be developed for assessing and quantifying the performance of cement-based materials. However, it was recognized that prescriptive criteria will continue to play an important role far into the future. Regarding prescriptive specifications for C_3A contents of cements, Bryant Mather in a keynote presentation at the workshop, commented that while performance standards are preferred, if satisfactory performance can be ensured more quickly and simply by placing a limit on C_3A calculated from chemical analysis, then such specifications should be invoked. In addition, however, he commented that performance tests are needed so that cements and concretes with novel compositions can compete and innovated materials or systems can be evaluated.

Prescriptive specifications for cements pose barriers to innovation because they put rigid limits on the range of permitted compositions. However prescriptive specifications can reduce risks that may arise from using cements with an unrestricted range of compositions, especially if there are not reliable performance tests. It has been suggested [35] that performance-based specifications be developed with optional prescriptive criteria. Assuming that prescriptive specifications will continue to be used in the future, we suggest that to be reliable and widely applicable, they should be based on a materials science perspective, rather than being based upon empiricism. An example of such a materials science approach is the computer modeling of microstructural and associated properties of hydrating portland cement deveoped by Bentz [36].

SUMMARY

The purpose of this paper is to discuss: 1) the current status of standards for controlling or predicting the durability or service life of portland cement-based materials exposed to aqueous sulfate solutions; and 2) improvements needed in the standards. Both standard performance tests and criteria, and prescriptive specifications, are available for selecting cement-based materials when sulfate attack is of concern. Some of the deficiencies in performance tests include lengthy testing periods, the insensitivity of the measurement tools to the progression of sulfate attack, and failure to simulate field degradation mechanisms. Ways to overcome these deficiencies are discussed based on the application of the methodology outlined in ASTM E 632, "Standard Practice for Developing Accelerated Tests to Aid Prediction of the Service Life of Building Components and Materials" [1]. Prescriptive specifications for sulfate resistance usually pose restrictions on the C_3A content of cements. Also, depending on the severity of the sulfate exposure, the use of blended cements or mineral admixtures may be specified.

Three major types of exposure conditions are associated with sulfate attack: 1) immersion in sulfate-containing water; 2) wetting and drying cycles; and 3) partial immersion with evaporation. Standard test methods, e.g., ASTM C 452 and C 1012 only address the immersion process, while no standard tests exist for exposure conditions 2 and 3, even though they may be more destructive than the immersion exposure condition. It is generally assumed that the destructive process in exposure conditions 2 and 3 involves precipitation/crystallization processes; however, this assumption does not appear to have been rigorously validated.

Based on our analysis, two rather distinct standards/specifications appear to be needed: 1) a rapid standard test for determining the potential reactivity of cements to sulfates; and 2) a standard method for the service life design of concrete in which sulfate attack is a one of the major degradation processes. Recommended approaches for developing such standards include modification of existing test methods and application of modeling based on material science principles.

ACKNOWLEDGEMENTS

The authors acknowledge the support by the National Institute of Standards and Technology's Partnership for High-Performance Concrete Technology (PHPCT) program for the preparation of this paper.

REFERENCES

[1] ASTM Designation E 632, "Standard Practice for Developing Accelerated Tests to Aid Prediction of the Service Life of Building Components and Materials" (ASTM, Philadelphia, PA).

[2] P.K. Mehta, "Mechanisms of Sulfate Attack on Portland Cement Concrete-Another Look," Cement and Concrete Research, V.13, pp. 401-406, 1983.

[3] A. Atkinson and J.A. Hearne, "Mechanistic Model for the Durability of Concrete Barriers Exposed to Sulphate-Bearing Groundwaters," Materials Research Society Symposium Proceeding, V. 176, pp. 149-156, 1990.

[4] C.F. Ferraris, J.R. Clifton, P.E. Stutzman, and E.J. Garboczi, "Mechanisms of Degradation of Portland Cement-Based Systems by Sulfate Attack," in Mechanisms of Chemical Degradation of Cement-based Systems, eds. K.L. Scrivener and J.F. Young, E&FN Spon (London, 1997).

[5] J.M. Pommersheim and J.R. Clifton, "Expansion of Cementitous Materials Exposed to Sulfate Solutions," Materials Research Society Symposium Proceedings,V. 333, pp. 363-368, 1994.

[6] J.R. Clifton and J. Pommersheim, "Sulfate Attack of Cementitous Materials: Volumetric Relationships and Expansions," NISTIR 5390, National Institute of Standards and Technology (Gaithersburg, MD 1994).

[7] J.R. Clifton and L.I. Knab, "Service Life of Concrete," Report No. NUREG CR-5466, U.S. Nuclear Regulatory Commission (Washington, D.C., 1989).

[8] G.L. Kalousek, L.C. Porter, and E.J. Benton, "Concrete for Long-Term Service in Sulfate Environment," Cement and Concrete Research, V.2 (1), pp. 79-90 (1972).

[9] G.C. Price and R. Peterson, "Experience with Concrete in Sulphate Environments in Western Canada," pp. 93-112, in Performance of Concrete, ed. E.G. Swenson, University of Toronto Press (1968).

[10] J.J. Hamilton and G.O. Handegord, "The Performance of Ordinary Portland Cement Concrete in Prairie Soils of High Sulphate Content," pp135- 158, *ibid.* 9.

[11] ASTM Designation C 150, "Standard Specification for Portland Cement" (ASTM, Philadelphia, PA).

[12] K. Mather, "Tests and Evaluation of Portland and Blended Cements for Resistance to Sulfate Attack," pp. 74-86, in Cement Standards Evolution and Trends, ed. P.K. Mehta, ASTM STP 663 (ASTM, Philadelphia, PA, 1977).

[13] ASTM Designation C 1012, "Standard Test Method for Length Change of Hydraulic-Cement Mortar Exposed to a Sulfate Solution," (ASTM, Philadelphia, PA).

[14] ACI 201 "Guide to Durable Concrete," AC I Manual of Concrete Practice: Part 1-1998 (American Concrete Institute, Farmington Hill, MI).

[15] Minimum Requirements for Durable Concrete: Carbonation- and Chloride-induced Corrosion, Freeze-Thaw Attack and Chemical Attack, ed. by D.W. Hobbs, British Cement Association (1998).

[16] P.K. Mehta, "Effect of Fly Ash Composition on Sulfate Resistance of Cement," Journal of the American Concrete Institute, V83 (6), pp. 994-1000 (1986).

[17] ASTM Designation C 1157M, "Standard Performance Specification for Blended Hydraulic Cement [Metric] " (ASTM, Philadelphia, PA).

[18] P.K. Mehta, "Sulfate Attack on Concrete-A Critical Review," pp. 105-130, Material Science of Concrete III, ed. by J. Skalny, American Ceramic Society (1992).
[19] ASTM Designation C 452, "Standard Test Method for Potential Expansion of Portland-Cement Mortars Exposed to Sulfate" (ASTM, Philadelphia, PA).
[20] M.D. Cohen and B. Mather, "Sulfate Attack on Concrete - Research Needs," AC I Materials Journal, V.88 (1), pp. 62-69 (1991).
[21] I. Biczok, *Concrete Corrosion and Concrete Protection*, Chemical Publishing Company, Inc., NY (1967).
[22] T. Patzias, "Evaluation of Sulfate Resistance of Hydraulic-Cement Mortars by the ASTM C 1012 Test Method," pp. 92-99 in Concrete Durability, ACI SP-100, American Concrete Institute (Farmington Hills, MI, 1987).
[23] M. Romanoff, "Underground Corrosion," National Bureau of Standards Circular 579 (1957): [Now the National Institute of Standards and Technology].
[24] Handbook of Chemistry and Physics, 73rd edition, CRC Press (1992).
[25] E.J. Reardon, "An Ion Interaction Model for Determination of Chemical Equilibra in Cement-Water Systems," Cement and Concrete Research, V.20, pp. 175-192 (1990).
[26] P.K. Mehta and O.E. Gjorv, "A New Test for Sulfate Resistance of Cements," Journal of Testing and Evaluation, V.2 (6), pp.510-514 (1974).
[27] P.K. Mehta, "Evaluation of Sulfate Resistance of Cements by a New Test Method," Journal of the American Concrete Institute, V.72 (10) pp. 573-75 (1975).
[28] P. Brown, "An Evaluation of the Sulfate Resistance of Cements in a Controlled Environment," Cement and Concrete Research, V.11, pp. 719-727 (1981).
[29] I.R. de Almeida, "Resistance of High Strength Concrete to Sulfate Attack: Soaking and Drying Test," pp. 1073-1092 in Concrete Durability, ACI SP-100, American Concrete Institute (Farmington Hills, MI, 1987).
[30] D.C. Hughes and T. Grounds, "The Use of Beams with a Single Edge Notch to Study the Sulphate Resistance of OPC and OPC/PFA Pastes," Magazine of Concrete Research, V.37 (131), pp. 67-74 (1985).
[31] K.A. Snyder and J.R. Clifton, "4SIGHT Manual: A Computer Program for Modelling Degradation of Underground Low Level Waste Concrete Vaults," NISTIR 5612, National Institute of Standards and Technology (Gaithersburg, MD, 1995).
[32] J.R. Clifton, D.P. Bentz, and J.M. Pommersheim, "Sulfate Diffusion in Concrete," NISTIR 5361, National Institute of Standards and Technology (Gaithersburg, MD, 1994).

[33] D.P. Bentz, E.J. Garboczi, and E.S. Lagergren, "Multi-Scale Microstructural Modeling of Concrete Diffusivity: Identification of Significant Variables," Cement, Concrete, and Aggregates, V.20 (1) pp. 129-139 (1998).
[34] G. Frohnsdorff and J. Clifton, "Cement and Concrete Standards of the Future: Report from the Workshop on Cement and Concrete Standards of the Future - October 1995," NISTIR 5933, National Institute of Standards and Technology (Gaithersburg, MD, 1997).
[35] G. Frohnsdorff, P.W. Brown, and J.H. Pielert, "Standard Specifications for Cements and the Role in their Development of Quality Assurance Systems for Laboratories," *Proceeding*, 8th International Congress on the Chemistry of Cement, Vol. VI, pp. 316-320 (Rio de Janeiro, Brazil, 1986).
[36] D.P. Bentz, "Three Dimensional Computer Simulation of Portland Cement Hydration and Microstructure Development," Journal of the American Ceramic Society, V.80 (1), pp. 3-21 (1997).

ARE SULFATE RESISTANCE STANDARDS ADEQUATE?

R.D. Hooton
Dept. of Civil Engineering
University of Toronto
Toronto, Ontario M5S 1A4 Canada

ABSTRACT

There are various types of standards related to sulfates in concrete and its components. There are standards which control internal sulfates in cementitious materials and also there are standards related to the resistance of cementitious materials to external sulfate attack. Also, there are standards for concrete which are used to provide durable concrete for various degrees of sulfate exposure. In spite of these standards, there are concerns that present standards may be inadequate for protection of the customer. In this contribution, current standards are reviewed. As well, it is postulated that standards cannot cover all situations and that the exposure conditions need to be understood and sound concrete practice needs to be exercised.

CEMENTITIOUS MATERIALS STANDARDS

Internal Sulfate Attack

Since the early days of portland cement production, calcium sulfate (typically gypsum and/or hemi-hydrate) has been interground with cement clinker to prevent flash set resulting from rapid reaction of the tricalcium aluminate (C_3A) phase. The amount of calcium sulfate needed is a function of both the C_3A content and the fineness of the ground cement. As will be discussed later, sulfate additions can have a beneficial effect on early strength and drying shrinkage, as long as excess sulfates that will cause deleterious expansion are avoided.

ASTM C150 for Portland cement was first adopted in 1941 with the five types of portland cement which still exist today. In 1941, C150 replaced ASTM C9 for Portland Cement and ASTM C74 for High-Early-Strength Portland Cement. The 1941 version of C150 contained limits on the SO_3 content of all types of cements but did not provide separate limits for high and low C_3A contents (see Table I). Except for a two year period from 1953 to 1955 when SO_3 limits were temporarily dropped (in favor of a ASTM C265 test limit of 0.50 g/ℓ dissolved SO_3 at 24 h), these limits have been raised over the years until 1971 as shown in Table I. In 1978, it was realized that the table limits in C150 were too conservative and that they prevented cement manufacturers from adding enough calcium sulfate to optimize the early strength of their cements. This is determined using the ASTM C563 Optimum SO_3 test in which Terra Alba or plant gypsum is interground at three addition rates and the 24 hour strengths are determined. Although, not explicitly covered in the standards, some manufacturers also optimize SO_3 content to optimize either later age strengths or to minimize drying shrinkage (using C596).

Table I. Evolution of Sulfate Limits in ASTM C150

		Type I	Type II	Type III	Type IV	Type V
1941	max SO_3 (%)	2.0	2.0	2.5	2.0	2.0
1946						
	if C_3A > 8%, max SO_3%	2.0	2.0	2.5	2.0	2.0
	if $C_3A \leq$ 8%, max SO_3 %	2.5	-	3.0	-	-
1953	all SO_3 limits dropped and C265 0.50 g/ℓ+ SO_3 limit adopted					
1955	SO_3 limits re-instituted but raised (C265 was dropped)					
	If $C_3A\leq$ 8%, max SO_3%	2.5	2.5	3.0	2.3	2.3
	if C_3A > 8%, max SO_3%	3.0	-	3.0	-	-
1960						
	if $C_3A \leq$ 8%, max SO_3%	2.5	2.5	3.0	2.3	2.3
	if C_3A > 8%, max SO_3%	3.0	-	4.0	-	-
1971						
	if $C_3A \leq$ 8%, max SO_3%	3.0	3.0	3.5	2.3	2.3
	if C_3A > 8%, max SO_3%	3.5	-	4.5	-	-
1978a	- same Table limits but a note was added that if optimum SO_3 by C563 is higher than limits, the limits can be exceeded by up to 0.5% if C265 value is less than 0.50 g/ℓ at 24 h					
1989	- same Table limits but C265 was dropped and C1038 expansion of 0.020% at 14 days was adopted. Also the 0.5% extra SO_3 limit was dropped.					

In any case, in 1978, if the Optimum SO_3 determined by C563 was found to be above the C150 Table limits, an additional 0.5% SO_3 was allowed provided that the C265 24 hour limit of 0.50 g/ℓ soluble SO_3 was not exceeded.

In C265, duplicate mortar samples are sealed in bags and cured for 24 hours at 23°C. The mortars are pulverized until they pass a 2.36 mm (#8) sieve, then slurried with distilled water. The slurry is filtered and the filtrate is analysed for soluble SO_3. (The dissolved sulfates are really SO_4^{2-} but the values are converted to SO_3 to be consistent with the normal oxide notation). The concept is that if there is excess SO_3 (defined by 0.50 g/ℓ) left in solution at 24 h of age, then when the ettringite (AFt) formed in the first 24 h normally decomposes to monosulfate (Afm), the sulfate left in solution will drive the reaction back to ettringite. This can lead to deleterious expansion or other damage due to internal sulfate attack.

The problems with C265 are that (1) the 24 h storage temperature of the mortars is critical to the test result, and (2) the multi-laboratory test variation is high around the failure limit of 0.50 g/ℓ. According to the precision statement, two lab results could differ by as much as 0.39 g/ℓ and still be acceptable when the results are in the range 0.30 to 0.85 g/ℓ SO_3!

Therefore, in the 1980's, tests were conducted by the ASTM sulfate content subcommittee C01.28 to evaluate replacement of the C265 test with the Canadian (CSA) 14 day mortar bar expansion test (now ASTM C1038). In this test, the cement is cast into mortar bars and the length change is monitored from 1 to 14 days of storage in saturated lime water at 23°C.

The concept behind C1038 is that the performance issue related to excess sulfates is deleterious expansion. The concerns raised about this test are (1) that the 14 day period is too short and (2) the 0.020% expansion limit at 14 days is not as conservative as the C265 limit. In defence of the first point, the accepted C452 external sulfate resistance test for portland cement is also a 14 day expansion test and one year expansion data by the author show an excellent correlation to the 14 day results for both C1038 and C452.

In 1989, the C150 Table footnote was altered to allow any amount of extra SO_3 as long as the C1038 14 day expansion was less than 0.020% (the C265 test was dropped). This still exists in the 1998 ASTM Book of Standards but there is a move being considered to re-instate C265 and the 0.5% extra SO_3 restriction due to recent concerns by some parties about delayed ettringite formation (DEF). DEF is a special case of internal sulfate attack which may occur when some concretes are exposed to elevated temperatures (in excess of 70°C) at an early age. Temperatures in excess of 70°C can destabilize ettringite and cause it to decompose. If it reforms at a later age, it can cause cracking and expansion. The conditions that cause DEF are controversial and there is much disinformation and secrecy resulting from factual information being tied up in litigations. Currently, there are no North American standards dealing with DEF except that the CSA precast concrete standard A23.4 has recently limited maximum concrete temperatures to 70°C. Because of the vested interests involved in current litigations, it is likely wise for the ASTM and CSA cementitious standards committee to move cautiously and not adopt any new test or test limits until more scientific results become known. Interim measures should be directed to controlling maximum concrete temperatures.

External Sulfate Attack

The portland cement aluminates were implicated in sulfate attack for many years but it was Thorvaldson[3] who explained how C_3A caused ettringite formation which caused the expansion. As a result of his work in the 1920's and 1930's, the Canada Cement Company patented the first Sulfate Resistant Portland Cement (SRPC) in North America. Since then, in ASTM C150, control of C_3A content has been used to manufacture Type V SRPC. The current C_3A limit of 5% has not changed since the 1941 version of C150. In Canada, the CSA A5 limit has been 3.5%.

In 1949, C150 also added a limit on the sum of $2(C_3A) + C_4AF$ of 20% maximum based on the work of Lerch[4] which showed that the ferrite phases may also contribute ettringite-forming alumina at slower rate. This limit was raised to 25% in 1985 which better fit Lerch's data.

The standard performance test for evaluating an SRPC has been to ASTM C452 14 day mortar bar expansion test developed by Lerch[4]. In this test, SO_3 content of a cement is raised to 7.0% using Terra Alba gypsum and mortar bars are cast. These bars are stored in water from 1 to 14 days at 23°C and the expansion is measured. The C150 expansion limit for SRPC (Type V) is 0.040% while in CSA A5, it is 0.035% (Type 50) and 0.045% for moderate sulfate resisting, Type 20 cement.

C452 was first published in 1960 but was not adopted in the C150 specification until 1971. The original SRPC optional test limit of 0.045% was reduced to 0.040% in 1985. (There was an attempt in the 1950's to develop a lean mortar bar test[5] but it was dropped due to poor productivity).

While this test has proven useful for testing sulfate resisting portland cement it was found not to be valid for evaluation of blended cements or combinations of portland cement and supplementary cementing materials. The reason for this is that with the excess SO_3 mixed into the mortars, the sulfate attack would start immediately and certainly before the pozzolanic and/or hydraulic reactions of the fly ashes, slags, silica fumes and natural pozzolans had initiated. The admixed gypsum greatly accelerates the test since it eliminates the slow diffusion stage of external sulfate attack, but in reality, the bulk of cementitious materials in concrete have a chance to hydrate before being exposed to external sulfates in service. Therefore, to address this problem, the ASTM C1.29 sulfate resistance committee, guided by K. Mather and T. Patzias, developed a new test method, C1012 in the 1970's which was first published in 1984. In this test, mortar bars are cured until a strength of 20 MPa is achieved, then they are exposed to 50 g/ℓ Na_2SO_4 solution at 23°C and expansion is measured. This test method had been adopted for blended cements (ASTM C595, C1157 and CSA A362) and for evaluating mixtures of portland cement with supplementary cementing materials (ASTM C989, C1240 and CSA A23.5). The expansion test limits that have evolved[1,2] are 0.10% at 6 months for high sulfate resistance (this can be superseded by meeting a limit of 0.10% at 12 months). For moderate sulfate resistance the limit is 0.10% at 6 months.

One of the obvious problems with C1012 is the length of time required to obtain results. However, it is difficult to avoid this since sulfates must diffuse to reaction sites before any expansion can occur. During development of this test, the original strength of 25 MPa before sulfate exposure was reduced to 20 MPa in order to shorten the test. In hindsight, this was counter-productive in terms of balancing the concept of allowing these blended materials to react with respect to the saving of only a few days in a six month test. As well, the test is not necessarily directly relevant to other types of sulfate salts than sodium sulfate. For example, magnesium sulfate typically results in softening and cracking of the bars without as much expansion. Therefore, visual indications of damage need to be recorded as well as expansion if other sulfate salts are substituted in this test. The original test allowed for a mixture of sodium and magnesium sulfates but it was found that this caused problems in some cases, so sodium sulfate was adopted.

While the pH of the solutions in C1012 are not controlled, in interlaboratory testing, it was found that the pH of the Na_2SO_4 storage solutions rapidly rose to 12.8-13.0 whereas the sodium-magnesium sulfate mixture only resulted in pH of about 10. Brown[6] and more recently Clifton et al[7] have advocated pH controlled tests, this may not model the situation of stagnant sulfates in contact with concrete foundations but it may better model flowing water situations, as in pipes.

None of the sulfate resistance tests including the C1012 model the situation of evaporative transport of sulfates into the specimens. This situation, which can occur in slabs on grade or in foundations subject to wetting/drying cycles, can pull sulfates in much faster than by diffusion. There may be some scope to modify C1012 to model this scenario in general, but not all possible situations could be modelled. This issue may better be addressed in concrete standards.

The C1012 test was developed to show the susceptibility (or resistance) of a cementitious system to attack by external sulfate solutions. It was not intended to model all or any field conditions for concrete nor concrete quality.

It is interesting to note that C1012 is more severe than C452 when used to evaluate SRPC. As shown in Figure 1, there are several cases where SRPC has passed the 0.040% 14 day limit in C452 but failed the 12 month C1012 limit of 0.10%. For all SRPC's tested by the author since 1977, all these have exceeded 0.10% expansion after exposure periods of between 7 and 20 months (some of the author's data is shown in Table II) and the bars start to crack and ravel at the edges. This illustrates the point that SRPC's are only sulfate resistant and not immune to attack. This may be due to the role of the ferrite phase. Therefore, steps must also be taken to make good quality concrete to slow sulfate penetration.

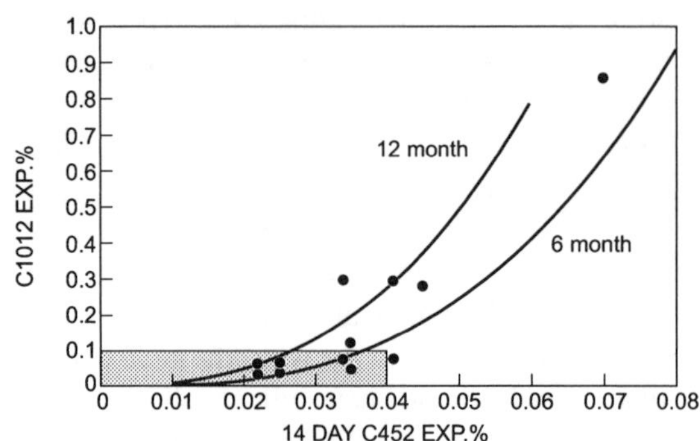

Figure 1: Comparison of 14 day C452 Expansions to 6 and 12 Month C1012 Expansions of Portland Cements.

This is shown in Figure 2, where 150 x 300 mm SRPC concrete (w/cm = 0.45) cylinders exposed to Na_2SO_4 or $MgSO_4$ for 22 years are showing microcracking and formation of ettringite and thaumasite (these concretes are described by Hooton and Emery[1]).

Since the field performance of Types II and V cements in concrete have been documented for over 50 years, by adopting the concept of "equivalence" the C1012 test can be used to evaluate blended cement systems versus Type II or Type V performance (assuming equal quality concretes). This is the concept used in ASTM C1157 and in the 1998 versions of CSA A362 and A23.5. It is interesting to note that the author has tested several portland-SCM combinations which have not exceeded 0.10% expansion in over 10 years of tests.[1,8]

CONCRETE STANDARDS

Both the ACI and CSA limits on concrete quality for sulfate exposures have evolved from the work performed by the U.S. Bureau of Reclamation. Both the ACI 318 and CSA A23.1 concrete standards limit the cement types and the maximum water to cementitious materials ratio (w/cm) based on the sulfate exposure. These limits are detailed in Table III. These w/cm limits are used to obtain relatively impermeable concretes which have a discontinuous capillary pore structure. This is regarded as the first line of defence with the cement type being a secondary issue.

Table II. Performance in C1012 Tests

C_3A	% at 6 months	% at 12 months	Time to >0.10%
(A) Moderate PC Performance in C1012 Tests			
5.9	0.074	0.294	~7 months
5.5	0.044	0.119	~11 months
7.4	0.117	0.517	5 months
7.3	0.072	0.235	8 months
7.9	0.076	0.291	~7 months
Limit	0.050	-	-
(B) SRPC Performance in C1012 Tests			
2.0	0.037	0.063	18 months
2.1	0.032	0.061	18 months
~2.0	0.052	0.113	11 months
3.8	0.060	0.273	7 months
1.4	0.037	0.061	20 months
Limit	0.050	0.100	-

Table III. Concrete Limit for Sulfate Exposure

Exposure	ACI 318		CSA A23.1		
	w/cm	cement type	w/cm	min strength (MPa)	cement type
Moderate 150-1500 mg/ℓ SO_4	0.50	II, IP, IS	0.50	30	20E,40E* 50E
severe 1,500-10,000 mg/ℓ	0.45	V	0.45	32	50E
very severe >10,000 mg/ℓ	0.45	V + SCM	0.40	35	50E

* E = Equivalent

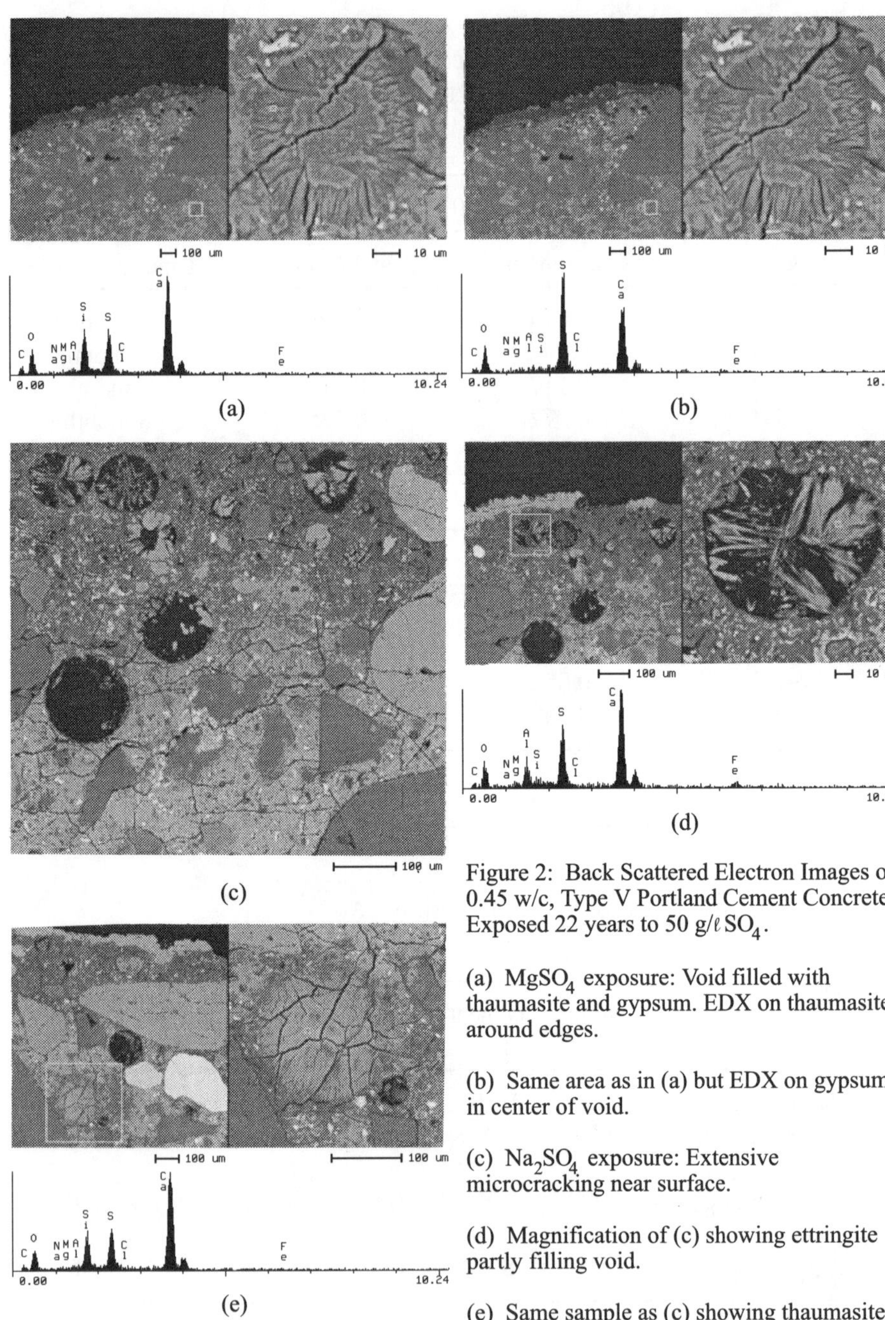

Figure 2: Back Scattered Electron Images of 0.45 w/c, Type V Portland Cement Concrete Exposed 22 years to 50 g/ℓ SO_4.

(a) $MgSO_4$ exposure: Void filled with thaumasite and gypsum. EDX on thaumasite around edges.

(b) Same area as in (a) but EDX on gypsum in center of void.

(c) Na_2SO_4 exposure: Extensive microcracking near surface.

(d) Magnification of (c) showing ettringite partly filling void.

(e) Same sample as (c) showing thaumasite completely filling void.

The ACI C201 Durability committee is currently balloting a revision to lower the w/cm to 0.40 for the very severe category which would then match the CSA limits.

With regard to evaporative transport of sulfates, such as in slabs on grade, there appears to be a consensus that this can also be minimized by provision of a low w/cm concrete with a discontinuous capillary pore structure (Haynes and O'Neill). Some have argued that the form of attack in these cases is physical, due to salt crystallization and related hydration/dehydration reactions (similar to the ASTM C88 sulfate soundness test for aggregates) but there is almost certain that if sulfates are carried up through the concrete, that chemical sulfate attacks will also occur. Since obtaining a low w/cm concrete is the solution to evaporative transport, it seems a moot point and the w/cm limits in the concrete standards should not be relaxed in these situations.

Another form of attack that has been found has resulted in the formation of thaumasite. This is a carbonate sulfo silicate hydrate which destroys the silicate matrix of concrete. Use of SRPC does not prevent the attack either. While originally thought to be rare and associated with cold temperatures, it has been found with increasing frequency,[10,11,12,13] often associated with carbonate aggregates and not always exposed to cold temperatures (see Figure 2). The condition under which thaumasite is a concern is not well known and as yet, there are no standards which address this issue. It is an area that may need to be addressed in concrete standard when more is known.

SUMMARY

There are standards for sulfate resistance of both cementing systems and concretes. The standards do not model all exposures or transport mechanisms so provision of concrete quality standards for impermeability need to be followed. Relying solely on cement type is not adequate for provision of sulfate resistance.

Standards are not perfect and new standards will undoubtedly be developed over time. Regardless, there is a need for the practitioner to understand the exposure conditions of concrete and to take appropriate precautions in specifying concrete since good quality concrete is the primary defence against sulfate attack.

REFERENCES

[1]R.D. Hooton and J.J. Emery, "Sulfate Resistance of a Canadian Slag Cement," *ACI Materials Journal*, **87** [6] 547-555 (1990).

[2]T. Patzias, "The Development of ASTM C1012 with Recommended Acceptance Limits for Sulfate Resistance of Hydraulic Cement," *Cement, Concrete, and Aggregates*, CCA GDP, **13** [1] 50-57 (1991).

[3]T. Thorvaldson, V.A. Vigfusson and R.K. Larmour, "The Action of Sulfates on the Component of Portland Cement," *Trans. Royal Soc. Canada*, 3rd Series, **21** Section III, p. 295 (1927).

[4]W.L. Lerch, "The Influence of Gypsum on the Hydration and Properties of Portland Cement Pastes," *Proceedings, ASTM*, **46** 1252-1292 (1946).

[5]D. Wolochow, "A Lean Mortar Bar Expansion Test for Sulfate Resistance of Portland Cements," Appendix A, *Proceedings, ASTM*, **52** 264-265 (1952).

[6]P.W. Brown, "An Evaluation of Sulfate Resistance of Cements in a Controlled Environment," *Cement and Concrete Research*, **11** 719-727 (1981).

[7]J.R. Clifton, G. Frohnschoff and C. Ferraris, "Standards for Evaluating the Susceptibility of Cement-Based Materials to External Sulfate Attack," in Quebec Seminar on Sulfate Attack Mechanisms, Materials Science of Concrete, Special Volume, American Ceramic Society, 1998.

[8]R.D. Hooton, "Influence of Silica Fume Replacement of Cement on Physical Properties and Resistance to Sulfate Attack, Freezing and Thawing, and Alkali-Silica Reactivity," *ACI Materials Journal*, **90** [2] 143-151 (1993).

[9]H. Haynes, R. O'Neill and P.K. Mehta, "Concrete Deterioration from Physical Attack by Salts," *Concrete International*, **18** 63-68 (1996).

[10]J.A. Bickley, R.T. Hemmings, R.D. Hooton, and J. Balinski, "Thaumasite Related Deterioration of Concrete Structures," pp. 159-175 in *Concrete Technology, Past, Present and Future*, ACI SP-144, American Concrete Institute, 1994.

[11]C. Rogers, M. Thomas and H. Lohse, "Thaumasite from Manitoba and Ontario, Canada," *Proceedings of 19th International Conference on Concrete Microscopy*, Cincinnati, 306-319 (1997).

[12] N.J. Crammond and M.A. Halliwell, "The Thaumasite Form of Sulfate Attack in Concretes Containing a Source of Carbonate Ions - A Microstructural Overview," pp. 357-380 in *Advances in Concrete Technology*, ACI SP-154, American Concrete Institute, 1995.

[13] M. Jones, "Industry Ignored Warnings on Thaumasite Threat, " *New Civil Engineer*, U.K., 9/16 April, p. 3, (1998).

ACKNOWLEDGEMENTS

Some of the C452/C1012 data was developed by Ms. M.-C. Lanctot at Lafarge's CTS laboratory in Montreal. The assistance of Prof. P.W. Brown and the staff at R.J. Lee Inc. are gratefully acknowledged for the SEM data. The concretes shown in Figure 2 were cast with support of Lafarge to Dr. J.J. Emery in 1976.

Participants

CANADA

Abdelkrim AMMOUCHE	Université Laval, Cité Universitaire, QC
Julie ARSENAULT	Université Laval, Cité Universitaire, QC
Remi BARBARULO	Université Laval, Cité Universitaire, QC
James BEAUDOIN	National Research Council of Canada, Ottawa, ON
Denis BEAUPRÉ	Service díExpertise en Matériaux, Québec, QC
Sandrine CATINAUD	Université Laval, Cité Universitaire, QC
Francois COUTURIER	SNC Lavalin, Montreal, QC
Anik DELAGRAVE	Université Laval, Cité Universitaire, QC
HélËne DESROSIERS	Université Laval, Cité Universitaire, QC
Benoit DURAND	Hydro-Québec, Varennes, QC
Sonia FLAMAND	Service d'Expertise en Matériaux, Québec, QC
Martin GENDREAU	Service d'Expertise en Matériaux, Québec, QC
Anik GINGRAS-GENOIS	Université Laval, Cité Universitaire, QC
Natalyia HEARN	University of Windsor, Windsor, ON
Doug HOOTON	University of Toronto, Toronto, ON
Olivier HOUDUSSE	Université Laval, Cité Universitaire, QC
Francois LAPLANTE	Hydro-Québec, Varennes, QC
Richard LARIVIRE	Hydro-Québec, Montréal, QC
Guillaume LEMAIRE	Université Laval, Cité Universitaire, QC
Zheng LIU	National Research Council of Canada, Ottawa, ON
Yannick MALTAIS	SIMCO Technologies, Québec, QC
Mel C. MARCHALL	M.C.Marchall Industrial Consultants, Delta, BC
Bruno MARCHAND	Hydro-Québec, Rouyn-Noranda, QC
Jacques MARCHAND	Service d'Expertise en Matériaux, Québec, QC
Valérie MICHAUD	Lafarge Canada Inc., Montreal, QC
Sidney MINDESS	University of British Columbia, Vancouver, BC
Fabien PEREZ	Université Laval, Cité Universitaire, QC
Stacey PERRON	National Research Council of Canada, Ottawa, ON
Michel PIGEON	Service d'Expertise en Matériaux, Quebec, QC
Philippe PINSONNEAULT	Ciment St-Laurent, Longueuil, QC
Elisabeth REID	Service d'Expertise en Matériaux, Québec, QC
Elaine ROBICHAUD	Hydro-Québec, Rouyn-Noranda, QC
Eric SAMSON	SIMCO Technologies, Québec, QC
Robert SUDERMAN	Lafarge Canada Inc., Montreal, QC
Basil TAMTSIA	National Research Council of Canada, Ottawa, ON

Julie THERRIEN	Université Laval, Cité Universitaire, QC
Michael THOMAS	University of Toronto, Toronto, ON
Guodong XU	National Research Council of Canada, Ottawa, ON
Tiewei ZHANG	Université Laval, Cité Universitaire, QC
Bruno ZUBER	Université Laval, Cité Universitaire, QC

DENMARK

Duncan HERTFORD	Aalborg Portland Cement, Aalborg
Gunnar M. IDORN	International Consultant, Naerum

FRANCE

Frédéric ADENOT	CEA Saclay, Gif-sur-Yvette
Pascal FAUCON	CEA Saclay, Gif-sur-Yvette
Gilles MARTINET	LERM, Paris
Micheline MORANVILLE	ENS Cachan, Cachan
Stéfanie PRENÉ	Électricité de France, Moret-sur-Loing

ISRAEL

Arnon BENTUR	Israel Institute of Technology (TECHNION), Haifa

JAPAN

Mitsunori KAWAMURA	Kanazawa University, Kanazawa, Ishikawa

NORWAY

Terje F. ROENNING	Norcem, R&D Cement and Concrete Laboratory, Brevik

SPAIN

Esperanza MENENDEZ	CSIC, Instituto Eduardo Torroja, Madrid

SWEDEN

Björn LAGERBLAD	Swedish Cement & Concrete Research Institute, Stockholm

SWITZERLAND

Juraj GEBAUER	Holderbank Management & Consulting Ltd., Holderbank

UNITED KINGDOM

Frederic P. GLASSER	University of Aberdeen, Old Aberdeen, Scotland
Stephen KELHAM	Blue Circle Technical Center, Greenhite, Kent
Hal F.W.TAYOR	Consultant, Coniston, Cumbria

U.S.A.

Gregory S. BARGER	Ash Grove Cement, Kansas City, KS
Ronald BERLINER	University of Missouri, Columbia, MO
David BONEN	USG Corporation Research Center, Libertyville, IL
Paul W. BROWN	Pennsylvania State University, State College, PA
Jeannette CHRISTENSEN	R.J.Lee Group, Monroeville, PA
Boyd CLARK	R.J.Lee Group, Monroeville, PA
Paul CLAYPOOL	R.J.Lee Group, Monroeville, PA
James CLIFTON	NIST, Gaithersburg, MD
Daniel CONSTANTINER	Master Builders Technologies, Cleveland, OH
Sidney DIAMOND	Purdue University, West Lafayette, IN
Bernard ERLIN	The Erlin Company, Latrobe, PA
Peter HAWKINS	California Portland Cement Co., Colton, CA
Harvey HAYNES	Haynes & Associates, Oakland, CA
Eugene D. HILL, Jr.	Openaka Corporation, Denver, CO
Felek JACHIMOWICZ	W.R.Grace & Co., Cambridge, MA
Vagn JOHANSEN	Construction Technology Laboratories, Skokie, IL
Jiann-Wen Woody JU	University of Calfornia, Los Angeles, CA
Kenneth S. KASDAN	Kasdan, Simonds, McIntyre, Epstein & Martin, Irvine, CA
Richard J. LEE	R.J. Lee Group, Monroeville, PA
Kumar P. MEHTA	University of California, Berkeley, CA
James S. PIERCE	U.S. Bureau of Reclamation, Denver, CO
Jan OLEK	Purdue University, West Lafayette, IN
Laura POWERS-COUCHE	Construction Technology Laboratories, Skokie, IL
Sadananda SAHU	R.J. Lee Group, Monroeville, PA
Jan P. SKALNY	J.P. Skalny Consulting, Holmes Beach, FL
William J. SUOJANEN	Kasdan, Simonds, McIntyre, Epstein & Martin, Irvine, CA
Oscar TAVARES	Lafarge Corporation, Southfield, MI
J. Francis YOUNG	University of Illinois, Urbana, IL

Index